Biology Unit 1
for CAPE® Examinations

Myda Ramesar, Mary Jones and Geoff Jones

T0320844

CAPE® is a registered trade mark of the **Caribbean Examinations Council (CXC)**. *Biology Unit 1 for CAPE® Examinations* is an independent publication and has not been authorised, sponsored, or otherwise approved by CXC.

CAMBRIDGE
UNIVERSITY PRESS

CAMBRIDGE
UNIVERSITY PRESS

Shaftesbury Road, Cambridge CB2 8EA, United Kingdom

One Liberty Plaza, 20th Floor, New York, NY 10006, USA

477 Williamstown Road, Port Melbourne, VIC 3207, Australia

314–321, 3rd Floor, Plot 3, Splendor Forum, Jasola District Centre,
New Delhi – 110025, India

103 Penang Road, #05-06/07, Visioncrest Commercial, Singapore 238467

Cambridge University Press is part of the University of Cambridge.

It furthers the University's mission by disseminating knowledge in the pursuit of
education, learning and reasearch at the highest international levels of excellence.

www.cambridge.org
Information on this title: www.cambridge.org/9780521176903

© Cambridge University Press & Assessment 2011

First published 2011

20 19 18 17 16 15 14

Printed in Great Britain by Ashford Colour Press Ltd.

A catalogue record for this publication is available from the British Library

ISBN 978-0-521-17690-3 Paperback

Additional resources for this publication are available through Cambridge GO.
Visit cambridge.org/go

Contents

Introduction

The new *Biology for CAPE® Examinations* course provides complete coverage of the CAPE® Biology syllabus. There are two books, one covering Unit 1 and one covering Unit 2. Some of the material is based on the *Cambridge OCR Advanced Sciences Biology 1 for OCR* and *Biology 2 for OCR*, but much is new. It has been brought up to date with new findings in numerous topics. Examples have generally been drawn from Caribbean contexts – for example, ecosystems are dealt with in the context of coral reefs, and plant reproduction is illustrated using Caribbean species.

The books address each of the learning outcomes in the CAPE® Biology syllabus. The material is organised in the same sequence as in the syllabus. It is written to ensure it will be accessible to students who have studied the Caribbean Secondary Education Certificate (CSEC®) Biology course, or other Biology courses at a similar level.

The depth and breadth of treatment of each topic is pitched at the appropriate level for CAPE® Biology students. Most chapters also include boxes containing material that goes a little beyond the requirements of the syllabus, to interest and stretch more able students.

The illustrations include numerous micrographs, some of which have accompanying interpretive diagrams. The diagrams have all been drawn by a biologist, and students should find them very helpful in developing their understanding of structures and processes.

Each chapter also includes self assessment questions (SAQs). These provide opportunities to check understanding and sometimes to make links back to earlier work. They often address misunderstandings that commonly appear in examination answers, and will help students to avoid such errors. Answers to the SAQs can be found at the back of the book.

Each chapter ends with numerous questions in the style of those that students will meet in the CAPE® examination papers, written by an experienced CAPE® examiner. These include three styles of question – multiple choice, structured and essay questions. The answers to these questions are available at **cambridge.org/go**

There is a glossary at the end of the book, where brief definitions of important biological terms are given.

Chapter 1
Aspects of biochemistry

By the end of this chapter you should be able to:

a   discuss how the structure and properties of water relate to the role that water plays as a medium of life;

b   explain the relationship between the structure and function of glucose;

c   explain the relationship between the structure and function of sucrose;

d   discuss how the molecular structure of starch, glycogen and cellulose relate to their functions in living organisms;

e   describe the generalised structure of an amino acid and the formation and breakage of a peptide bond;

f   explain the meaning of the terms primary, secondary, tertiary and quaternary structure of proteins, and describe the types of bonding (hydrogen, ionic, disulphide) and hydrophobic interactions which hold the molecule in shape;

g   outline the molecular structure of haemoglobin, as an example of a globular protein, and of collagen, as an example of a fibrous protein, ensuring that the relationships between their structures and functions are clearly established;

h   know how to carry out tests for reducing and non-reducing sugars (including quantitative use of the Benedict's test), for starch and lipids and (the biuret test) for proteins;

i   know how to investigate and compare quantitatively reducing sugars and starch;

j   describe the molecular structure of a triglyceride and its role as a source of energy;

k   describe the structure of phospholipids and their role in membrane structure and function.

The human body is made of many different types of molecule. Most of your body is water, which has small molecules made of two atoms of hydrogen combined with one atom of oxygen, formula H_2O. The other main types of molecule in organisms are proteins, carbohydrates, fats and nucleic acids. These molecules make up your structure, and they also undergo chemical reactions – known as **metabolic reactions** – that make things happen in and around your cells.

Up until the mid 1950s, we did not know very much about the structures of these molecules or how their structures might relate to the ways in which they behave. Since then, there has been an explosion of knowledge and understanding, and today a major industry has been built on the many applications to which we can put this knowledge. Biotechnology makes use of molecules and reactions in living organisms, and research continues to find new information about how molecules behave, and new uses for this technology in fields including medicine, agriculture, mining and food production.

In this chapter, we will look at the structures of the main types of molecule found in our bodies – and in those of every other living organism (Table 1.1). We will also see how these structures relate to the functions of the molecules.

Molecule	Percentage of total mass
water	60
protein	19
fat	15
carbohydrate	4
other	2

Table 1.1 Human body composition.

Water

Water is by far the most abundant molecule in our bodies. It is also one of the smallest and one of the simplest. Yet water is an amazing substance, with a collection of properties that is shared by no other. It is difficult to imagine how life of any kind could exist without water. The search for other life in the universe begins by searching for other planets that may have liquid water on them.

Water is a major component of all cells, often making up between 70% and 90% of the cell's mass. Your body is about 60% water. Water is also the environment of many living organisms, and life must have first evolved in water.

Water has very simple molecules, yet has some surprising properties. Other substances made of such small molecules tend to be gases at the temperatures found on most of the Earth, whereas water may be found as a solid, liquid or gas. The fact that water is often in liquid form allows it to act as the solvent in which metabolic reactions (the chemical reactions that take place in living organisms) can happen. Water provides a means of transporting molecules and ions from one place to another, because many of them can dissolve in it.

The structure of a water molecule

Figure 1.1a shows the structure of a water molecule. It is made up of two hydrogen atoms covalently bonded to one oxygen atom. The bonds are single covalent bonds, in which the single electron of each hydrogen atom is shared along with one of the six outer shell electrons of the oxygen, leaving four electrons which are organised into two non-bonding pairs. The bonds are very strong, and it is very difficult to split the hydrogen and oxygen atoms apart.

A covalent bond is an electron-sharing bond, and in this case the sharing is not equal. The oxygen atom gets slightly more than its fair share, and this gives it a very small negative charge. This is written δ− (delta minus). The hydrogen atoms have a very small positive charge, δ+ (delta plus). So, although a water molecule has a neutral electrical charge overall, different parts of it do have small positive and negative charges. This unequal distribution of charge in the molecule is called a **dipole**.

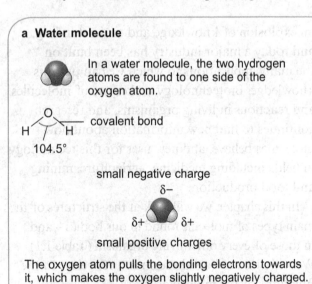

a Water molecule

In a water molecule, the two hydrogen atoms are found to one side of the oxygen atom.

covalent bond

104.5°

small negative charge

δ−

δ+   δ+

small positive charges

The oxygen atom pulls the bonding electrons towards it, which makes the oxygen slightly negatively charged. The hydrogen atoms have small positive charges. This unequal distribution of charge is called a dipole.

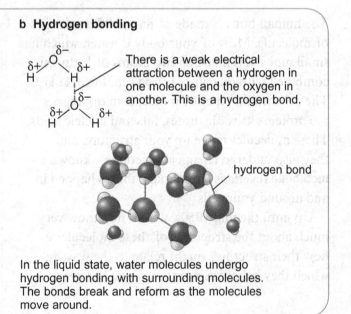

b Hydrogen bonding

There is a weak electrical attraction between a hydrogen in one molecule and the oxygen in another. This is a hydrogen bond.

hydrogen bond

In the liquid state, water molecules undergo hydrogen bonding with surrounding molecules. The bonds break and reform as the molecules move around.

Figure 1.1   a The structure of a water molecule; **b** hydrogen bonding.

These tiny charges mean that water molecules are attracted to each other – the positively charged hydrogen atoms in one molecule are attracted to the negatively charged oxygen atoms in other molecules. The attraction is called a **hydrogen bond** (Figure 1.1b). Every water molecule is hydrogen bonded to its four nearest neighbours.

Hydrogen bonds are weak, long-distance bonds. As we will see later in this chapter, they are very common and very important in biological molecules.

In the next few pages, we will look in detail at some of the special properties of water, and how these make water so important to all living organisms.

Ice, liquid water and water vapour

In a solid substance, such as ice, the molecules have relatively little kinetic energy. They vibrate continuously, but remain in fixed positions. In liquid water, the molecules have more kinetic energy, moving around past each other, forming fleeting hydrogen bonds with each other. In water vapour, the molecules are far apart, scarcely interacting with each other at all.

In solid water – ice – the hydrogen bonds hold the water molecules in a rigid lattice formation. As in all solids, the molecules vibrate, but they do not move around. As the temperature of liquid water decreases, the water molecules have less and less kinetic energy and they slow down. This allows each molecule to form the maximum number of hydrogen bonds (four) with other water molecules. When this happens, the water molecules spread out to form these bonds. This produces a rigid lattice which holds the water molecules further apart than in liquid water (Figure 1.2). Ice is therefore less dense than liquid water, and so it floats. Hydrogen bonding is responsible for this unique property of water.

The fact that ice floats on liquid water is important for aquatic living organisms. As the temperature of the air above the water falls, the water at the surface also cools. The water freezes from the surface, forming a layer of ice with liquid water underneath. This happens because the maximum density of water occurs at 4 °C.

As the surface water gets colder it gets denser and sinks to the bottom, displacing warmer water which rises to the surface and cools. However, when water gets colder than 4 °C, it becomes less dense than the warmer water below it and so it stays on the surface where it forms ice. The layer of ice insulates the water below it from further temperature changes. This water remains at 4 °C and stays liquid, providing an environment in which living organisms can continue to survive.

Changes in density of water as its temperature changes are the main cause of ocean currents and upwellings, which help to maintain the circulation of nutrients in seas and oceans.

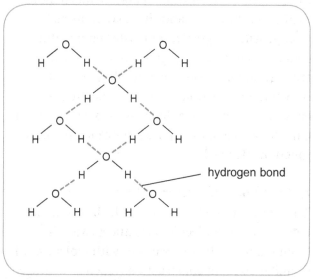

Figure 1.2 The structure of ice. Water molecules are held in a regular, fixed arrangement or lattice (shown in only two dimensions here).

Water, heat and temperature

Specific heat capacity

If you heat water, the temperature of the water rises. Temperature relates to the amount of kinetic energy that the water molecules have. As heat energy is added to the water, a lot of the energy is used to break the hydrogen bonds between the water molecules. Because so much heat energy is used for this, there is less heat energy available to raise the temperature. Water therefore requires a lot of heating in order to increase its temperature by very much. We say that it has a **high specific heat capacity**.

The specific heat capacity of water is $4.2\,J\,g^{-1}\,°C^{-1}$. This means that it takes 4.2 joules of energy to heat 1 g of water by 1 °C.

You – like all living organisms – make good use of this property. Being largely water, your body does not change its temperature easily. Large changes in the temperature of your external environment have relatively small effects on the temperature of your body. This is true for all living organisms.

For organisms that live in water, it means that the temperature of their external environment is relatively stable. It takes a lot of heat energy in order to change the temperature in, say, a lake or the sea. Air changes its temperature much more easily, so terrestrial (land-living) organisms have to cope with far greater and faster temperature changes than aquatic organisms. Changes in temperature inside cells can result in disruption of metabolic reactions, most of which require enzymes to make them happen. As you will see in Chapter 4, enzymes require a stable temperature to function effectively.

Latent heat of vaporisation

The energy needed to break the hydrogen bonds between water molecules also affects water's boiling point. Other substances with molecules of a similar size and construction – such as hydrogen sulphide, H_2S – form gases at room temperature. There are no hydrogen bonds holding the hydrogen sulphide molecules together, so they are free to fly off into the air. However, because of the hydrogen bonds, water at room temperature is liquid. It has to be heated to 100 °C before all the molecules have enough energy to break apart from one another so that the water turns from a liquid to a gas.

In liquid water, even well below its boiling point, some of the individual molecules have greater kinetic energy than others. These may have enough energy to escape from the surface of the liquid and fly off into the air to form water vapour. This is called **evaporation**. Like boiling, it requires a lot of energy because it involves breaking hydrogen bonds. As these energetic molecules leave the liquid, they reduce its average temperature (because the molecules that are left behind are the less energetic ones). The energy involved is called the **latent heat of vaporisation**. Evaporation therefore has a cooling effect.

Once again, our bodies make good use of this property of water. When liquid sweat lies on the surface of the skin, the water in the sweat absorbs heat energy from the body as it evaporates (Figure 1.3). It is our major cooling mechanism. It also helps to cool plant leaves in hot climates, as water evaporates from the surfaces of the mesophyll cells inside them.

Latent heat of fusion

Water also requires a lot of energy to change state from a solid to a liquid, or from a liquid to a solid. To change ice to water, 300 J of energy are required for each gram of ice. To change water to ice, 300 J must be lost from each gram of water. This is called the **latent heat of fusion**. It means that it is quite difficult to freeze water, so it tends to stay a liquid. This is very important to living organisms because cytoplasm contains a lot of water. The high latent heat of fusion reduces the likelihood of ice crystals forming inside cells. Once frozen, most cells are permanently damaged, because ice

Figure 1.3 This scanning electron micrograph (SEM) (page 35) shows sweat droplets emerging from pores on human skin (× 25).

crystals (which, you will remember, take up more space than liquid water) can pierce membranes.

Another factor that reduces the tendency of water to freeze is the presence of solutes. A salt solution has to be cooled well below 0 °C before it freezes. This is because the presence of the solute particles in between the water molecules disrupts the formation of hydrogen bonds between the water molecules, and therefore makes it more difficult for water molecules to assume the configuration they take up in ice (Figure 1.2).

Solvent properties of water

Water is an excellent **solvent**. The tiny charges on its molecules attract other molecules or ions that have charges on them (Figure 1.4 and Figure 1.5). The molecules and ions spread around in between the water molecules. This is called **dissolving**. A substance that dissolves in water is a **solute**, and the mixture of water and solute is called a **solution**.

Ionic compounds, such as sodium chloride, will generally dissolve in water. The ions have electrical charges on them, and these are attracted to the dipoles on the water molecules. The ions disperse among the water molecules.

Some covalent compounds can also dissolve in water. These tend to be ones that have small molecules and dipoles. Glucose and amino acids are examples of such compounds. (You can find out about the structures of glucose and amino acids on pages 8 and 14.) Even quite large covalent molecules can dissolve in water if they have plenty of small electrical charges on them. Many protein molecules come into this category, for example most enzymes and haemoglobin.

However, many other large covalent compounds are not soluble in water. These include starch, cellulose and other polysaccharides (pages 11 to 13) and many types of protein, especially fibrous proteins such as collagen (pages 20 to 21). Lipids are also insoluble in water (page 24).

The fact that water is such a good solvent is of huge importance to living organisms. When an ionic compound such as sodium chloride dissolves in water, the sodium ions and the chloride ions become separated from each other. This makes it easy for them to react with other ions or molecules. Many reactions, including most metabolic reactions, will only take place in solution, because this makes it possible for the ions or molecules to come into contact with each other. Cytoplasm contains many different substances dissolved in water, as do cell organelles such as mitochondria and chloroplasts (page 45).

Water can flow, and therefore it can carry dissolved substances from one place to another. This happens in our blood, and in the xylem

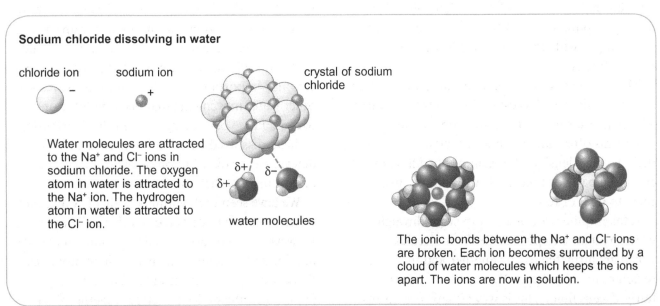

Sodium chloride dissolving in water

chloride ion   sodium ion   crystal of sodium chloride

$-$ $+$

Water molecules are attracted to the Na⁺ and Cl⁻ ions in sodium chloride. The oxygen atom in water is attracted to the Na⁺ ion. The hydrogen atom in water is attracted to the Cl⁻ ion.

$\delta+$ $\delta-$
$\delta+$

water molecules

The ionic bonds between the Na⁺ and Cl⁻ ions are broken. Each ion becomes surrounded by a cloud of water molecules which keeps the ions apart. The ions are now in solution.

Figure 1.4 Water as a solvent. Sodium chloride readily dissolves in water.

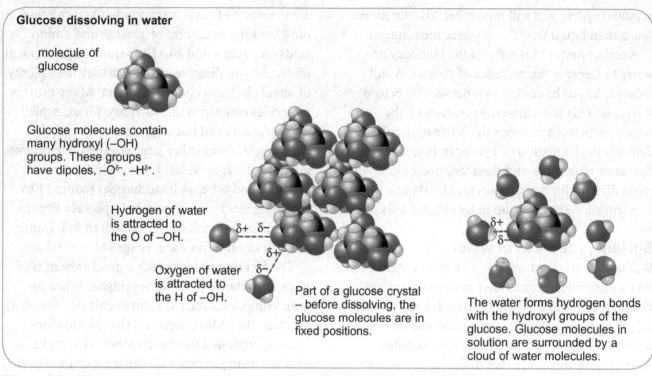

Glucose dissolving in water

molecule of glucose

Glucose molecules contain many hydroxyl (–OH) groups. These groups have dipoles, $-O^{\delta-}$, $-H^{\delta+}$.

Hydrogen of water is attracted to the O of –OH.

$\delta+$ $\delta-$

$\delta+$

Oxygen of water is attracted to the H of –OH.

$\delta-$

Part of a glucose crystal – before dissolving, the glucose molecules are in fixed positions.

$\delta+$
$\delta-$

The water forms hydrogen bonds with the hydroxyl groups of the glucose. Glucose molecules in solution are surrounded by a cloud of water molecules.

Figure 1.5   Water as a solvent. Glucose is a covalent molecule that dissolves in water.

vessels and phloem sieve tubes of a plant. Urea, the main nitrogenous excretory product of mammals, is removed from the body dissolved in water as urine.

Density and viscosity

Water molecules in liquid water are pulled closely together by the hydrogen bonds between them, and this makes water a relatively dense liquid. The density of pure water is $1.0\,\text{g cm}^{-3}$. Compare this, for example, with ethanol, which has a density of only $0.79\,\text{g cm}^{-3}$.

Most living organisms, containing a lot of water, have a density which is quite close to that of water. This makes it easy for them to swim. Aquatic organisms often have methods of slightly changing their average density – for example, by filling or emptying parts of their body with air – to help them to float or to sink.

It takes quite a lot of effort to swim through water. You have to push aside the molecules, which are attracted to one another and therefore reluctant to move apart. We say that water is a fairly **viscous** fluid. This is why aquatic organisms are often streamlined; their shape helps them to cut through water more easily.

Cohesion and surface tension

Water molecules tend to stick together, held by the hydrogen bonds that form between them. In liquid water, these bonds constantly form and break, each lasting for only a fraction of a second. The attractive force produced by these hydrogen bonds is called **cohesion**.

Cohesion makes it easy for water to move by **mass flow** – that is, a large body of water can flow in the same direction without breaking apart. This is important in the flow of blood in animals, and in the flow of water within xylem vessels in plants. It is also important in the formation of waves and other water movements that occur in lakes and oceans. These play a large part in the distribution of heat, dissolved gases and nutrients. They also determine the distribution of plankton (small organisms that drift in water).

We have seen that, within a body of water, each water molecule is attracted to others all around it. However, on the surface, the uppermost molecules have other molecules only below them, not above. So they are pulled downwards. These pulling forces draw the molecules closer together than in other parts of the water. This phenomenon is called **surface tension**. It forms a strong layer on the

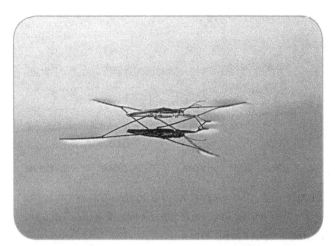

Figure 1.6 The water strider hunts by running over the surface of the water.

surface of the water – so strong that small animals are able to walk on it without difficulty (Figure 1.6).

pH

pH is a measure of the concentration of hydrogen ions in a solution. The more hydrogen ions, the lower the pH. A solution with a pH of 7 is neutral. A pH below 7 is acidic; a pH above 7 is alkaline.

Water itself is partially dissociated into hydrogen and hydroxide ions:

$$H_2O \rightleftharpoons H^+ + OH^-$$

As there are equal amounts of hydrogen ions (protons) and hydroxide ions in pure water, it has a pH of 7.

Some substances that dissolve in water tend to produce either hydrogen ions or hydroxide ions. Some proteins are able to 'soak up' these hydrogen or hydroxide ions, preventing the pH of the solution from changing. They are called **buffers**, and they are very important in living organisms because they help to maintain an ideal pH for enzymes to work. Cytoplasm and tissue fluids in living organisms are usually well buffered, so their pH is generally somewhere between 7 and 8.

Transparency

Pure water is **transparent** to the visible wavelengths of light. This allows aquatic photosynthetic organisms, such as phytoplankton and algae, to obtain enough light for photosynthesis.

However, light only travels a certain depth through water, and so the depths of the oceans are totally dark. Blue light is able to pass through greater depths of water than red and green light.

Reactivity

Water takes part in many metabolic reactions. For example, when large polymer molecules (such as starch or proteins) are broken down into their simpler units, water is involved. These are **hydrolysis** reactions, and they occur throughout all living organisms. The hydrolysis of carbohydrates and proteins is described on pages 9 and 15.

Water is also used in **photosynthesis**, where it provides hydrogen ions that are combined with carbon dioxide to make sugars. The oxygen atoms from the water molecules are released as oxygen gas. This is the source of the oxygen that we breathe.

SAQ

1 Explain each of these statements.
   a Sweating can cool the body to below the temperature of the air surrounding it.
   b A long column of water moves up through xylem vessels without breaking.
   c Small covalent molecules such as glucose can dissolve in water.
   d Reactions can occur in solution that would not occur if the reactants were in a solid state.
   e Sea water will only freeze if the temperature drops well below the freezing point of fresh water, 0 °C.

Carbohydrates

Carbohydrates are substances whose molecules are made up of sugar units. The general formula for a sugar unit is $C_nH_{2n}O_n$. Carbohydrates include sugars, starches and cellulose. Sugars are always soluble in water and taste sweet. Starches and cellulose, which are both examples of polysaccharides, are insoluble in water and do not taste sweet (page 11).

Monosaccharides

A carbohydrate molecule containing just one sugar unit is called a **monosaccharide**. Monosaccharides cannot be further hydrolysed. A monosaccharide is the simplest form of sugar.

Monosaccharides have the general formula $(CH_2O)_n$. They can be classified according to the number of carbon atoms in each molecule. Trioses have 3C, pentoses 5C and hexoses 6C (Table 1.2). Glyceraldehyde, ribose, glucose, fructose and galactose are examples of monosaccharides. **Glyceraldehyde** is a triose and is the first carbohydrate formed during photosynthesis. **Ribose** (a pentose) is a fundamental constituent of RNA, and a similar pentose, **deoxyribose**, is found in DNA. **Glucose** (a hexose) is the main respiratory substrate in many cells, providing energy when it is oxidised. All sugars taste sweet, and **fructose** is often found in fruits and in nectar – here, it attracts insects and other animals, which might then inadvertently disperse the plant's seeds, or transfer pollen from one flower to another.

In Figure 1.7, the sugars are all shown in their straight chain forms. However, pentoses and hexoses are usually also able to flip into a ring form, in which the chain links up with itself. Figure 1.8 shows the structure of two different ring forms of glucose. You can see that, in both of them, carbon atom 1 joins to the oxygen on carbon atom 5. The ring therefore contains oxygen, but carbon atom 6 is not part of the ring. In solution, glucose (and other monosaccharide) molecules continually flip between their ring and chain forms.

This ring can exist in two forms, known as α-**glucose** (alpha glucose) and β-**glucose** (beta

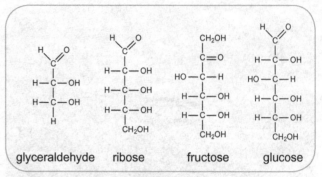

Figure 1.7 The straight chain structures of some common monosaccharides. Diagrams like this are called structural formulae.

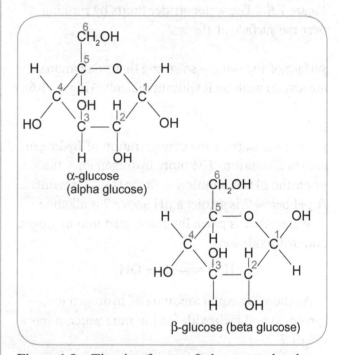

Figure 1.8 The ring forms of glucose molecules.

glucose). Both of them have a six-membered ring, made up of five carbons and one oxygen. They differ only in the orientation of the −H and −OH groups on carbon 1. The alpha position is defined

Type of sugar	Example	Functions
triose	glyceraldehyde	mainly as intermediates in metabolic pathways such as glycolysis (respiration) and in photosynthesis
pentose	ribose, deoxyribose	constituents of macromolecules (e.g. RNA, DNA, ATP)
hexose	glucose, fructose, galactose	soluble form of chemical energy that can be released by respiration; stored in some fruits to aid dispersal of seeds; found in nectar and honey (fructose and glucose)

Table 1.2 Some examples of monosaccharides.

as the −OH being on the opposite side of the ring from carbon 6. The beta position is defined as the −OH being on the same side as the ring as carbon 6.

Figure 1.9 shows some different forms that can be taken up by a molecule of fructose. Unlike glucose, it forms a five-membered ring, containing four carbon atoms and one oxygen atom. The closure of the ring occurs between carbon 2 and carbon 5. As with glucose, alpha and beta forms exist.

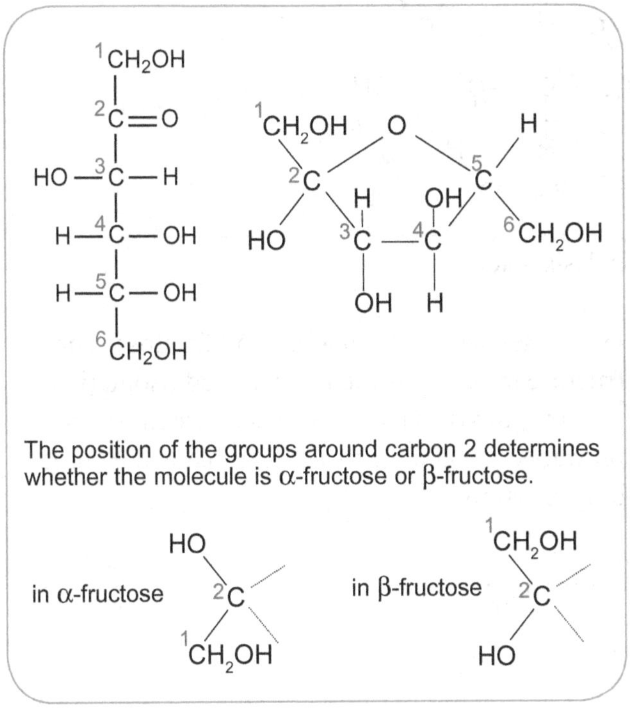

The position of the groups around carbon 2 determines whether the molecule is α-fructose or β-fructose.

Figure 1.9 Structural formulae of straight chain and ring forms of fructose. In the diagram of the ring, fructose is shown in a non-standard orientation with the numbers running anticlockwise.

SAQ

2 The molecular formula for a hexose sugar is $C_6H_{12}O_6$. Write the molecular formulae for
  a a triose
  b a pentose.

Disaccharides

Two monosaccharide molecules can link together to form a sugar called a **disaccharide**. Like monosaccharides, disaccharides are soluble in water and taste sweet.

For example, two α-glucose molecules can react to form **maltose** (Figure 1.10). This is a

condensation reaction, in which a water molecule is removed. The linkage formed between the two monosaccharides is called a **glycosidic bond**. It involves covalent bonds, and is very strong. In maltose, the glycosidic bond is formed between carbon atom 1 of one molecule and carbon atom 4 of the other. Both molecules are in the α form. We can describe this as an **α 1–4 glycosidic bond**.

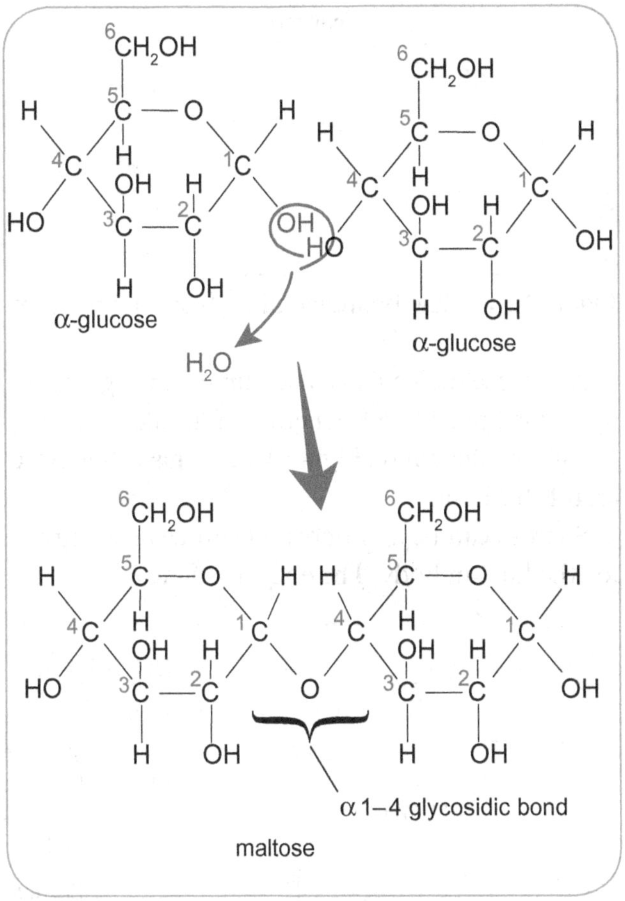

Figure 1.10 The formation of a glycosidic bond by a condensation reaction.

When carbohydrates are digested, glycosidic bonds are broken down by carbohydrase enzymes. The enzyme that breaks maltose apart is called maltase. This is a **hydrolysis reaction** (Figure 1.11).

Sucrose, ordinary table sugar, is a disaccharide that occurs naturally in the plant kingdom, especially in sugar cane, where it makes up 20% by weight, and in beets, where it makes up 15% by weight. As a result, sucrose is often known as cane sugar or beet sugar. Sucrose is the main form in which carbohydrates are transported in plants.

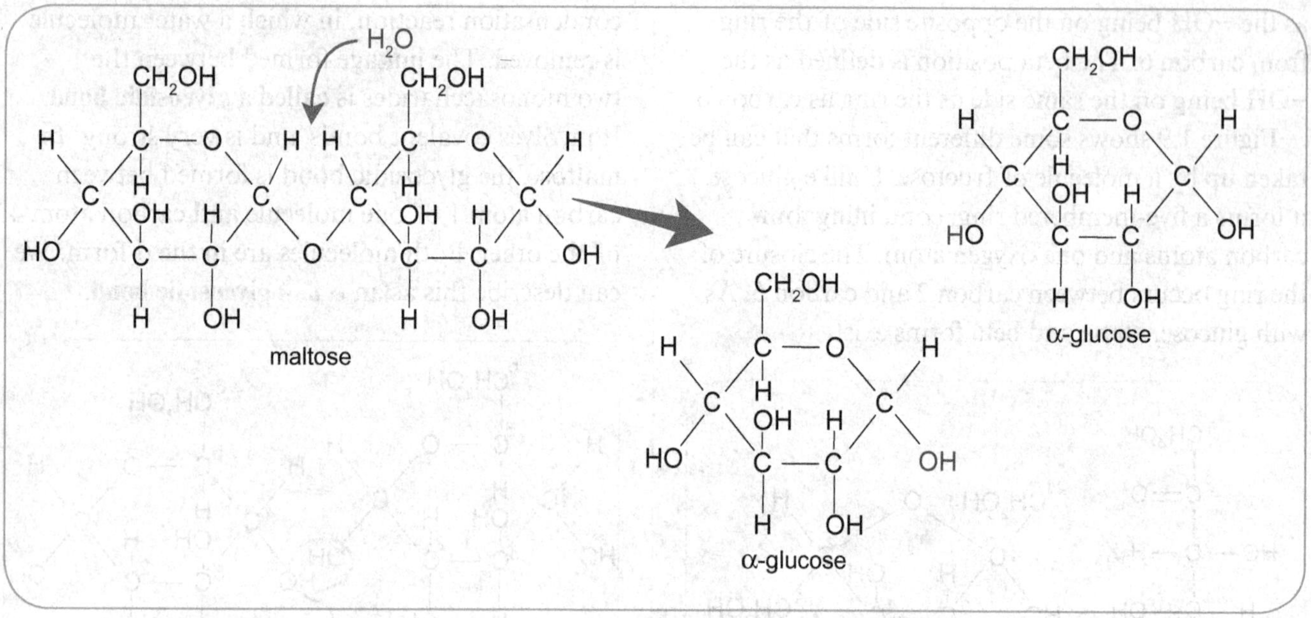

Figure 1.11 The breakage of a glycosidic bond by a hydrolysis reaction.

A sucrose molecule is made up of an α-glucose molecule linked to a β-fructose molecule by a 1–2 glycosidic bond (Figure 1.12). This is therefore an α1–β2 bond.

Sucrose can be an important source of energy for the human body. The enzyme sucrase hydrolyses sucrose to glucose and fructose. The fructose molecule is then rearranged (isomerised) to form glucose. Hence, every sucrose molecule produces two glucose molecules that can be used in respiration.

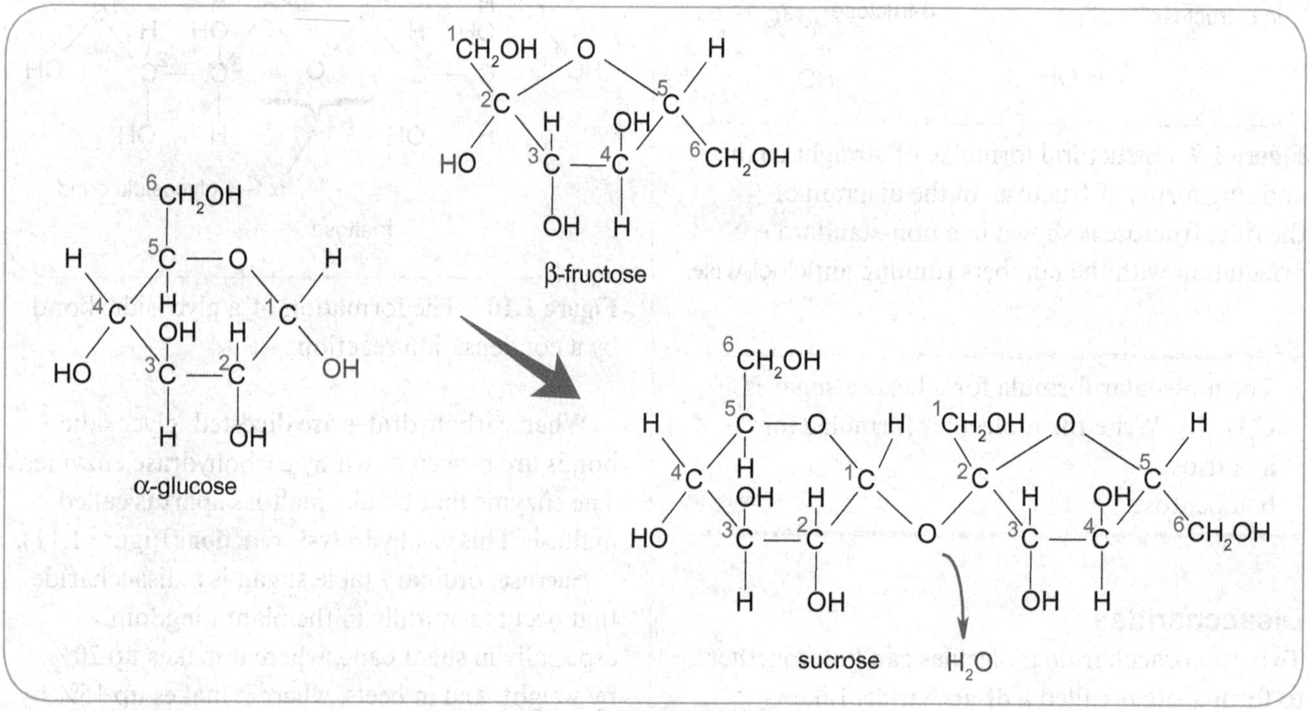

Figure 1.12 Formation of sucrose from glucose and fructose.

Reducing and non-reducing sugars

If you have a solution that you suspect contains sugars, you can use **Benedict's reagent** (or Fehling's solution) to test for it. Benedict's reagent contains copper(II) sulphate in alkaline solution. It is blue. When heated with some types of sugar, the Cu^{2+} ions are reduced to Cu^+ ions. They form insoluble copper(I) oxide, which forms a brick red precipitate.

Glucose, fructose and maltose all give a positive result with this test, and they are said to be **reducing sugars**.

The disaccharide sucrose, on the other hand, does not give a positive result with the Benedict's test, and it is said to be a **non-reducing sugar**. This is because part of the molecule that would react with the Benedict's reagent is involved in the glycosidic bond.

This is not true of all disaccharides. Maltose is a reducing sugar. In this case, the two glucose units are linked through carbon 1 and carbon 4, leaving carbon 1 on the second glucose unit free to form a reducing group.

You can still use the Benedict's test to find out if a solution contains a non-reducing disaccharide such as sucrose. First, carry out the Benedict's test as normal. If it is negative, take a fresh sample of the solution and heat it with hydrochloric acid. This will break the glycosidic bond, separating the sucrose into glucose and fructose. Now add an alkali such as sodium hydroxide to neutralise any left-over acid, and then try the Benedict's test again. The glucose and fructose (produced from the sucrose) will now give a positive result.

SAQ

3 The quantity of copper oxide that is formed, as a result of the Benedict's test, depends on the amount of reducing sugar in a solution. Describe how you could use this test to determine the concentration of glucose in an unknown solution.

4 Imagine that you have three solutions, one containing reducing sugar, one containing non-reducing sugar and one containing both types. How you could determine which is which?

Polysaccharides

Linking together thousands of α-glucose molecules with α 1–4 glycosidic bonds produces the carbohydrate **amylose** (Figure 1.13). This is a **polysaccharide** – a substance whose molecules contain long chains of monosaccharides linked together. Polysaccharide molecules can be enormous, and so cannot dissolve in water. A polysaccharide is an example of a **polymer**. Polymers are molecules made up of many **monomers** – in this case monosaccharides – linked together in a long chain.

An amylose molecule can contain several thousand glucose units. The long chain coils up into a spiral, which makes this big molecule very compact. The coil is held in shape by hydrogen bonds that form between the −OH groups attached to the carbon 1 of each sugar unit.

A similar polysaccharide is made when some of the glucose units join via glycosidic links between carbon 1 and carbon 6. This produces a branching molecule, called **amylopectin**. This can still spiral, but to a lesser extent than the unbranched amylose.

Starch contains a mixture of amylose and amylopectin. This is how plants store the carbohydrate that they make in photosynthesis. Starch is insoluble and metabolically inactive, so it does not interfere with chemical reactions inside the cell, nor does it affect the water potential (page 63). However, it can easily be broken down by the enzyme amylase, which breaks the glycosidic bonds and releases glucose that can be used in respiration.

Starch is stored in organelles called plastids. The starch forms grains or granules, and these have distinctive forms in different plant species.

Animal cells never contain starch, but they may contain a substance that is very similar to amylopectin, called **glycogen**. This is also made of many glucose molecules linked through α 1–4 glycosidic bonds with α 1–6 branches. Glycogen tends to have even more branches than amylopectin, and so is even less able to form spirals, making it less dense than amylopectin. Like starch, glycogen forms an insoluble and unreactive energy reserve in animal cells. It is even easier to break down to glucose than amylose or amylopectin, because its branching shape allows many enzymes to act simultaneously on one glycogen molecule. This may be related to the fact that most animals have higher metabolic rates than plants, and so may need energy reserves such as glycogen to be mobilised more rapidly. The structures of amylose and glycogen are shown in Figure 1.13.

In the human body, glycogen stores are found in the liver and in muscles, where little dark granules of glycogen can often be seen in photomicrographs. They are broken down to form glucose by an enzyme called glycogen phosphorylase, which is activated by the hormone insulin when blood glucose levels are low.

The presence of starch can be indicated using iodine. Iodine molecules can fit neatly inside the coiled molecules of amylose, and this produces a blue-black compound. Glycogen and amylopectin give a faint red colour with iodine.

Cellulose is another polysaccharide made of thousands of glucose molecules, but this time they are β-glucose and they are linked with β 1–4 glycosidic bonds (Figure 1.14). In β-glucose , the −OH group on carbon 1 lies above the ring, so – in order to form a glycosidic bond with carbon 4, where the −OH group is below the ring – one glucose molecule must be 'upside down' relative to its neighbours. Cellulose molecules do not coil, but lie straight. This allows hydrogen bonds to be formed between −OH groups of neighbouring molecules. This produces bundles of molecules lying side by side, all held together by thousands of hydrogen bonds. Although each hydrogen bond is weak, there are so many of them that they collectively form a very strong structure. These bundles are called **fibrils**, and they themselves form larger bundles called **fibres**. You can see these fibres in micrographs of cellulose cell walls (page 46). They are structurally very strong, with high tensile

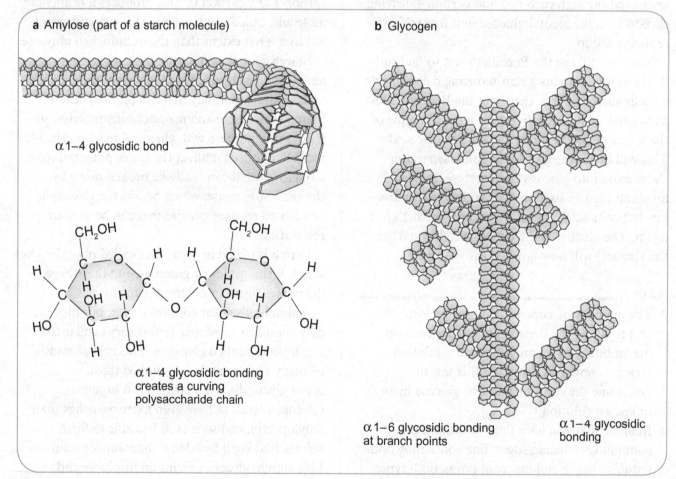

a Amylose (part of a starch molecule)

α 1–4 glycosidic bond

CH₂OH

CH₂OH

α 1–4 glycosidic bonding creates a curving polysaccharide chain

b Glycogen

α 1–6 glycosidic bonding at branch points

α 1–4 glycosidic bonding

Figure 1.13   Amylose (one of the two polysaccharides in starch) and glycogen.

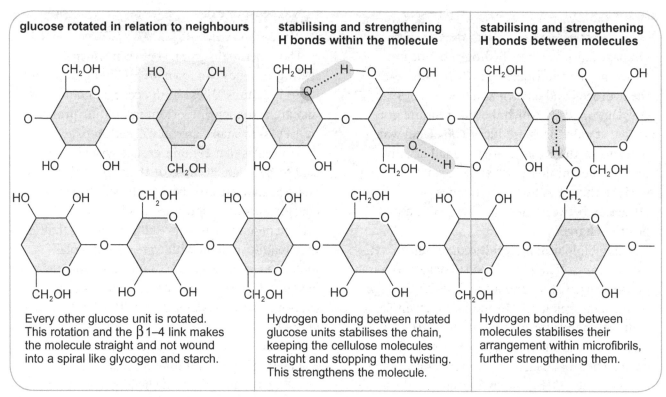

glucose rotated in relation to neighbours	stabilising and strengthening H bonds within the molecule	stabilising and strengthening H bonds between molecules
Every other glucose unit is rotated. This rotation and the β 1–4 link makes the molecule straight and not wound into a spiral like glycogen and starch.	Hydrogen bonding between rotated glucose units stabilises the chain, keeping the cellulose molecules straight and stopping them twisting. This strengthens the molecule.	Hydrogen bonding between molecules stabilises their arrangement within microfibrils, further strengthening them.

Figure 1.14 The structure of cellulose.

strength. This makes cell walls able to expand without breaking when a plant cell takes up water, enabling the cell to become turgid.

Cellulose is difficult to digest, because very few animals have an enzyme that can break its β 1–4 glycosidic bonds. It is therefore a good source of dietary fibre in the human diet, remaining undigested as it passes through the alimentary canal and stimulating peristalsis.

SAQ

5 Copy and complete this table to compare the structures and functions of amylose, cellulose and glycogen.

	Amylose	Cellulose	Glycogen
monosaccharide from which it is formed	α-glucose	β-glucose	α-glucose
type(s) of glycosidic bond			
overall shape of molecule			
hydrogen bonding within or between molecules			
solubility in water			
function			

Hydrogen bonding within biological molecules

On page 2, we saw how hydrogen bonds form between water molecules. They arise because the hydrogen and oxygen atoms in water form covalent bonds in which the electrons are not shared equally. This produces a **dipole**, in which one part of the group carries a small negative charge, and another part carries a small positive charge. A hydrogen bond is the attraction between these small negative and positive charges.

Many biological molecules contain hydroxyl, −OH, groups. These have a small negative charge on the oxygen atom and a small positive charge on the hydrogen atom. In starch, for example, hydrogen bonds form between the −OH groups on carbon 1 of different glucose units in the chain. The small negative charge on the oxygen atom of one −OH group is attracted to the small positive charge on the hydrogen atom of another −OH group. This pulls the molecule into a spiral.

The other main groups between which hydrogen bonds can form are −C=O groups and >N−H groups. These both occur in amino acids, and are extremely important in holding protein molecules in shape, as you will see on page 19.

Molecules that contain groups with dipoles are said to be **polar**. Because of their small electrical charges, they are attracted to the small electrical charges on water molecules. Such molecules can also be described as **hydrophilic**, meaning that they readily interact with water. Small polar molecules, such as glucose, tend to be soluble in water. However, large ones, such as starch, cannot dissolve in water despite the fact that they are polar.

Molecules that do not have dipoles are **nonpolar**. They do not interact readily with water, and so are said to be **hydrophobic**. Lipids (pages 22–24) are important examples of hydrophobic substances.

Proteins

Proteins are substances whose molecules are made of many **amino acids** linked together in long chains. Figure 1.15 shows the structure of an amino acid.

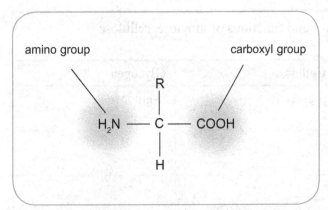

Figure 1.15   The basic structure of an amino acid.

All amino acids have a central carbon atom. Carbon atoms can form four covalent bonds with other atoms. In an amino acid, the central carbon forms one bond with a **carboxyl group**, −COOH.

There is another bond with a hydrogen atom, and a third bond with an **amino group**, −NH$_2$. The fourth bond, however, can be with any one of a whole range of different groups. The letter R is used to show this group. In animals such as humans, there are about 20 different amino acids, each with a different R group.

The amino acids that make up a protein are linked together during protein synthesis, which happens on the **ribosomes** in a cell (page 43). Here, separate amino acids are brought close to each other, and react together in a condensation reaction to form a linkage between them called a **peptide bond**. Figure 1.16 shows how this is done. Peptide bonds are very strong because they involve covalent bonds. As the peptide bond is formed, one hydrogen atom from one amino acid and one hydroxyl group from another amino acid join together to form a water molecule.

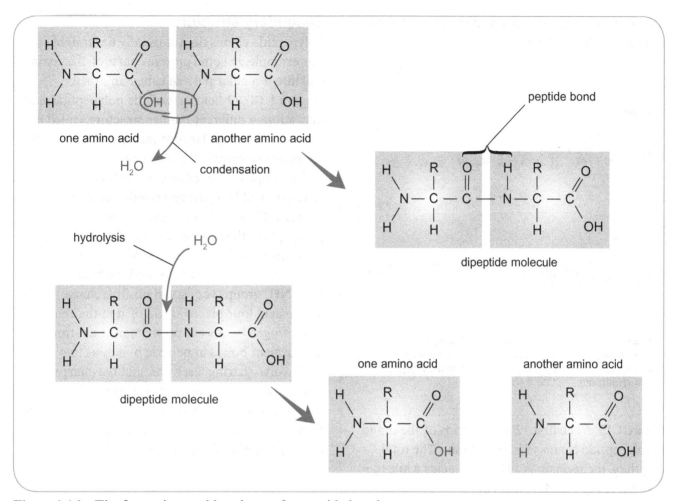

Figure 1.16 The formation and breakage of a peptide bond.

On the ribosome, a long chain of amino acids is formed, all linked together by peptide bonds. This chain is called a **polypeptide**. Protein molecules, as we shall see, contain one or more chains of polypeptides. Haemoglobin, for example, contains four polypeptides all coiled up with each other.

Polypeptides can be broken down by breaking their peptide bonds. This happens, for example, when protein molecules in the food you eat are digested in your stomach by protease enzymes. The reaction is a hydrolysis reaction. In this reaction, combination with a water molecule breaks the peptide bond between two amino acids and separates them (Figure 1.16).

Primary structure

If we look at the different amino acids in polypeptide chains, and the sequence in which they are linked together, we find that, for a particular polypeptide, they are almost exactly the same. This sequence is called the **primary structure** of the molecule.

As there are approximately 20 different amino acids, which can be linked together in any order, and as a polypeptide may contain hundreds of amino acids, there is a nearly infinite number of possible primary structures. A change in just one of the amino acids in the chain makes it a different polypeptide or protein, which may behave in a different way. In a haemoglobin molecule, there are two different types of polypeptide chain, which have slightly different primary structures. Figure 1.17 shows the amino acid sequence of part of each chain, called α- and β-chains.

The primary structure of a protein is determined by a **gene**. A gene is a length of DNA that carries a code determining the sequence in which amino acids will be linked together on a ribosome to form a polypeptide.

Each of the 20 different amino acids has its own name – for example, valine, leucine and cysteine.

Often, their names are abbreviated to three letters, such as val, leu and cys.

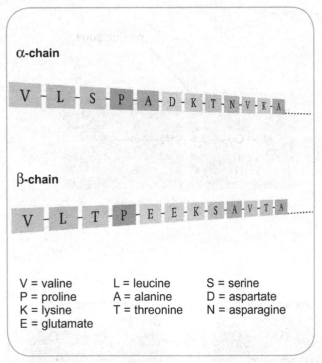

α-chain

V – L – S – P – A – D – K – T – N – V – K – A ……

β-chain

V – L – T – P – E – E – K – S – A – V – T – A ……

V = valine   L = leucine   S = serine
P = proline   A = alanine   D = aspartate
K = lysine   T = threonine   N = asparagine
E = glutamate

Figure 1.17 The first 12 amino acids of the α and β polypeptide chains in haemoglobin. The total length of the α-chain is 141 amino acids while the β-chain is 146 amino acids long.

Secondary structure

Polypeptide chains do not usually lie straight. For example, in some proteins parts of the chain coil into a regular pattern called an **α-helix** (Figure 1.18). Other parts of the polypeptide chain may adopt a different regular structure, called a **β-chain**. These regular arrangements make up the **secondary structure** of the protein.

The shape of the α-helix or the β-chain is maintained by hydrogen bonds that form between different amino acids. In an α-helix in a polypeptide, the hydrogen bonds form between the oxygen of the −CO group of one amino acid (very small negative charge) and the hydrogen of the −NH$_2$ group (very small positive charge) of the amino acid four places ahead of it in the chain.

Hydrogen bonds are nowhere near as strong as the covalent bonds in peptide bonds, as they do not involve sharing electrons, and the charges that attract one another are relatively small. However, if there are a lot of them, as there are in a protein, then between them they can be a major force in holding together the shape of a molecule.

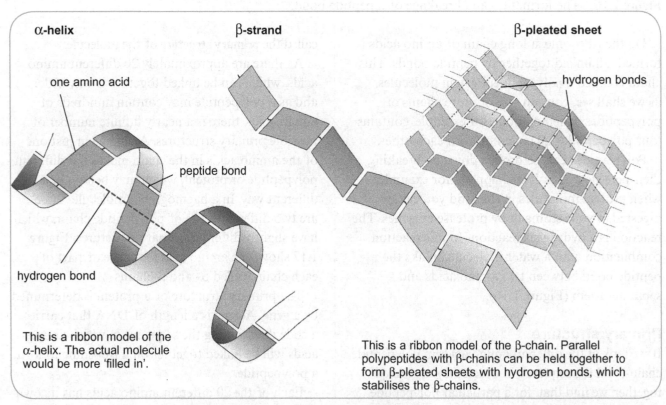

α-helix   β-strand   β-pleated sheet

one amino acid

hydrogen bonds

peptide bond

hydrogen bond

This is a ribbon model of the α-helix. The actual molecule would be more 'filled in'.

This is a ribbon model of the β-chain. Parallel polypeptides with β-chains can be held together to form β-pleated sheets with hydrogen bonds, which stabilises the β-chains.

Figure 1.18 Secondary structure in proteins – the α-helix and β-chain.

The 20 amino acids

Amino acids with side chains (R groups) containing $-NH_2$, $-COOH$ or $-OH$ groups are hydrophilic, and help to make a protein soluble in water. They may be involved in ionic bonding or hydrogen bonding within a protein molecule.

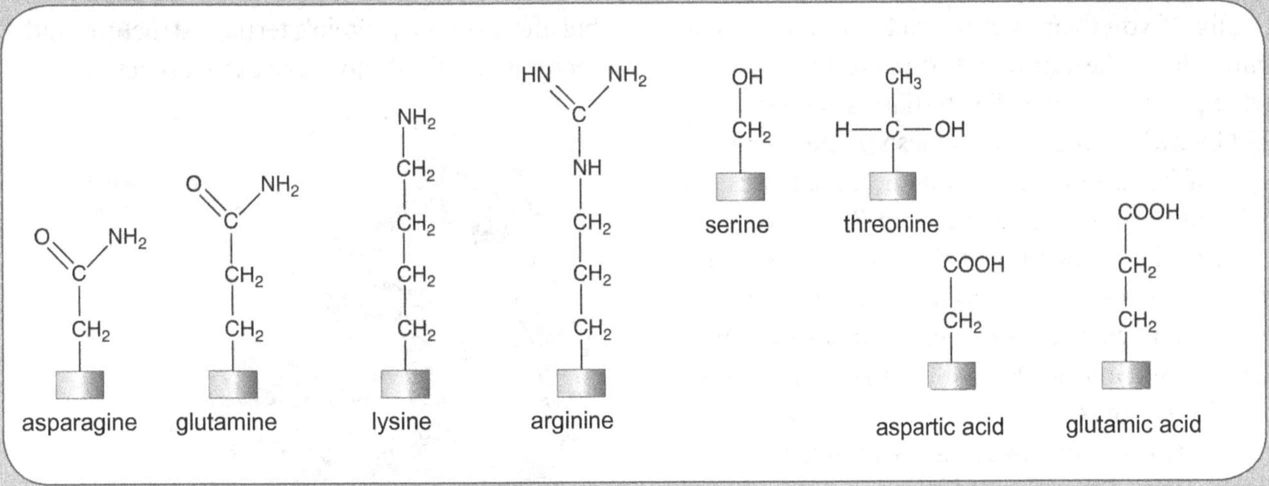

asparagine   glutamine   lysine   arginine   serine   threonine   aspartic acid   glutamic acid

Amino acids with side chains containing ring structures, or with $-CH_3$ groups, are hydrophobic. They may be involved in hydrophobic bonding within a protein molecule.

glycine   alanine   valine   isoleucine   leucine   methionine

phenylalanine   tyrosine   tryptophan   histidine   proline (the whole amino acid)

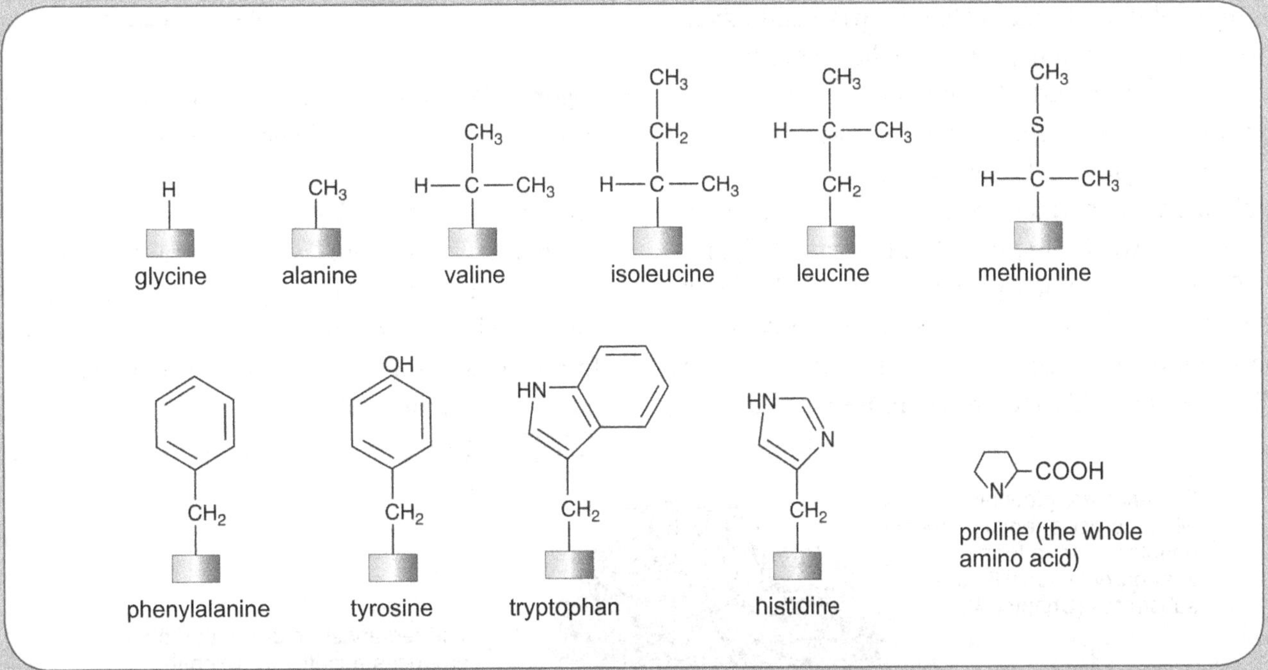

The amino acid cysteine, with an $-SH$ group in its side chain, is involved in forming disulphide bonds within a protein molecule.

cysteine

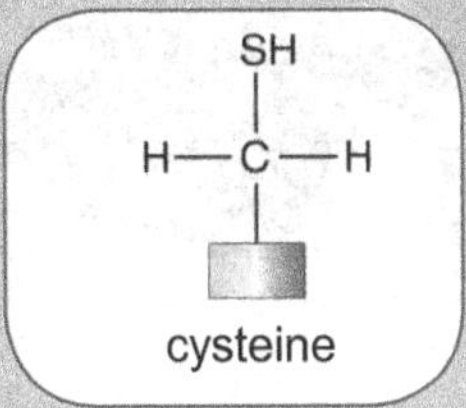

Tertiary structure

The chain can now fold round itself even more. The overall shape formed by this is called the **tertiary structure**. Imagine a curly cable attached to a kettle, for example. The regular coils are equivalent to the secondary structure – like an α-helix. If you then twist the coil into a particular shape, this is the equivalent of the tertiary structure of a protein. The tertiary structure is held by hydrogen bonds, and also by three other types of bond – disulphide bonds, ionic bonds and hydrophobic interactions between R groups.

Figure 1.19 shows the tertiary structure of one of the β polypeptide chains in a haemoglobin molecule. Each of the polypeptide chains winds itself around a little group of atoms with an iron ion (Fe^{2+}) at the centre. This group of atoms is called a **haem** group, and it is essential to the functioning of the haemoglobin molecule in its role as an oxygen transporter. The iron ion in each of the four haem groups in a haemoglobin molecule is able to bond with two oxygen atoms, so that one haemoglobin molecule is able to carry eight oxygen atoms in all. A group such as the haem group, which makes up part of a protein but is not an amino acid, is called a **prosthetic group**.

In haemoglobin, the two α-chains and two β-chains are curled into balls making haemoglobin a **globular protein** (Figure 1.19). Enzymes (Figure 1.20) are also globular proteins. Figure 1.21 shows the different kinds of bond that help to hold a polypeptide in its secondary and tertiary structures. These bonds hold each protein molecule in a very precise shape. This shape determines the function of the molecule.

The places in which the different kinds of bond can form are determined by the amino acid sequence in the protein – its primary structure. Even a change of a single amino acid can have a big effect on the protein's tertiary structure, and therefore its ability to carry out its function.

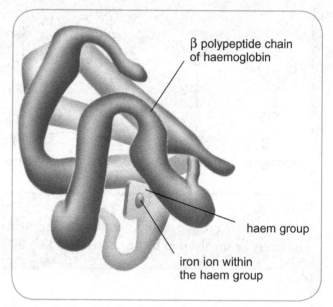

β polypeptide chain of haemoglobin

haem group

iron ion within the haem group

Figure 1.19 Tertiary structure of one β polypeptide chain of haemoglobin.

Quaternary structure

There is still one more level of complexity to come. We mentioned earlier that a haemoglobin molecule is made of four polypeptide chains. These four chains fit together to make the complete haemoglobin molecule.

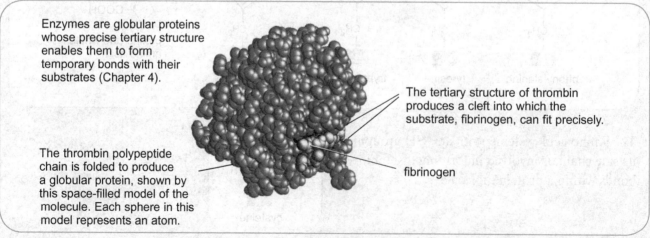

Enzymes are globular proteins whose precise tertiary structure enables them to form temporary bonds with their substrates (Chapter 4).

The tertiary structure of thrombin produces a cleft into which the substrate, fibrinogen, can fit precisely.

The thrombin polypeptide chain is folded to produce a globular protein, shown by this space-filled model of the molecule. Each sphere in this model represents an atom.

fibrinogen

Figure 1.20 The enzyme thrombin (blue) and its substrate fibrinogen (brown) attached to each other.

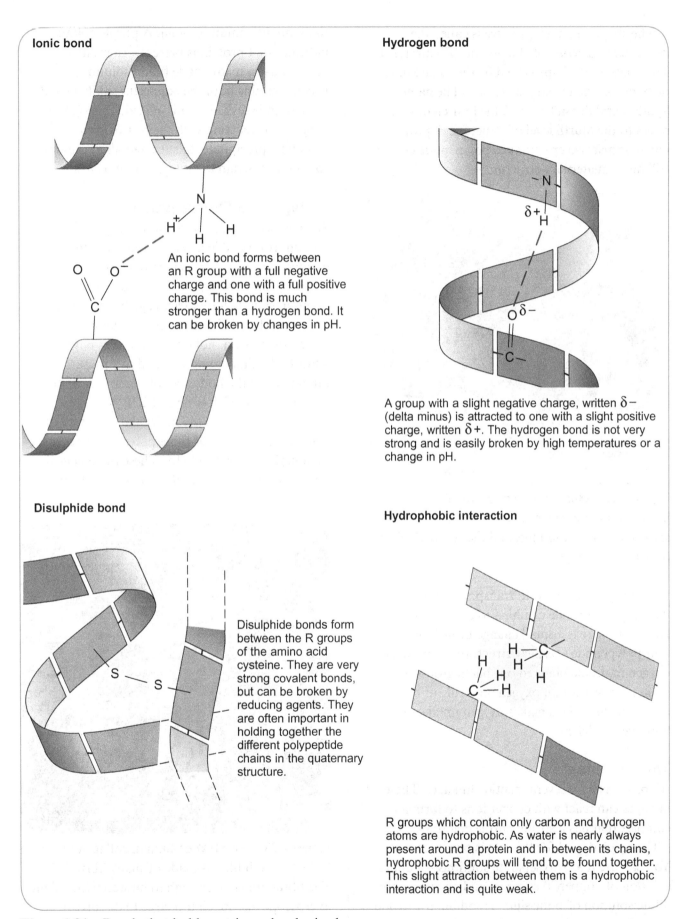

Ionic bond

An ionic bond forms between an R group with a full negative charge and one with a full positive charge. This bond is much stronger than a hydrogen bond. It can be broken by changes in pH.

Hydrogen bond

A group with a slight negative charge, written $\delta-$ (delta minus) is attracted to one with a slight positive charge, written $\delta+$. The hydrogen bond is not very strong and is easily broken by high temperatures or a change in pH.

Disulphide bond

Disulphide bonds form between the R groups of the amino acid cysteine. They are very strong covalent bonds, but can be broken by reducing agents. They are often important in holding together the different polypeptide chains in the quaternary structure.

Hydrophobic interaction

R groups which contain only carbon and hydrogen atoms are hydrophobic. As water is nearly always present around a protein and in between its chains, hydrophobic R groups will tend to be found together. This slight attraction between them is a hydrophobic interaction and is quite weak.

Figure 1.21 Bonds that hold protein molecules in shape.

The shape that they produce is called the **quaternary structure** of the protein (Figure 1.22). The quaternary shape is held by the same kinds of bonds as the tertiary structure. (The name 'quaternary' doesn't refer to the four chains – it refers to the fourth level of structure. A protein made of just two cross-linked polypeptide chains still has a quaternary structure.)

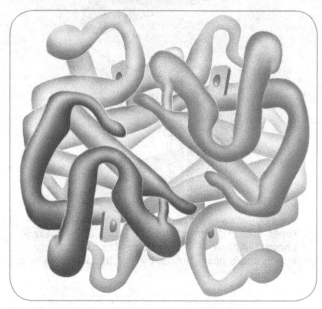

Figure 1.22 Quaternary structure of a haemoglobin molecule. The two α-chains are shown in purple and blue, and the two β-chains in brown and orange.

Haemoglobin is used to transport oxygen. An oxygen molecule can bind to each haem. However, there is a small change in each of the four polypeptide's tertiary structure when the first oxygen molecule binds to one haem group, which makes it easier for an oxygen molecule to bind to the other three. This makes oxygen uptake in the lungs very efficient.

The biuret test

All proteins have several peptide linkages. These linkages can react with copper ions to form a complex that has a strong purple colour.

The reagent used for this test is **biuret** reagent. You can use it as two separate soutions – a dilute solution of copper(II) sulphate and an even more dilute solution of potassium or sodium hydroxide – which you add in turn to the solution that you think might contain protein. A purple colour indicates that protein is present. Alternatively, you can use a ready-made biuret solution, which contains both the copper(II) sulphate and potassium hydroxide ready-mixed. To stop the copper ions reacting with the hydroxide ions and forming a precipitate, this reagent also contains sodium potassium tartrate or sodium citrate.

Collagen – a fibrous protein

We have seen that haemoglobin is a globular protein. It is soluble in water, and is found dissolved in the cytoplasm of red blood cells.

Collagen, however, is a fibrous protein. It is found in skin, tendons, cartilage, bones, teeth and the walls of blood vessels. It is an important structural protein, not only in humans but in almost all animals, and is found in structures ranging from the body wall of sea anemones to the egg cases of dogfish (Figure 1.23).

Figure 1.24 shows the molecular structure of collagen. It consists of three polypeptide chains, each in the shape of a helix. These three helical polypeptides wind around each other to form a 'rope'.

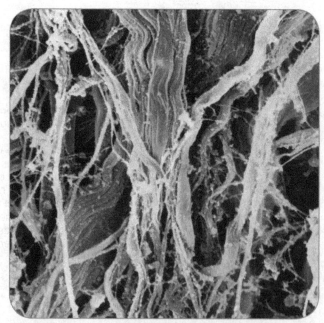

Figure 1.23 An SEM of human collagen fibres (×3000). Each fibre is made of many fibrils. The fibres are large enough to be seen with a light microscope and much too large to dissolve in water.

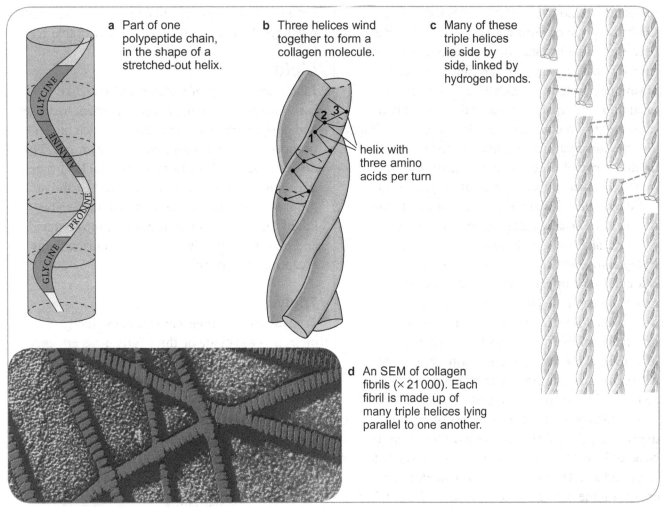

a Part of one polypeptide chain, in the shape of a stretched-out helix.

GLYCINE
ALANINE
PROLINE
GLYCINE

b Three helices wind together to form a collagen molecule.

helix with three amino acids per turn

c Many of these triple helices lie side by side, linked by hydrogen bonds.

d An SEM of collagen fibrils (× 21 000). Each fibril is made up of many triple helices lying parallel to one another.

Figure 1.24 The structure of collagen; the diagrams and photograph begin with the very small and work up to the not-so-small.

Almost every third amino acid in each polypeptide is glycine, the smallest amino acid. Its small size allows the three strands to lie close together and form a tight coil. Any other amino acid would be too large. The three strands are held together by hydrogen bonds.

Each complete, three-stranded molecule of collagen interacts with other collagen molecules running parallel to it. Bonds form between the R groups of lysines in molecules lying next to each other. These cross-links hold many collagen molecules side by side, forming **fibrils**. The ends of the parallel molecules are staggered – if they were not, there would be a weak spot running right across the collagen fibril. Fibrils associate to form bundles called **fibres**. Collagen has tremendous tensile strength – that is, it can resist strong pulling forces. The human Achilles tendon, which is

almost all collagen fibres, can withstand a pulling force of 300 N per mm² of cross-sectional area.

SAQ

6 Which aspects of the structure of collagen, shown in Figure 1.24, are
   a primary structure **b** secondary structure
   c tertiary structure **d** quaternary structure?

A comparison of globular and fibrous proteins

Many – but by no means all – globular proteins are soluble in water, and they are often metabolically active. Enzymes, for example, are globular proteins, as are antibodies and some hormones, including insulin and glucagon. Globular proteins that are insoluble in water include the transporter molecules found in cell membranes (page 59). Globular proteins must have a very precise shape, because this determines their function, and they therefore have a very precise primary structure, always being made of exactly the same sequence of amino acids making up a chain of exactly the same length. In contrast, fibrous proteins may have a rather more variable primary structure, with a limited range of different amino acids that can be joined together to form chains of varying lengths.

Fibrous proteins are not soluble in water, and they are not usually metabolically active. Most of them, like collagen, have a structural role. For example, keratin forms hair and nails, and is found in the upper layers of the skin, which it makes waterproof. The soluble globular protein fibrinogen, found in blood plasma, is converted into the insoluble fibrous protein fibrin when a blood vessel is damaged. The fibrin fibres form a network across the wound, in which platelets can be trapped to form a blood clot.

The solubility of these two types of protein depends on their structure. We have seen that, to be soluble in water, a molecule needs to be not too large and also to have groups with an electrical charge on the outside of the molecule. This makes them able to interact with water molecules (page 5). Globular proteins generally have these features. When their polypeptide chains fold, they do so with R groups carrying charges on the outside of the molecule, and R groups without charges on the inside. Fibrous proteins, in contrast, are generally much too large to be soluble in water (Table 1.3).

Lipids

Lipids are a group of substances that – like carbohydrates – are made up of carbon, hydrogen and oxygen. However, they have a much higher proportion of hydrogen than carbohydrates. They do not have dipoles and so are insoluble in water.

Lipids include fats, which tend to be solid at room temperatures, and oils, which tend to be liquid. In general, animals produce mostly fats and plants mostly oils, although there are many exceptions to this rule.

Triglycerides

Triglycerides get their name because their molecules are made of three **fatty acids** attached to a **glycerol** molecule (Figure 1.25). Fatty acids are acids because they contain a carboxyl group, −COOH.

The carboxyl groups of fatty acids are able to react with the −OH (hydroxyl) groups of glycerol, forming **ester bonds** (Figure 1.26). These linkages involve covalent bonds, and so are very strong. As with the formation of peptide bonds and glycosidic bonds, this is a condensation reaction. The breakage of an ester bond is a hydrolysis reaction.

Triglycerides are insoluble in water. This is because none of their atoms carries an electrical charge, and so they are not attracted to water

	Globular proteins	Fibrous proteins
Examples	haemoglobin, enzymes, antibodies, transporters in membranes, some hormones (e.g. insulin)	collagen, keratin, elastin
Primary structure	very precise, usually made up of a non-repeating sequence of amino acids forming a chain that is always the same length	often made up of a repeating sequence of amino acids, and the chain can be of varying length
Solubility	often soluble in water	insoluble in water
Functions	usually metabolically active, taking part in chemical reactions in and around cells	usually metabolically unreactive, with a structural role

Table 1.3 A comparison of globular and fibrous proteins.

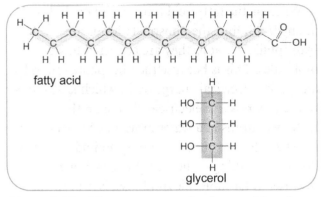

Figure 1.25 The structures of a fatty acid and glycerol.

molecules. They are said to be hydrophobic.

Fatty acids, and the fats in which they are found, can be classified as **saturated** or **unsaturated**. A saturated fat is one in which the fatty acids all contain as much hydrogen as they can. Each carbon atom in the fatty acid 'tail' is linked to its neighbouring carbon atoms by single bonds, while the other two bonds are linked to hydrogen atoms. An unsaturated fat, however, has one or more fatty acids in which at least one carbon atom is using two of its bonds to link to a neighbouring carbon atom, so it only has one bond spare to link to hydrogen (Figure 1.27). This double carbon–carbon bond forms a 'kink' in the chain.

Triglycerides are rich in energy, and they are often used as energy stores in living organisms. One gram of triglyceride can release twice as much energy as one gram of carbohydrate when it is respired, so they make compact and efficient stores. In humans, cells in a tissue called **adipose tissue** are almost filled with globules of triglycerides, and they make very good thermal insulators. Animals

SAQ

7 How might an ester bond be broken, and what is the name of the enzyme that catalyses this reaction?

8 **a** Use Figures 1.25 and 1.26 to work out the molecular formula of one triglyceride.
   b How do the proportions of carbon, hydrogen and oxygen in a triglyceride differ from their proportions in a carbohydrate such as glucose?

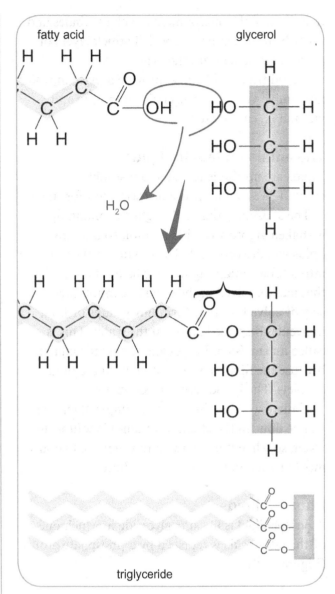

Figure 1.26 How a triglyceride is formed from glycerol and three fatty acids.

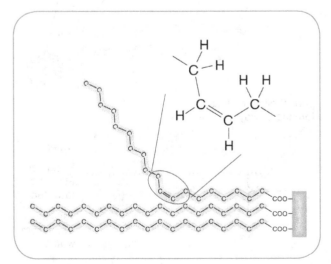

Figure 1.27 An unsaturated fat.

that live in cold environments, such as whales and polar bears, often have especially thick layers of adipose tissue beneath the skin.

Stored triglycerides also provide a place in which fat-soluble vitamins, especially vitamin D and vitamin A, can be stored.

The emulsion test for lipids

Lipids are insoluble in water, but soluble in ethanol. We can make use of this to test for them.

The substance that is thought to contain lipids is shaken vigorously with ethanol, so that any lipids will dissolve in it. The mixture is then poured into a tube containing water. The lipid molecules that had dissolved in the ethanol cannot now stay dissolved, so they form tiny groups of lipid molecules dispersed amongst the water. This is called an **emulsion**. Light cannot pass through it, because the light rays are reflected off the lipid droplets. The liquid looks milky white.

If there were no lipids in the sample, then there will be none in the ethanol. Ethanol is soluble in water, so when it is poured into water the two just mix together and make a clear solution.

Phospholipids

A **phospholipid** is like a triglyceride in which one of the fatty acids is replaced by a phosphate group (Figure 1.28).

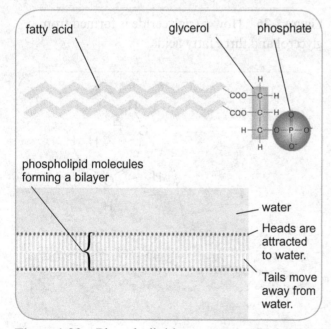

Figure 1.28   Phospholipids.

Whereas the fatty acid tails of a phospholipid are hydrophobic, the phosphate heads are hydrophilic – that is, they are attracted to water molecules. This is because the phosphate group has a negative electrical charge on it, which is attracted to the tiny positive electrical charge on the hydrogen atoms in a water molecule. So, when it is in water, the two ends of a phospholipid molecule do different things. The phosphate is drawn towards water molecules and dissolves in them. The fatty acids are repelled by water molecules and avoid them. In water, the phospholipid molecules arrange themselves in a sheet called a **bilayer**.

A phospholipid bilayer like this is the basic structure of a cell membrane. Phospholipids are one of the most important molecules in a cell. Without them, there could be no plasma membrane and the cell would simply cease to exist.

Cholesterol and steroids

Some people classify **cholesterol** as a lipid, whereas others do not. Here, we will include cholesterol within the group of compounds we call lipids. Figure 1.29 shows the structure of a cholesterol molecule. Cholesterol and other substances with similar structures, which are formed from it, are called **steroids**.

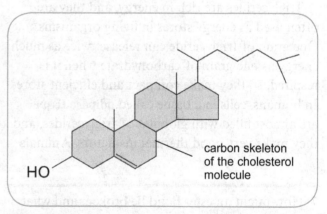

Figure 1.29   The structure of a cholesterol molecule.

There are a huge number of different kinds of steroids in the body. Many of them are hormones – for example, testosterone and oestrogen. Cholesterol itself is a major constituent of cell membranes (Chapter 3), where it helps to regulate the fluidity of the membrane.

Summary

- Water molecules have dipoles, which cause hydrogen bonds to form between them. This gives water a high specific heat capacity, a high latent heat of vaporisation, and a high latent heat of fusion.

- Water is most dense at 4 °C, which ensures that a body of water freezes from the top down, and that there is usually liquid water below the ice.

- Water is a good solvent for small molecules with electrical charges on them, which enables it to provide a suitable medium for metabolic reactions to take place, and to transport substances in solution from one part of an organism's body to another.

- Hydrogen bonding means that water is relatively dense and viscous, has high cohesion and high surface tension. Pure water has a pH of 7, and is transparent. Each of these properties has important implications for living organisms.

- Carbohydrates include sugars (monosaccharides and disaccharides) and polysaccharides.

- Monosaccharides have the general formula $C_nH_{2n}O_n$. They are soluble and taste sweet. All monosaccharides are reducing sugars. In the Benedict's test, they reduce Cu^{2+} ions to Cu^+ ions when heated, giving a brick-red precipitate of copper(I) oxide.

- Monosaccharides can exist in straight chain or ring forms. The ring forms can exist in α or β configurations.

- The monosaccharide glucose is used in respiration to provide energy. It is the form in which carbohydrates are transported in mammals.

- Disaccharides consist of two monosaccharides linked through a glycosidic bond, which is formed by a condensation reaction. Glycosidic bonds are broken by hydrolysis reactions. Disaccharides are soluble and taste sweet.

- The disaccharide sucrose is made up of fructose and glucose linked through an $\alpha 1 - \beta 2$ bond. It is a non-reducing sugar. It is the main form in which carbohydrates are transported in plants. It is a source of energy because it can easily be broken down to monosaccharides that can be used in respiration.

- Polysaccharides have molecules made up of large numbers of monosaccharides linked through glycosidic bonds. Their large size makes them insoluble in water.

- Starch is made up of a mixture of two polysaccharides: amylose and amylopectin. Both contain glucose linked through $\alpha 1–4$ bonds, but amylopectin also contains $\alpha 1–6$ branches. Amylose molecules coil up into tight spirals, held in place by hydrogen bonds. Starch is used as an energy storage compound in plant cells. Glycogen is very similar to amylopectin but with even more branches. It is used as an energy storage compound in animal cells.

- Cellulose is made up of glucose linked through $\beta 1–4$ bonds. This causes alternate glucose units to be inverted in relation to one another, forming straight chains that can link to neighbouring chains by hydrogen bonds. Cellulose therefore forms fibrils, which have high tensile strength and form strong cell walls around plant cells. The $\beta 1–4$ bonds make it difficult to digest.

continued ...

- Proteins are made up of one or more long chains of amino acids linked through peptide bonds. There are twenty different amino acids that occur naturally, and their sequence in a protein is known as its primary structure. This is determined by the genes in the nucleus of the cell.

- The chain of amino acids may coil into a regular shape such as an α-helix or β-pleated sheet, forming the secondary structure of the protein. This in turn may coil further, forming the tertiary structure. Sometimes more than one polypeptide chain is involved in the protein, forming its quaternary structure.

- Haemoglobin has a roughly spherical tertiary (and quaternary) structure, and is an example of a globular protein. Amino acids with dipoles on their R groups tend to be found on the outside of the molecule, making it soluble in water. Each haemoglobin molecule contains four polypeptide chains, each of which has a haem group that can bind reversibly with oxygen.

- Collagen forms long chains made up of three polypeptide chains twisted round one another. It is an insoluble fibrous protein, an important structural component of many animal tissues, such as tendons.

- Protein molecules are held in shape by hydrogen bonds, ionic bonds, disulphide bonds and hydrophobic interactions.

- Lipids include fats and oils. Triglycerides are lipids whose molecules contain glycerol to which three fatty acids are bonded through ester linkages. They are insoluble in water. They are good energy stores, containing twice as much energy per gram as polysaccharides.

- Phospholipids are like triglycerides with one of the fatty acids replaced by a phosphate group. This carries an electrical charge, so this end of the molecule is water-soluble. This causes phospholipids to form bilayers in water, which is the basis of cell membrane structure.

Questions

Multiple choice questions

1 Which statement is **not** a property of water?
   A It is a solvent for hydrophobic compounds.
   B In its pure form, it is transparent to visible wavelengths of light.
   C A relatively large amount of energy is needed to increase the temperature.
   D Water molecules have the ability to form hydrogen bonds with each other.

2 Which one of the following diagrams exemplifies the quaternary structure of a protein?

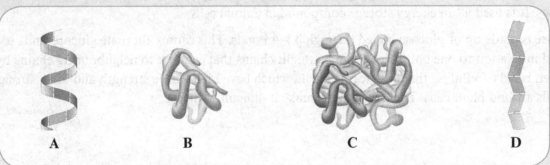

A   B   C   D

continued ...

3 The primary structure of a protein refers to the:

    A presence of alpha-helices or beta-sheets.

    B sequence and number of amino acids.

    C three dimensional folding of the molecule.

    D interactions of protein with prosthetic groups.

4 The diagram below shows a reaction resulting in the formation of a disaccharide.

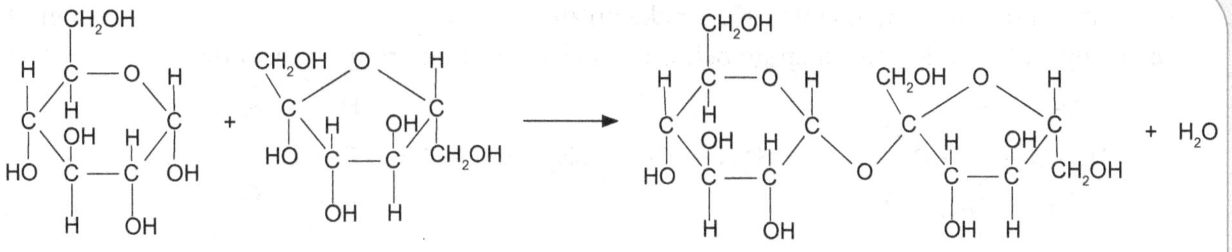

Which of the following correctly identifies molecules **I**, **II** and **III**?

	I	II	III
A	α-glucose	α-fructose	sucrose
B	α-glucose	β-fructose	sucrose
C	β-glucose	β-fructose	maltose
D	β-glucose	α-fructose	maltose

5 Four chemical tests are carried out on four solutions. Each solution contains two substances only. Which solution contains egg albumen and maltose?

	Heated with Benedict's solution	Heated with dilute HCl, neutralised, then heated with Benedict's solution	Mixed with biuret reagents	Mixed with iodine in potassium iodide
A	blue	brick red	blue	blue-black
B	blue	blue	purple	blue-black
C	brick red	brick red	purple	brown
D	blue	blue	purple	blue-black

6 An ester bond is formed between:

    A a carboxyl group and an amino group.

    B a hydroxyl group and a carboxyl group.

    C a hydroxyl group and amino group.

    D an aldehyde group and a carboxyl group.

7 How many molecules of oxygen can a molecule of haemoglobin transport?

    A 1   **B** 2   **C** 3   **D** 4

8 What is the general formula for a monosaccharide?

    A $(CH_2O)_n$ **B** $C(H_2O)_n$ **C** $(C_6H_{10}O_5)_n$ **D** $(C_5H_{10}O_5)_n$

continued ...

Structured questions

9 a The diagram below shows the molecular structure of the amino acid leucine.

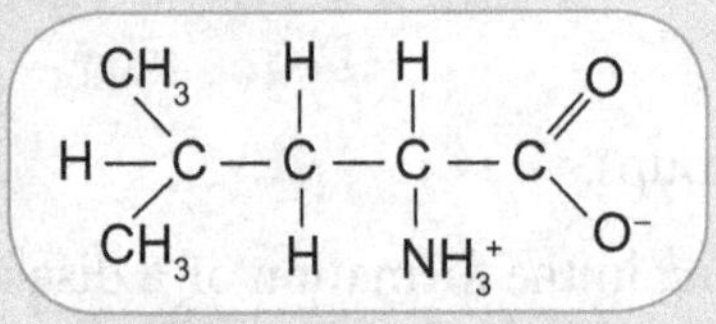

    i All amino acids have the same generalised structure. Copy the diagram above, and identify the components which make up amino acids. [4 marks]

    ii Copy and complete the diagram below to show the formation of a dipeptide.

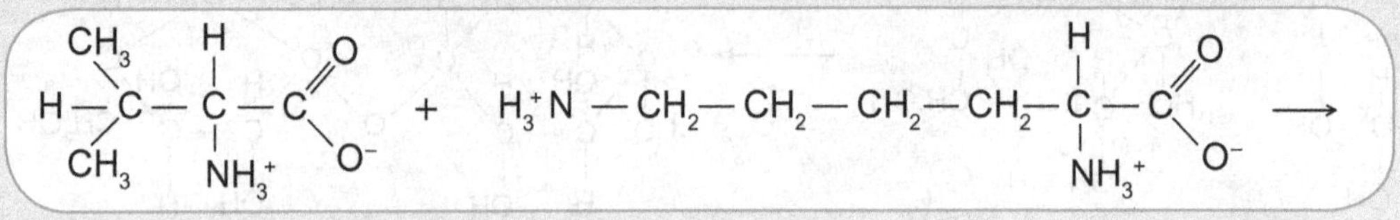

[3 marks]

  b i Proteins are made up of different levels of structure. Copy and complete the table below to explain the different levels of protein structure.

Level	Types of bond	Position of bonds	Effect of bond formation
Primary structure			
Secondary structure			
Tertiary structure			
Quaternary structure			

[3 marks]

    ii Explain the similarity and the difference in hydrogen bonding of the alpha-helix and the beta-pleated sheet secondary protein structure. [2 marks]

  c Proteins have many functions in living organisms. List three of these functions. [3 marks]

10 a i Draw the ring structure of an α-glucose molecule. [3 marks]

    ii How does the β form of this molecule differ in structure? [1 mark]

  b A student was asked to carry out certain food tests on a solution which contained a mixture of biological molecules. She wrote the following table before she started the tests, to help her with the procedure.

Food	Procedure and test reagents				
	Benedict's and heat	**Boil with dilute acid and neutralise**	**Biuret solution**	**Iodine in potassium iodide**	**Ethanol**
Reducing sugar					
Non-reducing sugar					
Starch					
Protein					
Lipid					

    i Copy the above table and complete to show which of these procedures and test reagents apply to the food test carried out on each of the foods listed. Fill in each box using a tick (✔) if it applies or a cross (✗) if it does not. [4 marks]

continued ...

ii The student conducted the tests and wrote the following in her lab report:

Reagent / procedure	Test results
Benedict's and heat	Brick-red precipitate
Biuret	Purple colour

What can she conclude from her results? [1 mark]

c In another experiment, students were given the following materials and apparatus:

Juices from three local fruits, 4% glucose solution, Benedict's solution, test tubes, beakers, syringes, Bunsen, tripod and gauze, stopwatch.

Describe the procedure that would be used to determine the amount of glucose present in each fruit juice in milligrams. [7 marks]

11 a i Sucrose is a disaccharide made up of glucose and fructose.

Show how these two molecules combine to form sucrose. [4 marks]

ii Identify the type of reaction involved in the formation of sucrose. [1 mark]

iii What is the relationship of the structure of sucrose to its function? [2 marks]

iv Sucrose is described as a non-reducing sugar. Explain this statement with reference to its structure. [2 marks]

b Starch (which is made up of amylose and amylopectin), cellulose and glycogen are all polymers of glucose. Copy and complete the table below.

	Starch	Cellulose	Glycogen
Form of glucose			
Bonds between monomers			
Features of molecule			
Function			

Essay questions [6 marks]

12 a 'Water is essential to life itself. Without water, life on Earth would not exist'.

Discuss this statement with reference to the structure and properties of water and explain how these relate to the role that water plays as a medium of life. [9 marks]

b i Haemoglobin is a globular protein. Explain what is meant by the term 'globular'. [2 marks]

ii Describe, with the aid of a diagram, how the structure of haemoglobin helps it to transport oxygen. [4 marks]

13 a i Describe the molecular structure of a starch molecule and identify its features, which make it suitable for storage in cells. [4 marks]

ii Explain how the molecular structure of cellulose differs from the starch molecule. [4 marks]

b Some natural fibres such as cotton and linen are made up of cellulose molecules.

Discuss how the structure of cellulose gives these natural fibres their strength. [3 marks]

c Collagen is a molecule with tensile strength as great as cellulose.

Explain how the structure of collagen relates to its strength. [4 marks]

14 a By means of an annotated diagram, illustrate the molecular structure of a triglyceride which consists of at least one unsaturated fatty acid. [5 marks]

b Explain how the structure of a phospholipid differs from a triglyceride. [3 marks]

c Discuss how the properties of phospholipids influence the formation and function of the cell membrane (see Chapter 3). [7 marks]

Chapter 2
Cell structure

By the end of this chapter you should be able to:

a   make drawings of typical animal and plant cells as seen under the light microscope;

b   describe and interpret drawings and electron micrographs of typical animal and plant cells, recognising the following membrane systems and organelles: rough and smooth endoplasmic reticulum, Golgi body, mitochondria, ribosomes, lysosomes, chloroplasts, cell membrane, nuclear envelope, centrioles, nucleus and nucleolus;

c   explain the differences between electron and light microscopy and between resolution and magnification;

d   outline the functions of the membrane systems and organelles listed in b above;

e   compare the structure of typical animal and plant cells, stressing similarities and differences;

f   describe the structure of a prokaryotic cell and compare the structure of prokaryotic cells with eukaryotic cells, including reference to the endosymbiotic development of eukaryotic cells;

g   explain the concepts of tissue and organ using as an example the dicotyledonous root;

h   make plan drawings to show the distribution of tissues within an organ, such as the dicotyledonous root, to include parenchyma, xylem and phloem.

All living organisms are made of cells. Cells are the basic units of living things, and most scientists would agree that anything that is not made of a cell or cells – for example, a virus – cannot be a living organism.

Some organisms, such as bacteria, have only one cell, and are said to be **unicellular**. Others have millions of cells. Any organism that is made up of more than one cell is said to be **multicellular**.

All cells are very small, but some are just large enough to be seen with the naked eye. The unicellular organism *Amoeba*, for example, can just be seen as a tiny white speck floating in liquid if you shake up a culture of them inside a glass vessel. These cells are about 0.1 mm across, which is unusually large. Human cells are usually somewhere between 10 μm and 30 μm in diameter (see Units of measurement box on page 32 for an explanation of μm). Bacterial cells are much smaller, often about 0.5 μm across. To see most cells, a microscope must be used.

Microscopes

The first microscopes were invented in the mid 17th century. They opened up a whole new world for biologists to study. Now biologists could see tiny, unicellular organisms whose existence had previously only been guessed at. They could also see, for the first time, that large organisms such as plants and animals are made up of cells.

Light microscopes

The early microscopes, like the microscopes that you will use in the laboratory, were **light microscopes**. Light microscopes use glass lenses to refract (bend) light rays and produce a magnified image of an object. Figure 2.1 shows how a light microscope works.

The specimen to be observed usually needs to be very thin, and also transparent. To keep it flat, it is usually placed on a glass slide with a very thin glass coverslip on top. For a temporary slide,

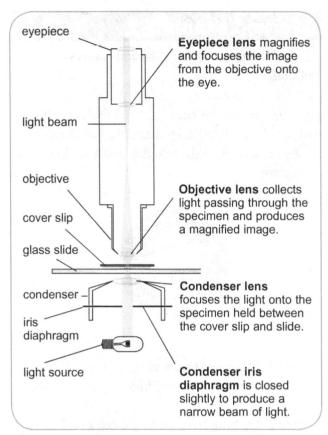

eyepiece

Eyepiece lens magnifies and focuses the image from the objective onto the eye.

light beam

objective

Objective lens collects light passing through the specimen and produces a magnified image.

cover slip

glass slide

condenser

Condenser lens focuses the light onto the specimen held between the cover slip and slide.

iris diaphragm

light source

Condenser iris diaphragm is closed slightly to produce a narrow beam of light.

Figure 2.1   How a light microscope works.

you can mount the specimen in a drop of water. To make a permanent slide, a liquid that solidifies to produce a clear solid is used to mount the specimen.

The slide is placed on a stage through which light shines from beneath. The light is focused onto the specimen using a **condenser lens**. The light then passes through the specimen and is captured and refracted by an **objective lens**. Most microscopes have three or four different objective lenses, which provide different fields of view and different magnifications. The greater the magnification, the smaller the field of view.

The light rays now travel up to the **eyepiece lens**. This produces the final image, which falls onto the retina of your eye. The image can also be captured using a digital camera or video camera, and viewed or projected onto a screen.

Many biological specimens are colourless when they have been cut into very thin sections, so a **stain** is often added to make structures within the specimen easier to see. Different parts of a cell, or different kinds of cells, may take up (absorb) a

stain more than others. For example, a stain called methylene blue is taken up more by nuclei than by cytoplasm, so it makes a nucleus look dark blue while the cytoplasm is pale blue. Methylene blue is taken up by living cells, but many other stains cannot get through the cell membrane of a living cell and can only be used on dead cells.

Magnification

Using a microscope, or even just a hand lens, we can see biological objects looking much larger than they really are. The object is **magnified**. We can define magnification as the size of the image divided by the real size of the object.

$$\text{magnification} = \frac{\text{size of image}}{\text{real size of object}}$$

For example, we can calculate the magnification of the drawing of a spider in Worked example 1.

Worked example 1

Calculation of the magnification of a drawing

$$\text{magnification} = \frac{\text{size of image}}{\text{real size of object}}$$

Below is a 'real' spider and a drawing of this spider.

Step 1 Measure the length of the 'real' spider. You should find that it is 10 mm long. The length of the spider in the drawing is 30 mm.

Step 2 Now, substitute these numbers into the equation above:

$$\text{magnification} = \frac{30}{10} = \times 3$$

Notice the '×' sign in front of the number 3. This stands for 'times'. We say that the magnification is 'times 3'.

When calculating magnifications, it is important to make sure all your measurements are in the same units. It is often best to convert everything into µm before you begin your calculation, as shown in Worked example 2.

Units of measurement

In biology, we often need to measure very small objects. When measuring cells or parts of cells, the most common (and useful) unit is the **micrometre**, written µm for short. The symbol µ is the Greek letter mu. One micrometre is one thousandth of a millimetre. (A micrometre is also known as a micron.)

$$1\,\mu m = \frac{1}{1000}\,mm$$

This can also be written $1 \times 10^{-3}\,mm$, or $1 \times 10^{-6}\,m$.

Even smaller structures, such as the organelles within cells, are measured using even smaller units. These are **nanometres**, written **nm** for short. One nanometre is one thousandth of a micrometre.

$$1\,nm = \frac{1}{1000}\,\mu m$$

This can also be written $1 \times 10^{-6}\,mm$, or $1 \times 10^{-9}\,m$.

SAQ

1 This is a photomicrograph – a photograph taken using a light microscope. The actual maximum diameter of the cell is 50 µm. Calculate the magnification of the photograph.

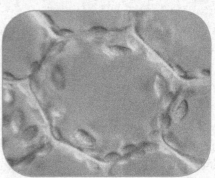

Worked example 2

Calculation of magnification and conversion of units

Let us say that we know that the real diameter of a red blood cell is 7 µm and we have been asked to calculate the magnification of this diagram.

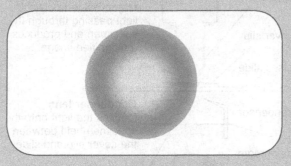

Step 1 Measure the diameter of the cell in the diagram. You should find that it is 30 mm.

Step 2 We have been given the cell's real size in µm, so we need to convert the 30 mm to µm. There are 1000 µm in 1 mm, so 30 mm is $30 \times 1000\,\mu m$.

Step 3 Now we can put the numbers into the equation:

$$magnification = \frac{size\ of\ image}{real\ size\ of\ object}$$

$$= \frac{30 \times 1000}{7}$$

$$= \times 4286$$

SAQ

2 A person makes a drawing of an incisor tooth. The width of the actual tooth is 5 mm. The width of the tooth in the drawing is 12 mm. Calculate the magnification of the drawing.

Worked example 3

Calculating magnification from a scale bar

This diagram shows a lymphocyte.

6 μm

We can calculate the magnification of the image of the lymphocyte without needing to measure it or know anything about its original size. We can simply use the **scale bar**. All you need to do is measure the length of the scale bar and then substitute its measured length and the length that it represents into the equation. (Remember to convert your measurement to μm.)

Step 1 Measure the scale bar. Here, it is 36 mm.

Step 2 Substitute into the equation.

$$\text{magnification} = \frac{\text{size of image}}{\text{real size of object}}$$

$$\text{magnification} = \frac{\text{the length of the scale bar}}{\text{the length the scale bar represents}}$$

$$= \frac{36 \times 1000 \, \mu m}{6 \, \mu m}$$

$$= \times 6000$$

SAQ

3 This is a photomicrograph of a transverse section through a leaf. Use the scale bar to calculate the magnification of the photomicrograph.

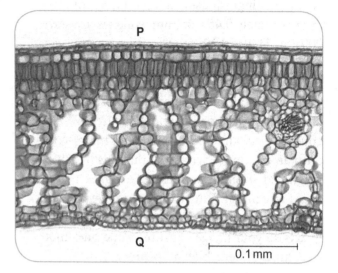

P

Q

0.1 mm

4 If we know the magnification, we can turn the equation around so that we can calculate the real size of something from its magnified image.

$$\text{real size of object} = \frac{\text{size of image}}{\text{magnification}}$$

Use your value for the magnification of the photomicrograph above to calculate the thickness of the leaf between **P** and **Q**.

Resolution

Light microscopes have one major disadvantage. They are unable to show objects that are smaller than about 200 nm across ($1 \, nm = \frac{1}{1000} \, \mu m$).

You might just be able to pick out such a structure, but it would appear only as a shapeless blur.

The degree of detail that can be seen in an image is known as the **resolution**. The tinier the individual points of information on an image – for example, the pixels on a monitor – the better the resolution. To see the very smallest objects, you need a microscope with very high resolution.

The absolute limit of resolution of a microscope is determined by the wavelength of the radiation that it uses. As a rule of thumb, the limit of resolution is about 0.45 times the wavelength. Shorter wavelengths give the best resolution.

The shortest wavelength of visible light is blue light, and it has a wavelength of about 450 nm. So the smallest objects we can expect to be able to distinguish using a light microscope are approximately 0.45 × 450 nm, which is around 200 nm. This is the best resolution we can ever expect to achieve using a light microscope. In practice, it is never quite as good as this.

It's important to understand that resolution is not the same as magnification (Figure 2.2). You could project an image from a light microscope onto an enormous screen, so that it is hugely magnified. There is no limit to how much you could magnify it. But your huge image will just look like a huge blur. There won't be any more 'pixels' in your image – just the same ones that were always there, blown up larger.

Electron microscopes

Light is part of the electromagnetic spectrum. To get around the limit of resolution imposed by the use of light rays, we can use a different type of wave with a shorter wavelength.

Electron microscopes use beams of electrons instead of light rays (Figure 2.3 and Figure 2.4). Electron beams have much shorter wavelengths than light rays. They therefore have much higher resolution, typically about 400 times better than a light microscope. Using an electron microscope, we can distinguish objects that are only 0.5 nm apart. This means that we can magnify things much more than with a light microscope and still obtain a clear image. With a light microscope, because of the relatively poor resolution, it is only useful to magnify an image up to about 1400 times. With an electron microscope, images remain clear up to a magnification of about 300 000 times.

Some electron microscopes work in a similar way to a light microscope, passing electrons

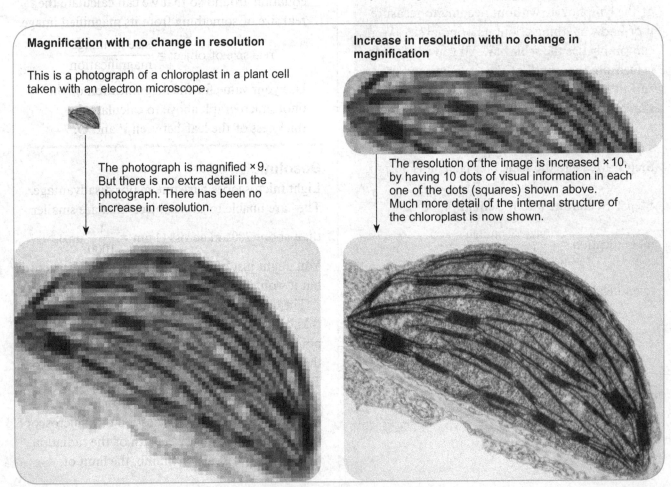

Magnification with no change in resolution

This is a photograph of a chloroplast in a plant cell taken with an electron microscope.

The photograph is magnified × 9. But there is no extra detail in the photograph. There has been no increase in resolution.

Increase in resolution with no change in magnification

The resolution of the image is increased × 10, by having 10 dots of visual information in each one of the dots (squares) shown above. Much more detail of the internal structure of the chloroplast is now shown.

Figure 2.2 The difference between magnification and resolution.

through a thin specimen. They are called **transmission electron microscopes** and produce images like the one in Figure 2.12 on page 40.

As with light microscopes, the specimens to be viewed need to be very thin, and to be stained so that the different parts show up clearly in the image that is produced. In electron microscopy, the 'stains' are usually heavy metals, such as lead or osmium. Ions of these metals are taken up by some parts of the cells more than others.

The ions are large and positively charged. The negatively charged electrons do not pass through them, and so do not arrive on the screen. The screen therefore stays dark in these areas, so the structures that have taken up the stains look darker than other areas.

Scanning electron microscopes work by bouncing electron beams off the surface of an object (Figure 2.4). They give a three-dimensional image, like the one in Figure 2.5. A scanning electron microscope can provide images that can be usefully magnified to almost the same extent as an image from a transmission electron microscope.

The original images produced by an electron microscope are in black, white and grey only, but false colours are often added using a computer, to make the images look more eye-catching and to help non-specialists to identify the different structures that are visible.

Table 2.1 compares light and electron microscopy.

Figure 2.4   These insects are being prepared for viewing in a scanning electron microscope, by having a thin, even layer of gold spattered over them. Gold has large atoms from which electrons will bounce off, giving a clear image of the surface of the insects' bodies.

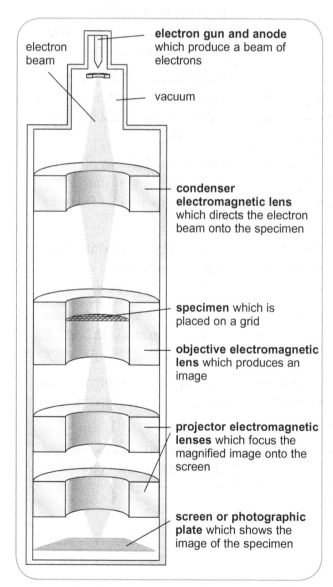

electron beam

electron gun and anode which produce a beam of electrons

vacuum

condenser electromagnetic lens which directs the electron beam onto the specimen

specimen which is placed on a grid

objective electromagnetic lens which produces an image

projector electromagnetic lenses which focus the magnified image onto the screen

screen or photographic plate which shows the image of the specimen

Figure 2.3   How an electron microscope works.

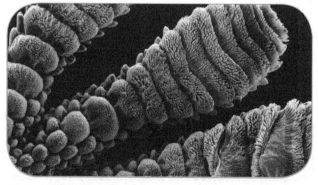

Figure 2.5   Coloured scanning electron micrograph (SEM) of the underside of a gecko's foot (*Tarentola mauritanica*). The foot is covered with ridges and microscopic hairs, which enable the gecko to cling to very smooth surfaces such as windows and ceilings.

Feature	Light microscopy	Electron microscopy
type of radiation used	visible light, wavelength 400 nm to 700 nm	electron beam, wavelength less than 1 nm
maximum resolution (approximate)	200 nm	0.5 nm
maximum useful magnification (approximate)	×1400	×300 000
type of lens (i.e. how the rays of radiation are bent)	glass lenses	electromagnets
how specimen is prepared	some types of light microscopy can use living cells, tissues or organisms; non-living cells and tissues can be cut very thinly and mounted in a transparent liquid on a slide	cells are always dead; for transmission electron microscopes they are cut extremely thinly; for scanning electron microscopy a surface is prepared; the specimen is dehydrated and then placed in a vacuum within the microscope
how specimen is stained	coloured stains that are absorbed by different parts of a cell, or by different types of cell (e.g. methylene blue, eosin)	heavy metals that are absorbed by different parts of a cell, or by different types of cell (e.g. gold, osmium)
how image is viewed	image can be viewed directly, using the eye, or directly projected onto a screen or fed to a monitor	image is produced by allowing the electrons to fall onto a fluorescent screen, or fed to a monitor
appearance of image	image is in colour, which can be the colour of the specimen, or the colours of the stains	image is black and white; false colours can be added using computer imaging software
distortion	material can be prepared for viewing without a high risk of distortion	staining techniques and effect of vacuum increase the risk of distortion
ease of use	relatively cheap to purchase, and can be used by a person after only a little training	very expensive to purchase and maintain; thorough training required
permanence of specimen	well-prepared slides can last for years	specimen deteriorates during viewing and cannot be kept

Table 2.1   Comparison of light and electron microscopy.

Cells

Appearance of cells seen with a light microscope

You are probably already familiar with the structure of animal and plant cells, as they are seen when we use a light microscope. Figure 2.6 is a photomicrograph of an animal cell, and Figure 2.8 is a photomicrograph of a plant cell.

Figure 2.7 is a diagram showing the structures that are visible in an animal cell using a light microscope, and Figure 2.9 is a similar diagram of a plant cell.

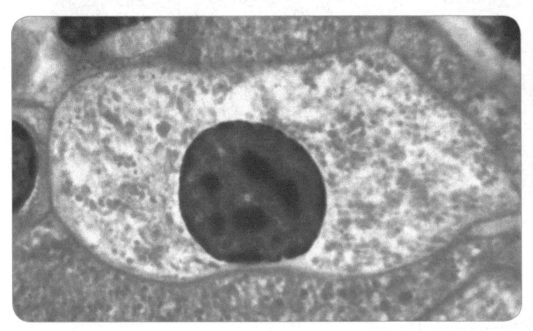

Figure 2.6 Photomicrograph of a stained animal cell (×1800).

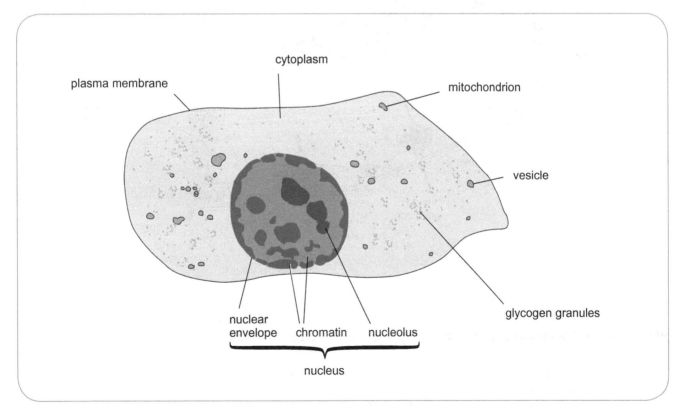

Figure 2.7 A diagram of an animal cell as it appears using a light microscope.

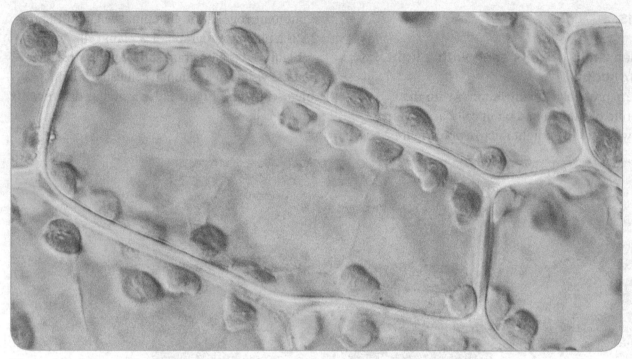

Figure 2.8 Photomicrograph of a cell in a moss leaf (×750).

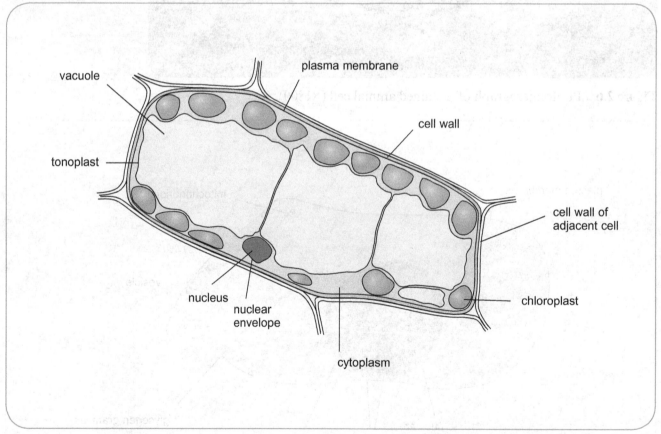

Figure 2.9 A diagram of a plant cell as it appears using a light microscope.

Appearance of cells seen with an electron microscope

As we have seen, electron microscopes are able to resolve much smaller structures than light microscopes. The structure that we can see when we use an electron microscope is called **ultrastructure**.

Figure 2.10 and Figure 2.11 are stylised diagrams summarising the ultrastructure of a typical animal cell and a typical plant cell. Figure 2.12 and Figure 2.14 are electron micrographs of an animal cell and a plant cell. Figure 2.13 and Figure 2.15 are diagrams based on these electron micrographs.

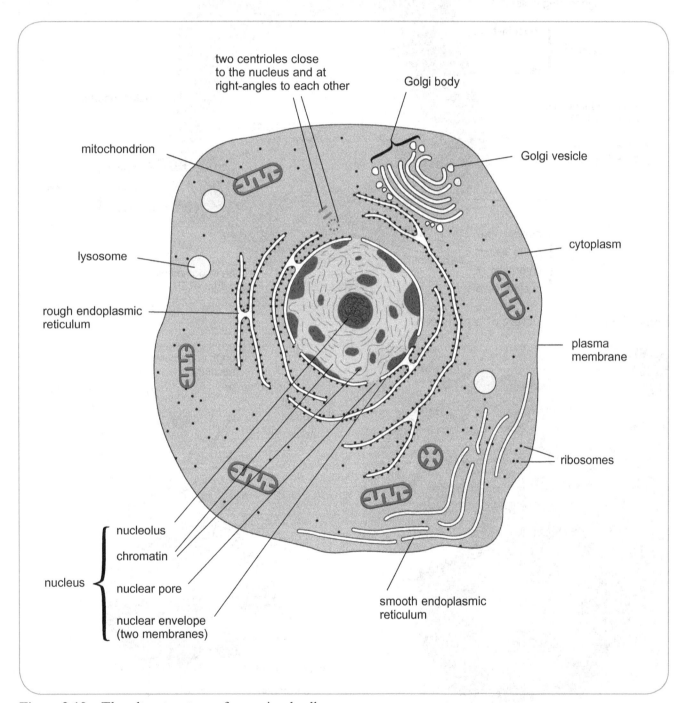

Figure 2.10   The ultrastructure of an animal cell.

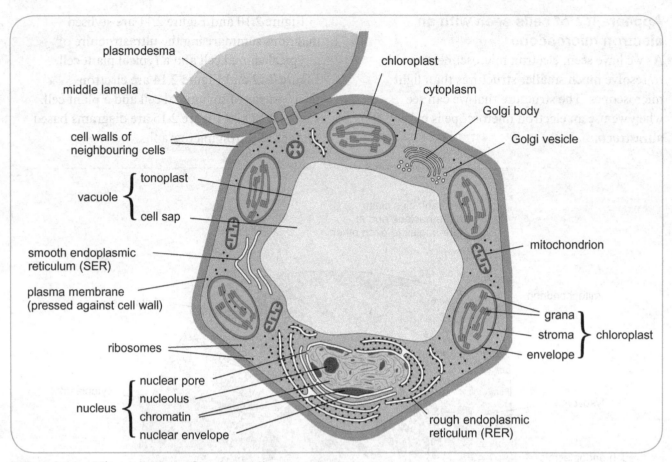

Figure 2.11 Ultrastructure of a plant cell.

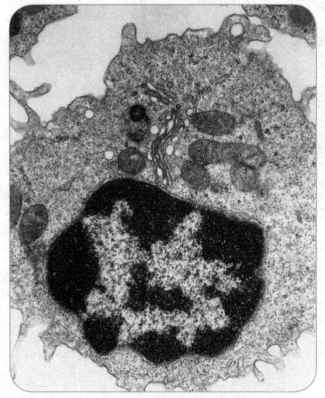

Figure 2.12 Transmission electron micrograph (TEM) of a white blood cell (×15 000).

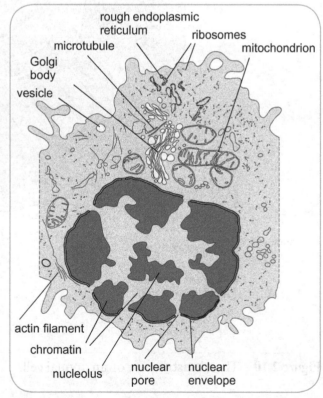

Figure 2.13 Drawing of an animal cell made from the electron micrograph in Figure 2.12.

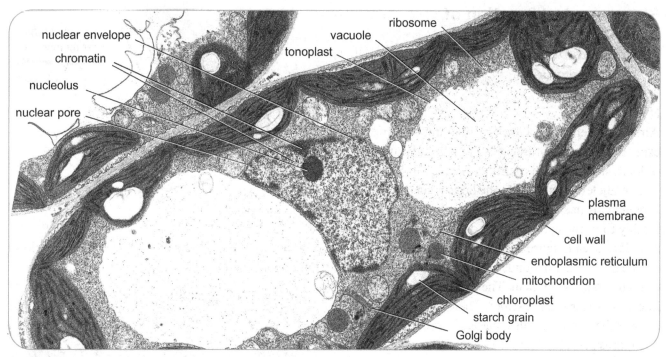

Figure 2.14 Electron micrograph of a plant cell (× 5600).

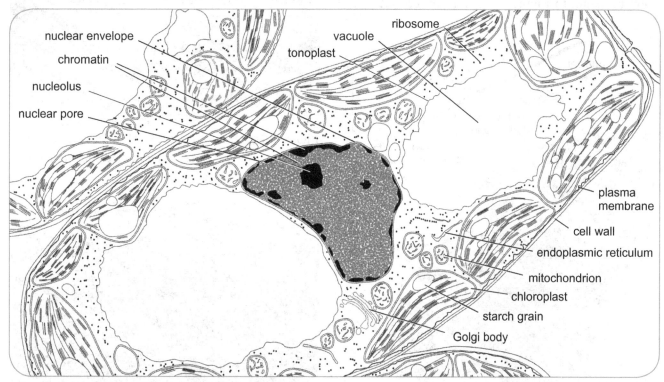

Figure 2.15 Drawing of a plant cell made from the electron micrograph in Figure 2.14.

5 State whether the electron micrograph in Figure 2.14 was made using a transmission electron microscope (TEM) or a scanning electron microscope (SEM). How can you tell?

6 Make a list of all the structures within a cell that are visible with an electron microscope but cannot be clearly seen with a light microscope.

Structure and function of organelles

The different structures that are found within a cell are known as **organelles**.

Nucleus

Almost all cells have a **nucleus**. Two important exceptions are red blood cells in mammals, and phloem sieve tubes in plants.

The nucleus is normally the largest cell organelle. It has a tendency to take up stains more readily than the cytoplasm, and so usually appears as a dark area (Figure 2.16).

The nucleus is surrounded by two membranes with a small gap between them. The pair of membranes is known as the **nuclear envelope**. There are small gaps all over the envelope, called **nuclear pores**.

Figures 2.17, 2.18 and 2.19 show a small part of the nuclear envelope and some nuclear pores in detail. In Figure 2.17, you can see that

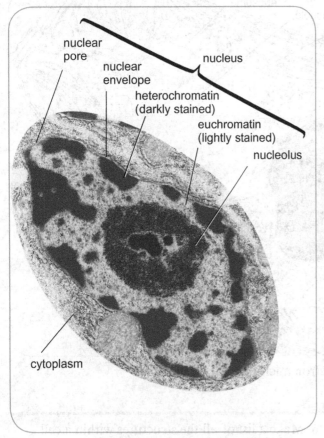

Figure 2.16   Coloured transmission electron micrograph (TEM) of a nucleus (× 10 000).

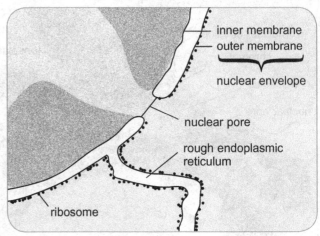

Figure 2.17   Detail of part of the nuclear envelope and a nuclear pore.

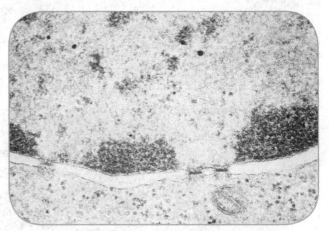

Figure 2.18   Coloured TEM of part of a nuclear envelope and two pores. The nucleus is above the envelope (× 18 700).

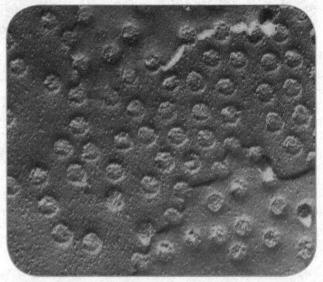

Figure 2.19   Coloured SEM of a surface view of a nuclear envelope, showing a cluster of nuclear pores (× 24 000).

the outer membrane of the nuclear envelope is directly joined to the rough endoplasmic reticulum (described in the next section) and has ribosomes attached to it. The inner and outer membranes of the nuclear envelope are connected at each nuclear pore. The pores prevent material moving freely between the nucleus and cytoplasm. They do allow the movement of some substances in each direction – for example, nucleotides and ATP from the cytoplasm into the nucleus, and messenger RNA from the nucleus into the cytoplasm.

The nucleus contains **chromosomes**. Chromosomes are long molecules of DNA. In a non-dividing cell, they are too thin to be visible as individual chromosomes, but form a tangle known as **chromatin**. Heterochromatin is usually very darkly stained whereas euchromatin is less darkly stained. Chromatin contains many proteins, such as histones (page 48), as well as DNA.

DNA carries a code that instructs the cell about making proteins, and the DNA in euchromatin can be used for transcription, the first stage of protein synthesis. During transcription, the information on DNA is copied onto molecules of messenger RNA (mRNA), which travel out of the nucleus, through the nuclear pores, into the cytoplasm.

An especially darkly staining area in the nucleus, the **nucleolus**, contains DNA that is being used to make ribosomal RNA (rRNA), a component of **ribosomes**, the tiny organelles where protein synthesis takes place.

Endoplasmic reticulum

Within the cytoplasm of every eukaryotic cell, there is a network of membranes, known as the **endoplasmic reticulum**. Some of these membranes have ribosomes attached to them, forming **rough endoplasmic reticulum**, RER for short (Figure 2.20). Some do not, and these form **smooth endoplasmic reticulum**, SER. The RER is usually continuous with the nuclear envelope.

The enclosed spaces formed by the membranes are called **cisternae**. The membranes keep these spaces isolated from the cytoplasm.

RER is where most protein synthesis takes place. Protein synthesis happens in the ribosomes that are attached to the membranes. As the

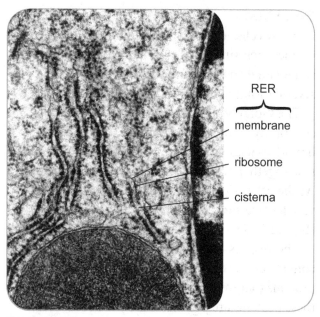

Figure 2.20 TEM showing endoplasmic reticulum (×40 000).

protein molecules are made, they collect inside the cisternae. From here, they can be transported to other areas in the cell – to the Golgi body, for example (Figure 2.21).

SER synthesises lipids, but has different roles in different cells. For example, in cells in the ovary and testis it is the site of production of steroid hormones such as oestrogen and testosterone. In liver cells, it is the place where toxins are broken down and made harmless.

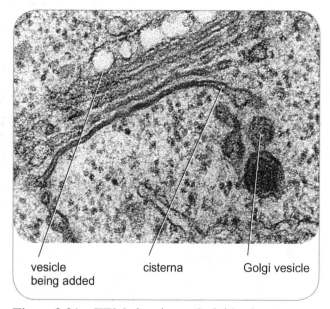

Figure 2.21 TEM showing a Golgi body (×35 000).

Golgi body

In many cells, a stack of curved membranes is visible, enclosing a series of flattened sacs. This is the **Golgi body** (Figure 2.21). Some cells have several Golgi bodies.

The Golgi body is not a stable structure; it is constantly changing. At one side, tiny membrane-bound **vesicles** move towards the Golgi body and fuse together, forming a new layer to the stack. At the other side, the sacs break down, forming vesicles that move away from the Golgi body (Figure 2.22).

The vesicles that fuse with the Golgi body have come from the endoplasmic reticulum. They contain proteins that were made there. In the Golgi body, these proteins are packaged and processed – for example, by adding sugars to a protein to make a glycoprotein.

Some of the processed proteins are then transported, in the vesicles that bud off from the Golgi body, to the plasma membrane. Here, the vesicles fuse with the membrane and deposit the proteins outside the cell, in a process called exocytosis. The production of useful substances in a cell and their subsequent release from it is called **secretion**.

Some vesicles, however, remain in the cell. Some of these contain proteins that function as digestive enzymes, and such vesicles are called **lysosomes**.

Lysosomes

Lysosomes are tiny bags of digestive enzymes. They are surrounded by a single membrane. They are usually about $0.5\,\mu m$ in diameter. Their main function is to fuse with other vesicles in the cell that contain something that needs to be digested – for example, a bacterium which has been brought into the cell by endocytosis (Chapter 3). They also help to destroy worn-out or unwanted organelles within the cell. The enzymes in the lysosome break down the large molecules in the bacterium or organelle, producing soluble substances that can disperse into the cytoplasm. The head of a sperm cell contains a special type of lysosome, called an **acrosome**, whose enzymes digest a pathway into an egg just before fertilisation takes place.

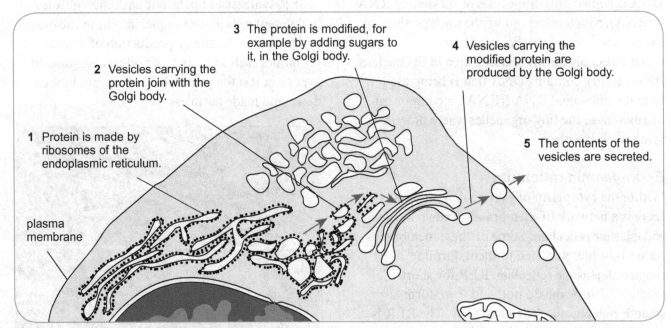

3 The protein is modified, for example by adding sugars to it, in the Golgi body.

2 Vesicles carrying the protein join with the Golgi body.

4 Vesicles carrying the modified protein are produced by the Golgi body.

1 Protein is made by ribosomes of the endoplasmic reticulum.

5 The contents of the vesicles are secreted.

plasma membrane

Figure 2.22   Function of the Golgi body.

Chloroplasts

Chloroplasts are found in some plant cells, but never in animal cells (Figure 2.23 and Figure 2.24). They are the site of photosynthesis.

A chloroplast has a double membrane, called an **envelope**, surrounding it. These membranes isolate the reactions that take place inside the chloroplast from the rest of the cell.

Inside the chloroplast, there are membrane-bound spaces called **thylakoids**. In places, the thylakoids form stacks called **grana** (singular: granum). The thylakoids contain chlorophyll, and this is where the light-dependent reactions of photosynthesis take place. In these reactions, light energy is captured by chlorophyll and used to split water molecules to provide hydrogen ions, which are then used to make ATP and a substance called reduced NADP. The ATP and reduced NADP are then used to make carbohydrates, using carbon dioxide from the air, in the light-independent reactions. The light-independent reactions take place in the 'background material' of the chloroplast, which is called the **stroma**.

Chloroplasts often contain **starch grains**. Starch is a carbohydrate that is used as an energy store in plants.

Mitochondria

Mitochondria are found in both plant and animal cells. Like chloroplasts, they are surrounded by a double membrane, also known as an envelope (Figure 2.25 and Figure 2.26).

Mitochondria are the site of **aerobic respiration** in a cell. Here, oxygen and energy-containing molecules produced from glucose are used to make **ATP**. ATP is the energy currency of a cell, necessary for every energy-using activity that it carries out. Each cell has to make its own ATP. Cells that use a lot of energy, such as muscle cells, therefore contain a lot of mitochondria.

The inner membrane of the mitochondrion is folded to form **cristae**. Here, ATP is made in a process that has many similarities with the production of ATP on the membranes inside a

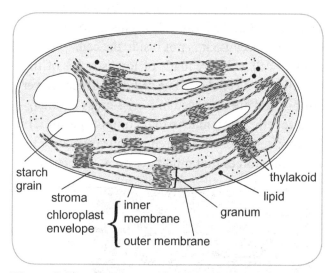

Figure 2.23 Structure of a chloroplast.

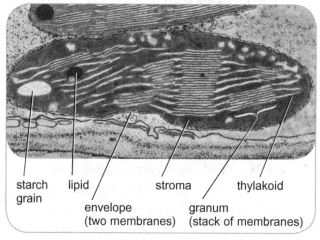

Figure 2.24 TEM of a chloroplast (×27 000).

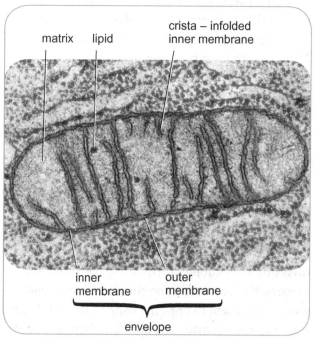

Figure 2.25 TEM of a section through a mitochondrion (×46 000).

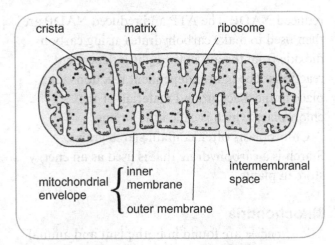

Figure 2.26 Diagram of a mitochondrion.

chloroplast. The 'background material' of the mitochondrion, called the **matrix**, is the site of the stages of aerobic respiration called the Krebs cycle.

Vacuole

A **vacuole** is a membrane-bound organelle that contains liquid. Mature plant cells often have large vacuoles that contain cell sap. The membrane surrounding the vacuole is known as the **tonoplast**. Cell sap contains a variety of substances in solution, especially sugars, pigments and also enzymes.

Plasma (cell surface) membrane

Every cell is surrounded by a **plasma membrane**, sometimes known as the **cell surface membrane**. This is a thin layer made up of lipid (fat) molecules and protein molecules. Its role is to control what enters the cell and what leaves it. You can read about the movement of substances through the plasma membrane in Chapter 3.

Centrioles

Centrioles are found in animal cells but not in plant cells (Figure 2.27). Centrioles make and organise tiny structures called **microtubules**, which are made of a protein called **tubulin**. During cell division, microtubules form the **spindle**, and are responsible for moving the chromosomes around, and pulling them to opposite ends of the cell. Plant cells also use microtubules during cell division, but they are not organised by centrioles.

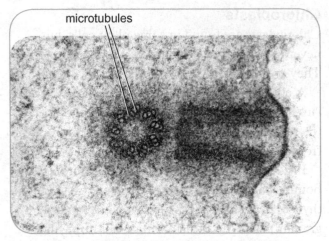

Figure 2.27 TEM showing the two centrioles of an animal cell (at right angles to each other) (×126 000).

Cell walls

Plant cells are always surrounded by a **cell wall** (Figure 2.28 and Figure 2.29). This is not an organelle, because it is not inside the cell.

Plant cell walls are made of long strands of the carbohydrate **cellulose**. The cellulose fibres are very strong, and are arranged in a criss-cross manner, held together by a matrix that contains **pectin**. This composite structure has tremendous resistance to stretching forces that might act on it – for example,

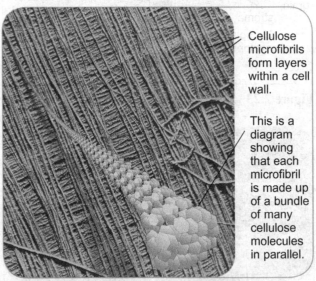

Cellulose microfibrils form layers within a cell wall.

This is a diagram showing that each microfibril is made up of a bundle of many cellulose molecules in parallel.

Figure 2.28 SEM and diagram showing the structure of a plant cell wall (background SEM ×600 000). Notice that the microfibrils lie in different directions in different layers, which greatly increases the mechanical strength of the cell wall.

if the cell has taken up a lot of water and is expanding, the cell wall holds firm, preventing the cell from bursting.

Pectin is also found in the **middle lamella** that cements one cell to another (Figure 2.29).

Adjacent plant cells are often linked together by **plasmodesmata** (singular: plasmodesma). These are passageways that cross the cell wall of each cell. The passageway is lined by the cell membrane, and contains other membranes that are part of the endoplasmic reticulum. The two cells are therefore in direct contact with one another.

Table 2.2 compares the structures found in animal cells and plant cells.

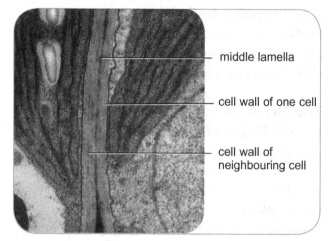

Figure 2.29 TEM of plant cell walls. Where two plant cells lie next to each other, a structure called the middle lamella holds the adjacent walls firmly together (×18 000).

Structure	Plant cells	Animal cells
cell wall	tough, elastic cell wall made of mainly cellulose and also hemicellulose, pectin and sometimes lignin	no cell wall
plasmodesmata	plasmodesmata link adjacent cells, providing a continuous membrane-lined passageway between them	no plasmodesmata
plastids (e.g. chloroplasts)	chloroplasts are present in some plant cells; leucoplasts, which contain stores of starch but no chlorophyll, are found in tissues such as potato tubers	no plastids
vacuole	large, permanent vacuole often present, surrounded by a single membrane called a tonoplast; the vacuole contains cell sap	no large permanent vacuole, but frequently contain numerous small, temporary vacuoles or vesicles (small vacuoles)
centrioles	no centrioles	a pair of centrioles usually present
cilia and flagella (long, thin motile extensions of a cell containing microtubules)	not present in most plant cells	may be present in some animal cells
lysosomes	not usually evident	usually present
carbohydrate storage	may contain starch grains (inside plastids)	may contain glycogen granules

Table 2.2 Structural differences between plant and animal cells.

Prokaryotic cells

Prokaryotic means 'before nucleus'. Prokaryotes are single-celled organisms that do not have nuclei. Cells that do have nuclei are said to be **eukaryotic**.

The structure of a prokaryotic cell

Figure 2.30 shows the structure of a typical prokaryotic cell. The most obvious difference between this cell and a eukaryotic cell is the lack of a nucleus. The prokaryote's DNA lies free in the cytoplasm.

In eukaryotic cells, the DNA is organised into several chromosomes, in which a long strand of DNA is associated with proteins called **histones**. This is not the case in prokaryotes. The DNA is not usually associated with histones, and it is circular rather than linear as in eukaryotes.

This arrangement of the DNA is so different that some people think we should not use the term 'chromosome' to describe it. However, it has now become common for scientists to talk about bacterial chromosomes, despite the fact that they are not the same as the chromosomes in eukaryotic cells.

Bacteria often have additional small, circular DNA molecules called **plasmids**.

Prokaryotes also lack complex membrane-bound organelles, such as mitochondria, chloroplasts and endoplasmic reticulum. They do have ribosomes, but these are smaller than in eukaryotic cells, and they are always free in the cytoplasm rather than attached to membranes.

Prokaryotes are surrounded by a cell wall, but its structure is not at all like that of plant cells. The prokaryote cell wall is made up of fibres of **peptidoglycan**. Like plant cell walls, this cell wall stops the cell bursting if it expands.

Table 2.3 summarises the differences and similarities between eukaryotic cells and prokaryotic cells.

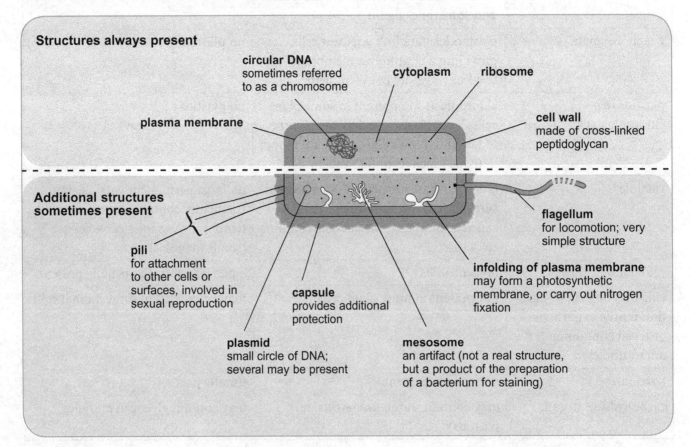

Figure 2.30 The structure of a typical prokaryotic cell.

Structure	Eukaryotic cell	Prokaryotic cell
nucleus	usually present, surrounded by a nuclear envelope and containing a nucleolus	no nucleus, and therefore no nuclear envelope or nucleolus
mitochondria	usually present	never present
chloroplasts	present in some plant cells	never present
endoplasmic reticulum	always present	never present
ribosomes	relatively large, about 30 nm in diameter (80 S)	relatively small, about 20 nm in diameter (70 S)
chromosomes	DNA arranged in several long strands, associated with histones	circular DNA, not associated with histones in bacteria
cell wall	cellulose cell walls present in plant cells	cell wall always present, made of peptidoglycan
cilia and flagella	sometimes present	some have flagella, but these have a different structure from those in eukaryotic cells

Table 2.3   Comparison of the ultrastructure of eukaryotic and prokaryotic cells.

The endosymbiont theory

Prokaryotic cells are older and more diverse than eukaryotic cells. Prokaryotic cells have probably been around for at least 3.5 billion years, while eukaryotic cells arose only about 1 billion years ago.

Eukaryotic cells, unlike prokaryotic cells, contain numerous organelles that are surrounded by membranes. Most of these, including the Golgi body and lysosomes, are derived from the membranes of the endoplasmic reticulum. However, mitochondria and chloroplasts have a more intriguing origin.

It is thought that mitochondria and chloroplasts were once prokaryotic cells that invaded a eukaryotic cell and lived inside it. Both cells gained from this arrangement. The prokaryotes brought with them the ability to carry out complex metabolic reactions (respiration and photosynthesis) that provided the eukaryote with energy. The eukaryote provided the prokaryotes with nutrients. Such an arrangement, whereby two organisms of different species live closely together, is called symbiosis. The prokaryotes are said to be **endosymbionts** – 'endo' means 'within'. This means that every cell in your body is actually a very close association of at least two different organisms – your 'own' eukaryotic cell, and a prokaryotic cell that became a mitochondrion.

Some researchers also believe that the nucleus of a eukaryotic cell arose in a similar way, but the evidence for this is not yet as clear as for the other organelles.

What evidence is there for this amazing theory for the origins of mitochondria and chloroplasts?

- Mitochondria and chloroplasts each have a pair of membranes (an envelope) surrounding them. The outer membrane is similar to other membranes of eukaryotic cells, while the inner one has similarities with membranes of prokaryotes.
- Mitochondria and chloroplasts each have their own small piece of DNA, which is circular like that of prokaryotes and is not enclosed in a nucleus.
- Mitochondria and chloroplasts each have their own ribosomes, which are the same size as those of prokaryotes (smaller than the ribosomes found elsewhere in eukaryotic cells).

- Mitochondria and chloroplasts reproduce by binary fission, like bacteria, before the whole eukaryotic cell divides by mitosis.
- Mitochondria and chloroplasts are often very similar in size to prokaryotic cells.

We can see other examples of symbiosis between prokaryotes and eukaryotes happening today. For example, nitrogen-fixing bacteria, such as *Rhizobium*, live inside the root cells of leguminous and other plants (Figure 2.31). They are still clearly bacteria, and we can also find these cells living freely in the soil. Perhaps one day they will have evolved to look less like cells and more like organelles within the eukaryotic cell.

Tissues and organs

In a multicellular organism such as an animal or plant, different cells are specialised to carry out particular functions. Each cell in the organism has an identical set of genes, but only some of these genes are expressed ('switched on') in any one cell. So, for example, in your body you have muscle cells that are specialised for contraction, bone cells that are specialised to provide support, and rod cells in the retina in your eye that are specialised to turn the energy from light rays into electrical energy in nerve impulses.

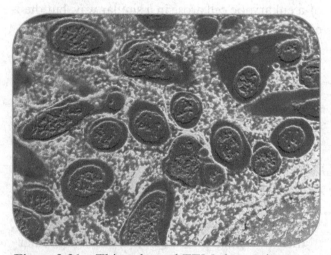

Figure 2.31 This coloured TEM shows nitrogen-fixing bacteria, *Rhizobium* (red), in the cytoplasm (blue) of a cell of a pea plant. There is a symbiotic relationship between the bacteria and plant. It is thought that mitochondria and chloroplasts may have evolved from a relationship similar to this one (× 7480).

These different types of cell are not scattered randomly about. Cells that carry out the same function are usually grouped together, forming a **tissue**. Tissues may be further grouped into **organs**. One example of an organ in a plant is a root.

A dicotyledonous root

The root of a dicotyledonous plant, such as buttercup, *Ranunculus*, is an example of an organ that contains many different kinds of tissue (Figure 2.32, Figure 2.33 and Figure 2.34).

A diagram that shows the distribution of the tissues in an organ is called a **plan drawing**. Figure 2.34 is a plan drawing made from the micrograph in Figure 2.32. Notice that it does not show any individual cells, just the areas in which the different cell types are found.

The tissues in a dicotyledonous root

The cells covering the outside of the root are **epidermal cells**, and they make up a tissue called the **epidermis**. Some of these cells have extensions that form root hairs, which provide a large surface area for the absorption of water and mineral salts. These are not visible in Figure 2.32, as the section has been taken in a part of the root where root hairs are not present.

Beneath the epidermis is the **cortex**. This is mostly made up of **parenchyma tissue**. The cells

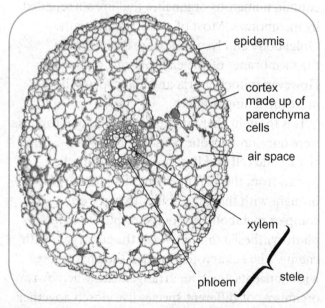

Figure 2.32 Light micrograph of a *Ranunculus* root (× 17).

in parenchyma are relatively unspecialised. They generally have fairly thin cellulose cell walls. Their functions are to allow water to move from the edge of the root to the centre, either by passing through the cells or through their cell walls (you will find out more about this in Chapter 6 in Unit 2). They may also store starch, as an energy reserve for the plant. When they are turgid (with their vacuoles full), they help to provide support.

In the root shown in Figure 2.32, there are quite large **air spaces** in between the cells in the parenchyma. These are sometimes found in plants that grow in wet soil. The cells in the root must have oxygen so that they can carry out aerobic respiration, and these spaces allow oxygen and other gases to diffuse easily from the stem down into the root. Plants that grow in normally aerated soil generally do not have air spaces in their roots, as they can get oxygen from the air in between the soil particles.

In the centre of the root is a group of **vascular tissues**, making up the **stele** (Figure 2.33). Surrounding the stele is a layer of cells, the **endodermis**. The endodermis often contains cells specially waterproofed by a deposit known as the **Casparian strip** (see Chapter 6 in Unit 2).

The stele has a layer of cells, the **pericycle**, which surrounds the vascular tissues that are involved in the transport of materials through the plant. The two main tissues in the stele are **xylem tissue** and **phloem tissue**.

Xylem tissue contains dead, empty cells with lignified side walls and no end walls, called **vessel elements**. Lignin is a very strong, waterproof substance – it is what gives wood its strength. The vessel elements are arranged end to end to form continuous tubes through which water flows in an unbroken column all the way from the roots to the top of the plant. As well as transporting water, vessel elements provide support to the plant, as their lignified walls are very strong.

You can often recognise xylem vessel elements in a transverse section (TS) of a root by their large, empty lumens (spaces inside them), surrounded by thick, angular, darkly stained walls.

Phloem tissue contains **phloem sieve elements**. These are living cells with cellulose cell walls, but their end walls are perforated with many tiny pores, allowing sucrose solution to pass through by mass flow. These cells do not contain a nucleus, and have only a thin layer of cytoplasm, with few

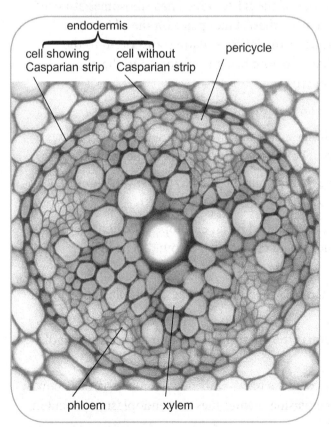

Figure 2.33 Detail of the stele of the root of *Ranunculus* (× 235).

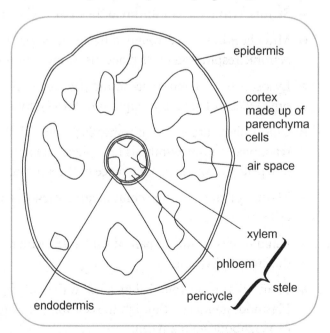

Figure 2.34 Low power plan of the *Ranunculus* root shown in Figure 2.33.

organelles. Each sieve element is connected by many plasmodesmata to a companion cell, which has a nucleus and many organelles.

In a TS of a root, look for the phloem tissue in between the 'arms' of the xylem tissue. Its cells have much thinner walls than those in xylem tissue, and will generally be stained differently because the cell walls contain cellulose and not lignin. Sometimes, if the section has gone through exactly the right place, you can pick out a sieve plate.

You will find out much more about xylem and phloem in Chapter 6 in Unit 2.

Summary

- Light microscopes use light rays, focused by glass lenses, to provide magnified images of specimens. Their best resolution is about 200 nm. Specimens are usually stained using coloured solutions, and it is possible to view living cells, tissues or whole organisms.

- Electron microscopes use electron beams, focused by electromagnets, to provide magnified images of specimens. Their best resolution is about 0.5 nm. Specimens are usually stained using heavy metals and must be placed in a vacuum, so they must be totally dehydrated and it is not possible to view living material.

- Magnification is the size of an object in an image divided by its real size.

- Both animal and plant cells contain a nucleus surrounded by a nuclear envelope, in which pores allow the transfer of mRNA from nucleus to cytoplasm. The nucleus contains a nucleolus, where rRNA is synthesised. Chromosomes are contained within the nucleus.

- The nuclear envelope is continuous with the membranes of the RER, which has ribosomes attached to the outer surface of the pairs of membranes. Protein synthesis takes place on the ribosomes. Proteins for export from the cell pass into the cisternae of the RER, and are transported to the Golgi body in vesicles. They are processed and packaged before being transported to the plasma (cell surface) membrane and secreted from the cell by exocytosis.

- Steroid synthesis and the breakdown of toxins take place on the SER.

- Mitochondria are surrounded by an envelope. The inner membrane is very folded, forming cristae. Aerobic respiration takes place inside mitochondria. The Krebs cycle happens in the matrix.

- Lysosomes are membrane-bound packets of enzymes, which may be used to digest worn-out organelles or bacteria brought into the cell by phagocytosis.

- Chloroplasts are found in some plant cells. They are surrounded by an envelope. The light-dependent reactions of photosynthesis take place on membranes where chlorophyll is found. The light-independent reactions take place in the stroma.

- Mature plant cells often contain large vacuoles, bound by a membrane called a tonoplast, filled with cell sap.

- Animal cells contain a pair of centrioles, which organise the microtubules that make up the spindle during cell division.

- Plant cells are surrounded by a cell wall made up of cellulose fibres embedded in a matrix of pectin. Plasmodesmata run through the cell wall, linking the plasma membranes and endoplasmic reticulum of neighbouring plant cells.

continued ...

- Prokaryotic cells (for example, bacteria) do not have a nucleus; their DNA forms a circular molecule that lies free in the cytoplasm. They are much smaller than eukaryotic cells, and do not contain membrane-bound organelles such as mitochondria or Golgi bodies. Their ribosomes are smaller than those of eukaryotes. Their cell walls contain peptidoglycans, not cellulose.

- There is strong evidence that mitochondria and chloroplasts were once prokaryotes, which came to live inside eukaryotic cells in a symbiotic relationship.

- A tissue is a group of cells of the same origin, which are specialised to carry out a particular function. Several different tissues are generally grouped together to form an organ. An example of an organ is a plant root, containing tissues including parenchyma, xylem and phloem.

Questions

Multiple choice questions

1 What are the advantages of using an electron microscope?
   I   very high magnification
   II very high resolution
   III can use living material
   IV natural colour of sample maintained

   A **I** only
   B **I** and **II** only
   C **I**, **II**, and **III** only
   D **I**, **II**, **III** and **IV**

2 Which of the following evidence supports the endosymbiotic theory for the origin of eukaryotes?
   A Mitochondria and chloroplasts have their own circular DNA.
   B Mitochondria are larger than bacterial cells.
   C Chloroplasts are surrounded by a single membrane.
   D Both the chloroplasts and mitochondria divide by budding.

3 The parts of a cell include mitochondria, a nucleus, ribosomes and smooth endoplasmic reticulum. This cell could **not** be:
   A from a mango plant.
   B a bacterium.
   C part of a housefly.
   D from a budding yeast cell.

4 The limit of resolution can best be defined as:
   A twice the wavelength of light.
   B the minimum distance that two objects must be apart in order to be distinguished as separate objects.
   C the size of the smallest object which can be seen using the microscope.
   D the degree of sharpness of the image produced by the microscope.

continued ...

5 Of the following organelles, which group is involved in manufacturing proteins and lipids by the cell?

A smooth ER, ribosomes, rough ER

B vacuole, rough ER, lysosome

C lysosome, vacuole, ribosome

D smooth ER, ribosome, vacuole

6 A chloroplast in a *Hibiscus* leaf measured 1.0 µm. On an electron micrograph the length of the organelle was 20 mm. What is the magnification of the electron micrograph?

A 200 times

B 2000 times

C 20 000 times

D 200 000 times

7 Which of the following pairs of organelles and functions are correctly matched?

A ribosome – manufacture of lipids

B central vacuole – storage

C mitochondrion – trapping of sunlight

D lysosomes – movement

8 In which of the following structures does aerobic respiration and ATP production occur?

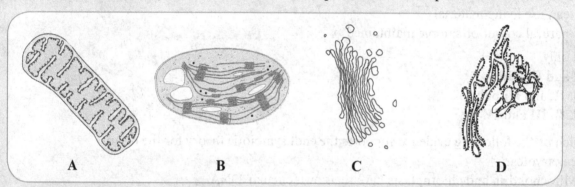

9 Substances enter and leave cells continuously. These substances must pass through the:

A nucleus.

B lysosomes.

C plasma membrane.

D microtubules.

10 Which of the following correctly describes a dicotyledonous root?

	Type of structure	Feature
A	tissue	a collection of organs with a particular function
B	tissue	a collection of cells with similar structure performing a particular function
C	organ	a collection of tissues which forms a structural and functional unit
D	organ	a collection of cells with similar structure performing a particular function

continued ...

Structured questions

11 The light micrograph shown below is a cross section a mature *Ranunculus* (buttercup) root.

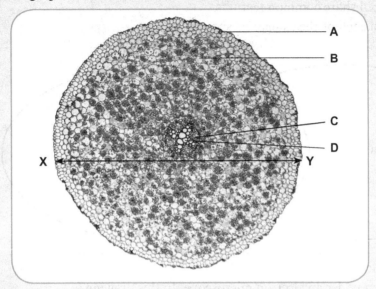

a Name the tissues **A** to **D**.   [4 marks]

b Draw a plan drawing to show the distribution of the different tissues shown in the
   micrograph of the root. Make your drawing the actual size. (No labels required)   [4 marks]

c If the micrograph is magnified 120 times, what is the actual width of the specimen
   from **X** to **Y**?   [2 marks]

d The root is described as an organ. Identify the **three** tissues which make up the root
   and state **one** function of each tissue.   [5 marks]

12 a The electron micrograph
      shown to the right is of
      a plant cell.

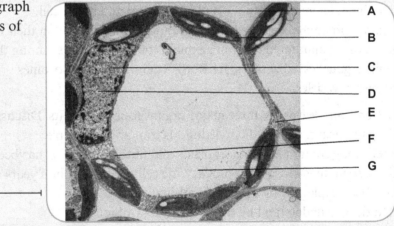

100 μm

i   Identify the structures labelled **A** to **G**.   [4 marks]

ii   Which structures shown above occur in plant cells only?   [1 mark]

iii An animal cell differs from a plant cell. Identify **two** ways an animal cell
     can be identified.   [2 marks]

b The scale bar represents an actual length of 100 μm.
   Calculate the magnification of the electron micrograph. Show your working.   [1 mark]

c What are the functions of the parts **A** to **D**?   [3 marks]

d What are the advantages and disadvantages of using an electron microscope
   instead of a light microscope to view this specimen?   [4 marks]

continued ...

13 a The diagram below represents the relationship between prokaryotic and eukaryotic cells. Some features are unique to either prokaryotes or eukaryotes while both types of cell share common features.
State **three** features in each of the categories shown. [6 marks]

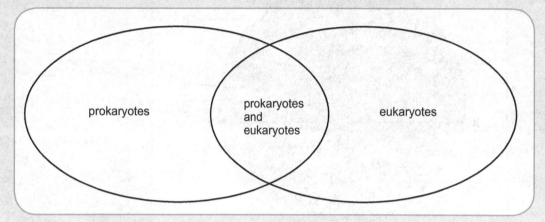

b Make a generalised drawing of:

  i a bacterium as seen under an electron microscope. In this drawing, label a site of protein synthesis. [3 marks]

  ii a mitochondrion as seen under an electron microscope. [3 marks]

c What is a common feature of both a bacterium and a mitochondrion? [1 mark]

Essay questions

14 a i With the use of a diagram, describe the features of a typical animal cell.

  ii State **three** differences between a plant cell and an animal cell. [9 marks]

b Both plant and animal cells have a membranous network within their cells. By means of an annotated diagram, explain the relationships among the following organelles: ribosomes, RER, SER, Golgi body, lysosomes and secretory vesicles. [6 marks]

15 a Roots, stems and leaves are three major organs found in plants. Discuss what the term 'organ' means, using a dicotyledonous root as an example. [8 marks]

b Plant cells may contain both chloroplasts and mitochondria. It has been postulated that these structures arose from prokaryotic cells about a billion years ago. This has been explained by the endosymbiont theory.

  i What do you understand by the term 'endosymbiont'? [2 marks]

  ii Discuss the evidence which suggests that both chloroplasts and mitochondria have prokaryotic origins. [5 marks]

16 a The electron microscope was used to elucidate the ultrastructure of both eukaryotic and prokaryotic cells. What are the advantages of using the electron microscope, rather than a light microscope, to study cells? [3 marks]

b Compare eukaryotic and prokaryotic cells in terms of: cell wall, size, packaging of DNA, structures involved in protein synthesis. [12 marks]

Chapter 3
Membrane structure and function

By the end of this chapter you should be able to:

a explain the fluid mosaic model of membrane structure;

b outline the roles of phospholipids, cholesterol, glycolipids, protein and glycoproteins;

c explain the processes of diffusion, facilitated diffusion, osmosis, active transport, endocytosis and exocytosis;

d know how to investigate the effects on plant cells of immersion in solutions of different water potentials.

Membranes

Every living cell is surrounded by a membrane. This is called the **plasma membrane**, or the **cell surface membrane**. The plasma membrane defines the limits of the cell. It separates the cell's contents from its external environment, and it controls what can pass from this environment into the cell, and from the cell into the external environment. It is **partially permeable**.

Membranes are also found inside cells. Some organelles are surrounded by a single membrane – for example, lysosomes. The nucleus, mitochondria and chloroplasts each have two membranes around them, making up an **envelope**. Most eukaryotic cells also have an extensive network of membranes within their cytoplasm, forming the rough endoplasmic reticulum, the smooth endoplasmic reticulum and the Golgi body (Figure 3.1). Like the plasma membrane, these membranes inside the cell are partially permeable, and therefore able to control what can pass through them. They separate what happens inside the organelle from what is happening in the rest of the cell.

Table 3.1 summarises the functions of membranes around and inside cells. You will find out more about some of these functions in this chapter.

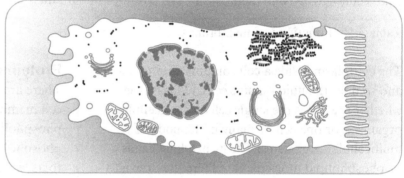

Figure 3.1 Membranes in an animal cell (shown in red).

Structure of cell membranes

All cell membranes have a similar structure. They are normally between 7 nm and 10 nm thick, which makes them invisible with a light microscope but visible using an electron microscope. They are formed from a double layer of molecules called **phospholipids** (Chapter 1), in which many different kinds of **proteins** are situated.

Phospholipid bilayer

Phospholipid molecules have an unusual property. Their heads have a tiny electrical charge, and this attracts them to water molecules. But their tails don't have a charge, and they are repelled from water molecules. We say that the heads of the phospholipids are **hydrophilic** (water-loving) and the tails are **hydrophobic** (water-hating) (Figure 3.2).

Function	Example
Membranes are partially permeable, controlling what passes through them.	The plasma membrane allows small or uncharged particles to pass through it; protein channels and transporters control the passage of larger or charged particles.
Membranes produce different compartments inside cells.	Mitochondria are surrounded by two membranes, which isolate the reactions taking place inside from the reactions taking place in the cytoplasm.
Membranes are important in cell signalling.	A substance produced by one cell docks into a receptor in the plasma membrane of another, causing something to happen in the second cell.
Membranes can allow electrical signals to pass along them.	The membrane of the axon of a motor neurone transmits action potentials from the central nervous system to a muscle.
Membranes provide attachment sites for enzymes and other molecules involved in metabolism.	The inner membrane of a mitochondrion contains molecules needed for the production of ATP. The inner membrane of a chloroplast contains chlorophyll needed for photosynthesis.

Table 3.1 Functions of membranes.

The cytoplasm inside a cell contains a lot of water, and so does the fluid outside cells. (This is true whether the cell is the single cell of a unicellular organism, or one cell of many in the body of a multicellular organism.) The hydrophilic heads of phospholipid molecules are therefore drawn to these watery fluids, while the hydrophobic tails are repelled by them. This causes the phospholipids to arrange themselves in a double layer, with heads facing outwards and tails facing inwards. This is called a **phospholipid bilayer** (Figure 3.3).

Protein components of cell membranes

There are many different protein molecules in cell membranes. They are much larger than phospholipid molecules. They float in the phospholipid bilayer, and are sometimes referred to as 'protein icebergs in a phospholipid sea'. Most plasma membranes consist of approximately 50% lipid and 50% protein by mass.

The proteins are sometimes classified into two main groups on the basis of their positions within the membrane.

- **Peripheral** or **extrinsic proteins**. These can be removed from the membrane by mild treatment with detergent, which breaks them free from the phospholipids. They do not penetrate the lipid bilayer at all, but simply associate with the hydrophilic surfaces of the phospholipid bilayer.

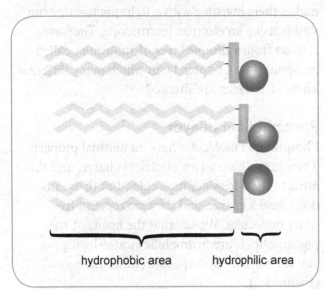

hydrophobic area   hydrophilic area

Figure 3.2 Phospholipid molecules.

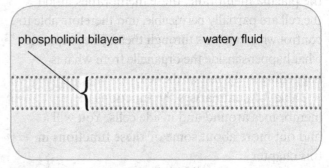

phospholipid bilayer   watery fluid

Figure 3.3 A phospholipid bilayer.

- **Integral** or **intrinsic proteins.** These can only be removed by disrupting the membrane structure. This is because they lie actually within the phospholipid bilayer. Some of them span the entire membrane, reaching from one side to the other, and these are known as **transmembrane** proteins. These proteins have areas of their molecules that are hydrophilic and other areas that are hydrophobic, and are therefore said to be **amphipathic**. They normally have hydrophilic areas on the parts of the molecule that reach into the watery fluids on both the outer surface and the inner surface of the membrane, and hydrophobic parts that lie in the phospholipid bilayer (Figure 3.4). As you will see, some of them also have hydrophilic parts lining a channel that runs right through the molecule, providing a hydrophilic passageway through which ions and other particles can move.

SAQ

1 Think back to what you know about protein structure. Suggest what makes some parts of a transmembrane protein hydrophilic, and some parts hydrophobic.

The proteins are held in place in the membrane in various ways.

- The hydroxyl groups of the amino acids, and other R groups that have small electrical charges, are attracted to the charged polar heads of the phospholipids.

- The hydrophobic regions of the protein are attracted to the lipid tails, by hydrophobic interactions.

Some proteins are firmly attached to protein structures called microfilaments, in the cytoplasm of the cell. These microfilaments make up a kind of 'scaffolding' in the cell, called the cytoskeleton.

You can see a summary of the roles of proteins in cell membranes in Figure 3.5 and Table 3.2. Their roles in helping substances to move through the membrane are explained on pages 61 and 70 to 71.

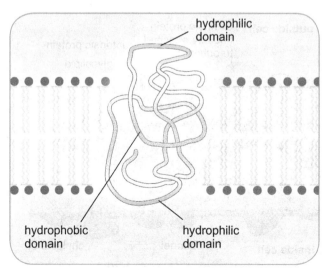

Figure 3.4   A transmembrane protein.

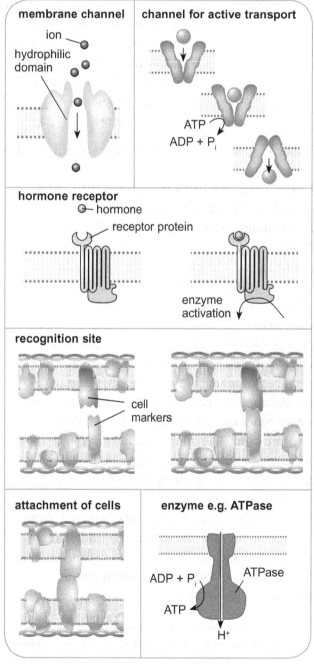

Figure 3.5   Roles of membrane proteins.

Component	Roles
phospholipid	• forms the bilayer which is the fundamental basis of the membrane in which all other components are embedded • provides a barrier to water-soluble (hydrophilic) substances, such as ions and molecules that carry a charge
cholesterol	• helps to maintain the fluidity of the membrane, preventing it from becoming too stiff when temperatures are low, or too fluid when temperatures are high
protein and glycoprotein	• form channels through which hydrophilic substances can pass; the channels can be opened and closed • act as transporters that can move substances across the membrane up their concentration gradients, with the use of energy from ATP • act as receptor sites, allowing specific molecules from outside the cell, such as hormones, to bind with them and then set up responses within the cell • act as recognition sites, because their precise structure may be specific to a particular type of cell or to a particular individual (note: glycolipids also have this role) • bind to proteins in neighbouring cell membranes, holding the cells together • act as enzymes

Table 3.2 Roles of the components of cell membranes.

Other components of cell membranes

As well as phospholipids, membranes also contain another type of lipid. This is **cholesterol**. Cholesterol molecules lie among the phospholipids, helping to make up the bilayer. Their function is outlined in Table 3.2.

Many of the phospholipid and protein molecules have short chains of sugar molecules attached to their surfaces and are known as **glycolipids** and **glycoproteins**. The carbohydrate (sugar) chains are always on the outer surface of the membrane. These chains often function as sites where other specific molecules, from outside the cell, can bind. They therefore act as **receptor sites**, which can be important in cell signalling, when molecules from outside the cell – for example, a hormone, a neurotransmitter or an antigen – bind with their specific receptors, bringing about changes within the cell. Often, the whole outer surface of a membrane is covered with a very large number of these carbohydrate chains, known as a **glycocalyx**.

SAQ

2 What is the difference between the outer surface and the inner surface of the plasma membrane?

The fluid mosaic model of membrane structure

Figure 3.6 shows the structure of a plasma (cell surface) membrane. This is called the **fluid mosaic model** of membrane structure. 'Fluid' refers to the fact that the molecules in the membrane are in constant motion, moving around within their own phospholipid monolayer (not normally swapping sides). The term 'mosaic' refers to the way the membrane would look if viewed from above, with a mosaic pattern of protein molecules.

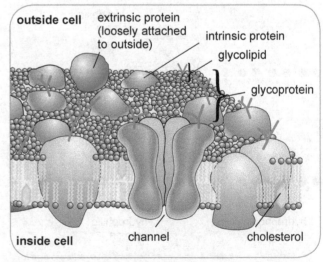

Figure 3.6 Part of a plasma membrane.

Movement across cell membranes

Some substances move into and out of cells using vesicles, by exocytosis and endocytosis (page 71). However, many substances pass through the plasma membrane itself. Some of these substances move **passively** – that is, the cell does not have to use energy to make them move. Passive processes include **diffusion, facilitated diffusion** and **osmosis**. Other substances are **actively** moved by the cell, which uses energy to make them move up their concentration gradients. This is called **active transport**.

Diffusion

Particles are constantly moving around randomly. They hit each other and bounce off in different directions. Gradually, this movement results in the particles spreading evenly throughout the space within which they can move. This is **diffusion**.

If there are initially more particles in one place than another, we say there is a **concentration gradient** for them. Diffusion is the net movement of molecules or ions down their concentration gradient – that is, from a place where they are in a high concentration to a place where they are in a lower concentration.

There are usually a large number of different kinds of particles bouncing around inside and outside a cell, on both sides of its plasma membrane. Some of these particles hit the plasma membrane. If they are small – like oxygen and carbon dioxide molecules – and do not have an electrical charge, they can easily slip through the phospholipid bilayer.

Oxygen enters a cell like this. Inside the cell, aerobic respiration constantly uses up oxygen, so the concentration of oxygen inside the cell is low. If there is more oxygen outside the cell, then there is a concentration gradient for oxygen. Oxygen molecules on both sides of the plasma membrane are moving freely around, and some of them hit the plasma membrane and pass through it. This happens in both directions, but because there are more oxygen molecules in a given volume *outside* the cell than *inside*, more of them will pass through the membrane from outside to inside rather than in the opposite direction. The overall effect is for oxygen to move from outside the cell, through the plasma membrane, into the cytoplasm.

The two features of a substance that affect its ability to diffuse across a cell membrane are:

- whether or not there is a charge on its molecules or ions – if there is a charge, it is less likely to be soluble in lipids and therefore less able to move through the phospholipid bilayer;
- its size – smaller molecules are more likely to be able to get through.

Ions (which all carry a big charge), molecules such as glucose that have small charges, and large molecules such as proteins are all unable to diffuse freely through the phospholipid bilayer. Water can get through, and so can small uncharged molecules such as oxygen and carbon dioxide, and lipid-soluble molecules no matter how large they are, such as glycerol, alcohol and steroids.

Facilitated diffusion

Cells provide special pathways through the plasma membrane which allow ions (or charged molecules) to pass through. Such pathways are provided by **channel proteins**. These proteins lie in the membrane, stretching from one side to the other, forming a hydrophilic channel through which ions can pass. The ions pass through by diffusion, down their concentration gradient. This process is called **facilitated diffusion**. It is just like ordinary diffusion, except that the molecules or ions only get through the membrane if they happen to bump into a channel (Figure 3.7).

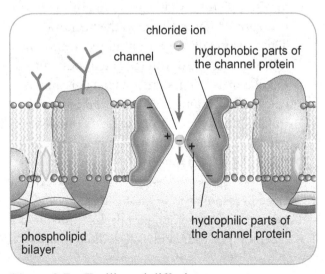

Figure 3.7 Facilitated diffusion.

Each channel formed by a protein allows only a specific ion or molecule to pass through. The protein can change its shape, making the channel either open or closed. Cells can therefore prevent or allow specific substances to pass through their membranes by facilitated diffusion.

Sometimes two different substances are transported through the same channel. A **symport** allows movement of the two substances in the same direction. An **antiport** allows movement in opposite directions. Symports and antiports can also be involved in active transport (page 70).

Comparing diffusion and facilitated diffusion

We have seen that diffusion and facilitated diffusion are essentially the same – they both involve the random movement of molecules or ions, which results in their net movement down a concentration gradient. Neither of them involves any energy input (as ATP) from the cell. The only difference between them is that, in facilitated diffusion, the molecules or ions have to move through protein carrier molecules whereas in simple diffusion, the molecules just move through the phospholipid bilayer.

But this difference does influence how the concentration gradient affects the *rate* at which diffusion takes place. Imagine, for example, a cell that has a limited number of sodium ion channels. Even when all of these are open, the sodium ions only have these channels through which they can diffuse, no matter how great the difference in the sodium ion concentration on each side of the membrane (the concentration gradient). This would mean that, if we were to keep increasing the concentration gradient for sodium ions, we would reach a point where the rate of diffusion cannot get any faster (Figure 3.8). The ions are already moving through all the channels as fast as they can, so the number of channels has limited the maximum rate at which facilitated diffusion can take place. Oxygen molecules, on the other hand, can diffuse through any part of the phospholipid bilayer. For oxygen, therefore, we would expect the rate of diffusion to keep on increasing if we increase the concentration gradient.

Channel control

The rate of facilitated diffusion can be controlled by opening or closing the channels. Most protein channels can be classified into the following three groups:

- **voltage-gated channels**. These open or close in response to tiny changes in the voltage (potential difference) across the membrane. This is especially important in the membranes of neurones, where voltage-gated sodium ion channels, potassium ion channels and calcium ion channels cause action potentials to pass along the neurone and pass into other neurones. You will find out much more about this in Unit 2 Chapter 10.
- **mechanically gated channels**. These open or close in response to mechanical changes. For example, Pacinian corpuscles (pressure receptors) in your skin have sodium ion channels that open in response to pressure, causing action potentials to be set up and to sweep along sensory neurones to your brain. In the cochlea of your ear, pressure changes in the fluid, caused by sound waves, affect sodium ion channels, causing action potentials to be sent along the auditory nerve to the brain.
- **ligand-gated channels**. These open or close in response to the attachment of a small signalling molecule – a ligand – that binds with the channel protein. For example, sodium ion channels on the membrane of a post-synaptic neurone open when the neurotransmitter acetylcholine binds with them, allowing sodium ions to flood into the post-synaptic neurone and set up an action potential there.

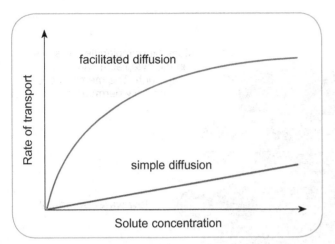

Figure 3.8 The effect of concentration on the rate of simple diffusion and facilitated diffusion.

In Figure 3.8, the rate for simple diffusion is shown as being much lower than that for facilitated diffusion. This is often the case, but not always. It depends on how permeable the phospholipid membrane is to the substance which is diffusing, and how many channel proteins there are. If the membrane is very permeable, and there are not many channel proteins, then simple diffusion could happen faster than facilitated diffusion.

SAQ

3 Explain under what circumstances carbon dioxide might diffuse into a palisade cell in a leaf, and how the process takes place.
4 Explain the roles of the hydrophobic and hydrophilic parts of the channel protein shown in Figure 3.7.
5 Explain why facilitated diffusion is said to be an example of *passive* transport.

Osmosis

Water molecules, although they carry charges (Chapter 1), are very small. They are therefore able to pass through the lipid bilayer by diffusion, slipping through tiny spaces between the phospholipids. They can also move through permanently open protein channels called aquaporins. This movement of water molecules down their diffusion gradient, through a partially permeable membrane, is called **osmosis**.

It is not correct to use the term 'concentration' to describe how much water there is in something. Concentration refers to the amount of solute present. Instead, the term **water potential** is used. The symbol Ψ (psi) can be used to mean water potential.

The water potential of a solution is a measure of how much water the solution contains in relation to other substances, how freely the water molecules can move and how much pressure is being applied to it. A solution containing a lot of water, and under pressure, is said to have a **high water potential**. A solution containing a lot of dissolved substances (solutes) and little water, and not under pressure, has a **low water potential**. You can think of water potential as being the tendency for water to leave a solution.

By definition, pure water at normal atmospheric pressure is given a water potential of 0. The more solute you dissolve in the water, the lower its water potential gets. The freedom of the water molecules is reduced because they are attracted to the solute molecules. Therefore, a solution of sugar has a water potential which is less than 0 – that is, it has a negative water potential.

Just as we don't normally talk about the 'concentration' of water, we don't normally use the term 'concentration gradient' either. Instead, we use the term **water potential gradient**. Water tends to move *down* a water potential gradient, from where there is a lot of water to where there is less of it (Figure 3.9). It diffuses out of a dilute solution (a lot of water – high water potential) and into a concentrated solution (a lot of solute – low water potential).

Why is this important? The cells in your body contain watery cytoplasm and are surrounded by watery fluids. Blood cells, for example, float in blood plasma. Water can move freely through the plasma membrane of the blood cells, but most of the substances dissolved in the water cannot. If there is a water potential gradient between the contents of a cell and the blood plasma, then water will move either into or out of the cell. If a lot of water moves like this, the cell can be damaged.

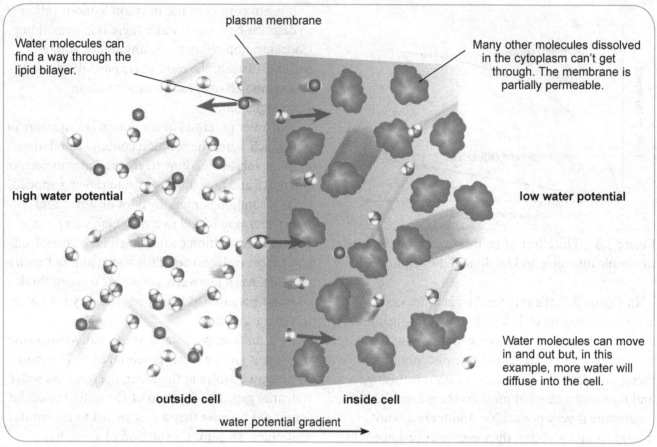

Figure 3.9 How osmosis occurs.

Osmosis and animal cells

Figure 3.10 shows what happens when animal cells are placed in solutions with water potentials higher or lower than the water potential of the cytoplasm inside the cells. If the solution outside the cell has a higher water potential than the cytoplasm, then water enters the cell by osmosis. If the water potential gradient is very steep, so much water may enter that the cell bursts.

If the water potential gradient is in the other direction, then water leaves the cell by osmosis. The cell may shrink, sometimes becoming 'star-shaped', described as being **crenated**. The concentration of the solutes in the cytoplasm increases, and this may adversely affect metabolic reactions taking place inside the cell.

If two solutions have the same osmotic potential they are said to be **isotonic**. A **hypertonic** solution will absorb water by osmosis from a **hypotonic** solution.

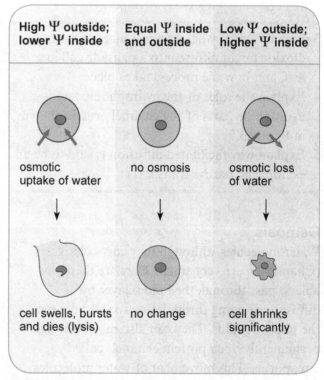

Figure 3.10 Osmosis and animal cells.

Osmosis and plant cells

Figure 3.11 shows what happens when plant cells are placed in solutions with water potentials higher or lower than the water potential of the cytoplasm in the cells.

Water moves into or out of the cell, down its water potential gradient, just as in an animal cell. The cell wall does not directly affect this movement, because it is fully permeable to water and to most of the solutes dissolved in it.

If the cell is put into water, then – just as in an animal cell – water enters by osmosis. But this time the cell does not burst. This is because, as it swells, it has to push out against the strong cell wall. The cell wall resists expansion of the cell, exerting a force called **pressure potential**. The cell becomes full and stiff, a state called **turgor**.

If a plant cell is put into a concentrated solution, then water leaves it by osmosis. The cell therefore shrinks. If a lot of water is lost, the contents no longer press outwards on the cell wall, and the cell loses its turgor. It is said to be **flaccid**.

The strong cell wall cannot cave in very much, so as the volume of the cell gets smaller and smaller, the plasma membrane may eventually pull away from the cell wall. The plasma membrane is often damaged in this process. A cell in this state is said to be **plasmolysed**. The cell usually dies.

SAQ

6 Look at Figure 3.11. Explain why, in the cell in diagram 3 in the bottom row, the space between the cell membrane and the cell wall is filled with the same solution that is outside the cell.

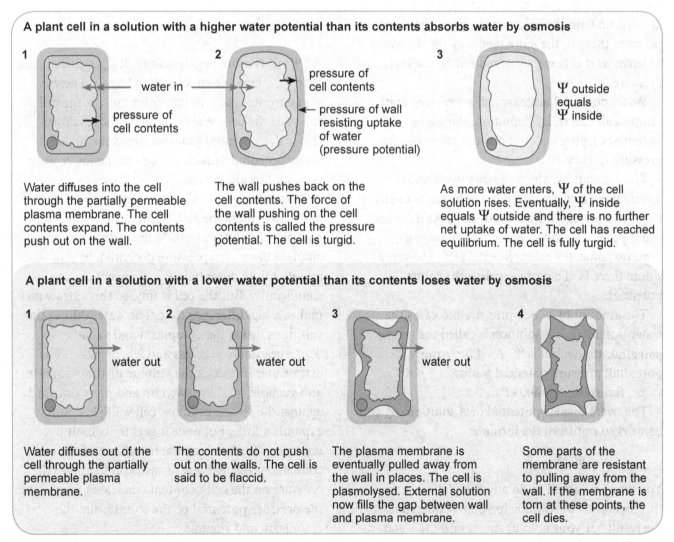

Figure 3.11 Osmosis and plant cells.

Investigating the effect of solutions on plant cells

Plant cells are often exposed to solutions of concentrations that differ from the concentration of their cytoplasm or vacuoles. This can have large effects on the cell, causing it to lose or gain water. We can use these effects to work out the water potential of the contents of the cell.

We have seen that the water potential of a solution is affected by:

- the concentration of solute in the solution – *the more solute, the lower the water potential*;
- the pressure that is applied to the solution – *the more pressure, the higher the water potential*.

It may help you to remember this if you think about the water potential of a solution as being the tendency for water to leave the solution. The more solute there is in it, the *less* tendency there is for the water to leave, because the water molecules are tied up with the solute molecules. The more pressure there is, the *more* tendency for the water to leave, as it is being squeezed out by the high pressure.

Water potential is measured in pressure units, kilopascals (kPa). By definition, the water potential of pure water at normal atmospheric pressure is 0 kPa.

The amount by which the dissolved solute lowers the water potential of a solution is called the **solute potential**. Its symbol is Ψ_s. As it *lowers* water potential, solute potential is always a *negative* value, for example −500 kPa. The more solute there is, the more negative the solute potential.

The amount by which pressure increases the water potential of a solution is called the **pressure potential**. Its symbol is Ψ_p. As it *increases* water potential, pressure potential is always a *positive* value, for example +500 kPa.

The overall water potential of a solution, Ψ, can be worked out from the formula:

$$\Psi = \Psi_p + \Psi_s$$

You won't be asked to do any calculations using this formula, but it will help you to understand the results of your investigations into the water potential of plant cells.

A plant cell in pure water

Let's think about what happens when a plant cell is placed in pure water at atmospheric pressure. There is no solute in the pure water, so its solute potential, $\Psi_{s\,outside}$, is 0. The concentration of solute in the cell is much more than the concentration outside, so its solute potential, $\Psi_{s\,inside}$, is more negative. If the pressure on the solutions inside the cell and outside the cell are the same, then the pressure potentials, $\Psi_{p\,outside}$ and $\Psi_{p\,inside}$, are the same.

What does this mean for the overall water potentials inside and outside the cell? Which one is larger? We know that:

$$\Psi = \Psi_p + \Psi_s$$

So:

$$\Psi_{outside} = \Psi_{p\,outside} + \Psi_{s\,outside}$$

$$\Psi_{inside} = \Psi_{p\,inside} + \Psi_{s\,inside}$$

As $\Psi_{s\,inside}$ is more negative than $\Psi_{s\,outside}$, this means that Ψ_{inside} must be smaller than $\Psi_{outside}$. There is a water potential gradient from outside the cell to inside the cell. Water therefore moves down this water potential gradient, from the higher water potential outside the cell to the lower water potential inside the cell.

Now think about what happens as the water gradually enters the cell. There is probably a large amount of solution outside the cell, so not much changes there – its pressure potential, $\Psi_{p\,outside}$, and its solute potential, $\Psi_{s\,outside}$, won't change significantly. But the cell is tiny, so this extra water makes a big difference to it. The water dilutes the solutions inside the cytoplasm and vacuole, so $\Psi_{s\,inside}$ gradually gets less and less negative. The extra water increases the volume of the cytoplasm and vacuole, so these swell up and press outwards against the strong cellulose cell wall. The wall expands a little, but once it gets to its limit it resists any further increase in volume of the cell's contents, pressing back inwards on them. So the pressure on the cell's contents increases, increasing the pressure potential of the solutions in the cytoplasm and vacuole.

Eventually, we get to a point where no more water can enter the cell. The cell becomes fully turgid. We have reached equilibrium. If there is no net water movement by osmosis, then we know that the water potential outside must equal the water potential inside the cell. At equilibrium:

$$\Psi_{outside} \; = \; \Psi_{inside}$$

We also know that $\Psi_{outside}$ was 0. So, at equilibrium, Ψ_{inside} must also be 0.

We can take this argument one step further. We know that:

$$\Psi_{inside} \; = \; \Psi_{p\,inside} \; + \; \Psi_{s\,inside} \; = \; 0$$

This means that, for a cell in pure water, at equilibrium with its surroundings, $\Psi_p = -\Psi_s$.

A plant cell in a concentrated solution

Now let's think about what will happen if we put our plant cell into a solution that is more concentrated than the cell contents. Now, $\Psi_{s\,inside}$ is less negative than $\Psi_{s\,outside}$. If the cell is fully turgid, with its cell wall pushing back and resisting the outward force of the cell contents pushing on it, then $\Psi_{p\,inside}$ is greater than $\Psi_{p\,outside}$.

Can you see that this makes the water potential inside the cell greater than the water potential outside the cell? You may just be able to see that straight away. The equations can show us why this is.

$$\Psi_{outside} \; = \; \Psi_{p\,outside} \; + \; \Psi_{s\,outside}$$

$$\Psi_{inside} \; = \; \Psi_{p\,inside} \; + \; \Psi_{s\,inside}$$

If $\Psi_{p\,inside}$ is *greater* than $\Psi_{p\,outside}$, and $\Psi_{s\,inside}$ is *less negative* than $\Psi_{s\,outside}$, then Ψ_{inside} is greater than $\Psi_{outside}$.

There is therefore a water potential gradient from inside the cell to outside the cell. Water leaves the cell by osmosis, down the water potential gradient.

Now think about what happens as water gradually leaves the cell. As the water leaves, the volume of the contents shrinks, and the pressure exerted by the cell wall on the contents decreases. $\Psi_{p\,inside}$ gets less and less. The loss of the water also increases the concentration of the solution inside

the cell, so $\Psi_{s\,inside}$ gets more and more negative. The overall water potential inside the cell, Ψ_{inside}, therefore gradually decreases.

As the contents of the cell steadily shrink, there comes a point at which they stop pushing outwards on the cell wall, and the cell wall therefore stops pushing back. At this point, pressure potential is 0. If the contents keep on shrinking, then the cell membrane begins to pull away from the cell wall, and plasmolysis takes place. So the point at which pressure potential just becomes 0 is called **incipient plasmolysis**. 'Incipient' means 'just about to begin'.

At incipient plasmolysis, because $\Psi_{p\,inside}$ is 0, then:

$$\Psi_{inside} \; = \; \Psi_{s\,inside}$$

The way in which $\Psi_{p\,inside}$, $\Psi_{s\,inside}$ and the overall Ψ_{inside} change when a plant cell is placed into solutions of different concentrations (water potentials) is shown in the graph in Figure 3.12.

Finding the water potential of the cells in a plant tissue

This can be done using pieces of plant tissue, such as potato tuber. We can do it by measuring changes in mass, or changes in volume, of pieces of tissue.

A good way of doing this is to cut cylinders of potato tuber (remember to peel it first) using a cork borer. Cut each cylinder to exactly the same length. Either measure the mass of each cylinder, or measure its length. Then place each cylinder in a sucrose solution of a different concentration, including pure water. Leave the cylinders long enough for equilibrium to be reached, then remove each one from the solution, pat it dry and measure its new mass or length.

You will find that the tissue placed in pure water will gain in mass or length, as the cells take up water by osmosis and swell. The tissue placed in the most concentrated solution will lose mass or length, as the cells lose water by osmosis and shrink.

Now you need to calculate the percentage change in mass or length of each cylinder. Plot this against the concentration of the sucrose solution. You will get a graph that looks something like the one in Figure 3.13.

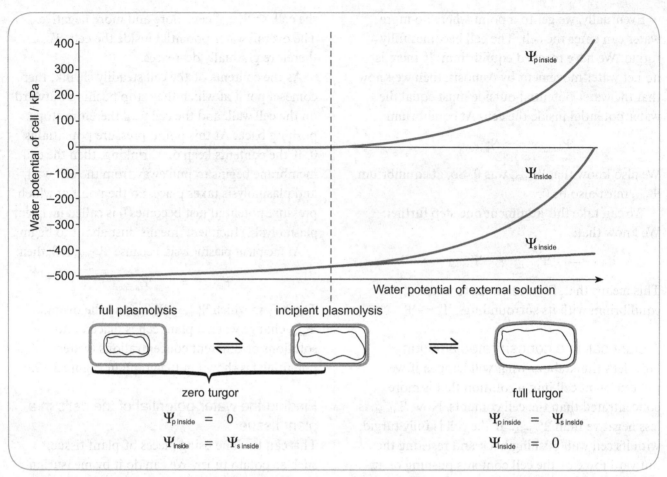

Figure 3.12 The effect of the concentration of the external solution on pressure potential, solute potential and overall water potential of a plant cell. What happens to the volume of the cell as you move from left to right on the graph?

The concentration of the sucrose solution in which we would expect there to be 0 change in the mass or length of the potato cylinders is $0.26\,\text{mol}\,\text{dm}^{-3}$. This means that there would no net movement of water either into or out of the potato tissue in this solution – there is no water potential gradient. The water potential of the potato cells must therefore be equal to the water potential of a $0.26\,\text{mol}\,\text{dm}^{-3}$ sucrose solution. The water potential of the sucrose solution is equal to its solute potential, as it is at atmospheric pressure, so $\Psi_{p\,\text{outside}}$ is 0.

We can look up the solute potential of sucrose solutions in Table 3.3.

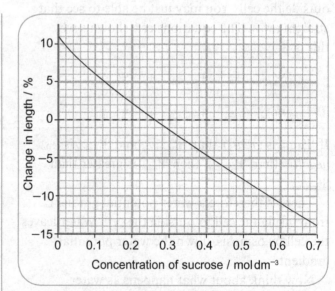

Figure 3.13 Results of an experiment to find the water potential of the cells in potato tuber tissue.

Concentration of sucrose solution / mol dm^{-3}	Solute potential / kPa
0.05	−130
0.10	−260
0.15	−410
0.20	−540
0.25	−680
0.30	−860
0.35	−970
0.40	−1120
0.45	−1280
0.50	−1450
0.55	−1620
0.60	−1800
0.65	−1980
0.70	−2180
0.75	−2370
0.80	−2580
0.85	−2790
0.90	−3000
0.95	−3250
1.00	−3500

Table 3.3 Solute potential of sucrose solutions.

Finding the solute potential of plant cells

You can do this using a thin tissue such as a piece of onion epidermis. This time, you can look at individual cells rather than having to measure changes in a big piece of tissue.

Once again, this involves placing the cells in sucrose solutions of different concentrations. Take several microscope slides, and place a drop of sucrose solution on each slide, a different concentration on each (remember to label the slides first). On one slide, the drop should be pure water. Then cut little pieces of onion epidermis and place one in each drop. Push them down gently with a seeker, so that they are fully immersed, and then carefully place a coverslip over each piece of epidermis. Leave them for 10 minutes or so, to give time for osmosis to take place and equibrium to be reached.

Now look at each slide in turn under the microscope. For all the cells in your field of view, decide whether they are plasmolysed or not. (Figure 3.14 shows what plasmolysed onion cells look like.) Write down the number that are plasmolysed and the number that are not. Then move the slide so that you can do the same in a different part of the epidermis. Keep on until you have counted at least 50 cells in all.

Repeat for each of the other tissues, in their different solutions. Record all your results in a table, and plot a graph of the percentage of cells that were plasmolysed against the concentration of the sucrose solution. You will probably get a graph that looks rather like the one in Figure 3.15.

Now look for the concentration at which 50% of cells are plasmolysed. We can take this as the point at which, on average, the onion cells were at incipient plasmolysis. If you look back at page 67, you will see that, at incipient plasmolysis:

$$\Psi_{inside} = \Psi_{s\,inside}$$

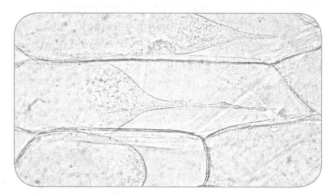

Figure 3.14 Plasmolysed cells in onion epidermis.

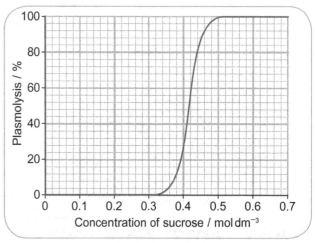

Figure 3.15 Results of an experiment to find the solute potential of the cells in onion epidermis tissue.

So we can say that at this point:

$$\Psi_{\text{inside}} = \Psi_{\text{s inside}}$$

$$\Psi_{\text{outside}} = \Psi_{\text{s outside}} \quad (\text{because } \Psi_{\text{p outside}} \text{ is } 0)$$

$$\Psi_{\text{inside}} = \Psi_{\text{outside}}$$

(because osmosis has stopped, so there is no water potential gradient).

Therefore,

$$\Psi_{\text{s inside}} = \Psi_{\text{s outside}}$$

We can therefore read off the concentration of the sucrose solution at which the curve is at 50% plasmolyis and use Table 3.3 to convert this to a solute potential. This gives us the mean solute potential of the cytoplasm and cell sap inside the onion cells.

Active transport

So far, we have looked at three ways in which substances can move down a concentration gradient (or a water potential gradient) from one side of the plasma membrane to the other. The cell does not have to do anything to make this happen, except perhaps to open a channel to allow facilitated diffusion to take place. These methods are all passive.

However, there are many instances where a cell needs to take up, or get rid of, substances whose concentration gradient is in the 'wrong' direction. This is usually the case with **sodium ions** (Na^+) and **potassium ions** (K^+). Most cells need to contain a higher concentration of potassium ions, and a lower concentration of sodium ions, than the concentration outside the cell. To achieve this, cells constantly pump sodium ions out and potassium ions in, up their concentration gradients. This requires energy input from the cell, so it is called **active transport** (Figure 3.16).

Active transport is carried out by **transporter / carrier proteins** in the plasma membrane, working in close association with ATP, which supplies the energy. The ATP is used to change the shape of the transporter proteins. The shape change moves three sodium ions out of the cell and two potassium ions in. This is going on all the time in most of your cells, and is called the **sodium–potassium pump**. It is estimated that more than a third of the ATP produced in your cells by respiration is used as fuel for the sodium–potassium pump.

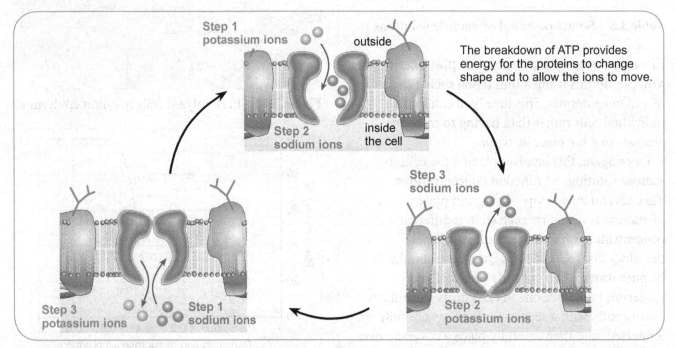

Figure 3.16 An example of active transport – the sodium–potassium pump. Start at step 1 for each ion in turn, and work your way round clockwise. Potassium ions are green and sodium ions are blue.

Direct and indirect active transport

The type of active transport involved in the sodium–potassium pump is **direct active transport**. The transporter protein uses energy from ATP to move sodium ions out of the cell and potassium ions in. This is an example of an **antiport**. However, there are also cases where the ion that is to be moved is not directly moved by the transporter protein. This is called **indirect** transport, and it is often used, for example, to move glucose molecules across cell membranes.

Figure 3.17 shows how indirect active transport works. At one side of the cell, the sodium–potassium pump pushes sodium ions out of the cell. This lowers the concentration of sodium ions inside the cell, producing a concentration gradient across the membrane on the opposite side of the cell. Here, sodium ions passively move down their concentration gradient into the cell, passing through channel proteins. These channel proteins allow glucose to pass through along with the sodium ions, so this is known as **cotransport** and is an example of a **symport**. The energy stored in the sodium ion concentration gradient is used to move the glucose molecules *up* their concentration gradient. The effect is the same as if the glucose molecules had been actively transported into the cell.

This process takes place in the ileum, where it is used to move glucose from the lumen of the ileum and into the cells covering the villi. It also happens in the cells making up the wall of the proximal convoluted tubule in the kidney, absorbing glucose from the glomerular filtrate so that it can be returned to the blood.

Exocytosis and endocytosis

All the mechanisms of movement across membranes that we have looked at so far involve individual ions or molecules moving. Cells can also move substances in bulk across the membrane.

Moving substances *out of* a cell in this way is called **exocytosis** (Figure 3.18). The substance to be released from the cell is contained in a tiny membrane-bound sac called a **vesicle**.

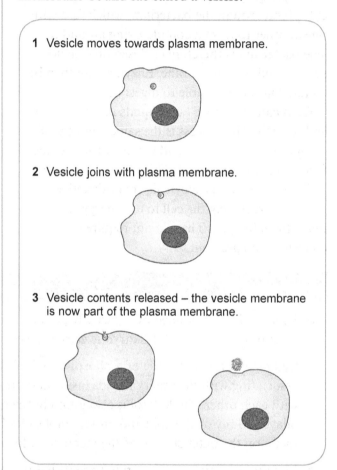

1 Vesicle moves towards plasma membrane.

2 Vesicle joins with plasma membrane.

3 Vesicle contents released – the vesicle membrane is now part of the plasma membrane.

Figure 3.18   Exocytosis.

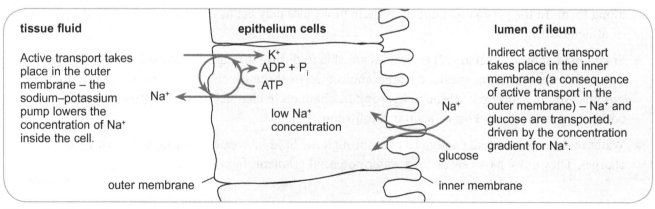

tissue fluid

Active transport takes place in the outer membrane – the sodium–potassium pump lowers the concentration of Na⁺ inside the cell.

epithelium cells

K⁺
ADP + P_i
ATP
Na⁺
low Na⁺ concentration

outer membrane

lumen of ileum

Indirect active transport takes place in the inner membrane (a consequence of active transport in the outer membrane) – Na⁺ and glucose are transported, driven by the concentration gradient for Na⁺.

Na⁺
glucose

inner membrane

Figure 3.17   Indirect active transport of glucose.

The vesicle is moved to the plasma membrane along microtubules. The membrane around the vesicle fuses with the plasma membrane, emptying the vesicle's contents outside the cell.

Moving substances *into* a cell in this way is called **endocytosis** (Figure 3.19). A good example is the way that a phagocyte (a type of white blood cell) engulfs a bacterium (also called phagocytosis). The cell puts out fingers of cytoplasm around the bacterium, which fuse with one another to form a complete ring around it. The bacterium is therefore enclosed in a vacuole, surrounded by a membrane. Enzymes can then be secreted into the vacuole to digest it.

Cells can also move bulk liquids into the cell by endocytosis. The process is the same – fingers of cytoplasm surround a small volume of liquid and form a vacuole around it. This is called **pinocytosis**.

Endocytosis and exocytosis are both active processes, requiring the cell to use energy to make them happen. They do not require a concentration gradient.

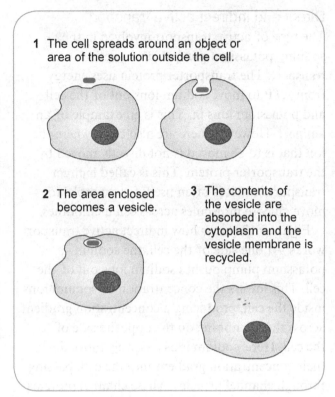

1 The cell spreads around an object or area of the solution outside the cell.

2 The area enclosed becomes a vesicle.

3 The contents of the vesicle are absorbed into the cytoplasm and the vesicle membrane is recycled.

Figure 3.19   Endocytosis.

Summary

- Every cell is surrounded by a partially permeable plasma (cell surface) membrane, which controls what passes through it. Eukaryotic cells also contain many internal membranes.

- Cell membranes are made of a phospholipid bilayer, with the hydrophilic phospholipid heads associating with the watery liquids inside and outside the cell, and the hydrophobic tails associating with each other. Cholesterol is also found in the bilayer. Proteins float within the phospholipid bilayer. Many of the lipid and protein molecules have short chains of carbohydrates attached to them, on the outer surface of the membrane, forming glycolipids and glycoproteins.

- Membranes help to form compartments within the cell within which particular metabolic pathways (series of metabolic reactions, such as aerobic respiration) can take place. They provide surfaces for the attachment of enzymes. In neurones and muscle cells, electrical impulses (action potentials) pass along them. In the plasma membrane, protein molecules may act as receptors or for cell recognition or attachment.

- Small, lipid-soluble (uncharged) molecules are able to diffuse through the phospholipid bilayer, down their concentration gradient. Larger molecules, or those which carry a charge and are therefore not lipid-soluble, must pass through hydrophilic channels in channel proteins, also down their concentration gradient. This is facilitated diffusion.

- Water molecules are small enough to pass through the lipid bilayer, even though they carry small charges. They move passively down a water potential gradient, by osmosis.

continued ...

- Water potential is affected by the concentration of the solute and the pressure exerted on the solution. The contribution of the solute is called the solute potential, and the contribution of the pressure is called the pressure potential. A high concentration of solute lowers the water potential, and a high pressure increases it. $\Psi = \Psi_p + \Psi_s$

- An animal cell placed in a solution of high water potential may absorb so much water by osmosis that it bursts. Plant cells do not burst in these circumstances, because the strong cell wall resists expansion of the cytoplasm, and the cell simply becomes turgid.

- An animal cell placed in a solution of low water potential loses water by osmosis, becoming shrunken or crenated. A plant cell in these circumstances may lose so much water that the cell membrane pulls away from the cell wall, so that the cell becomes plasmolysed.

- The water potential of the cells in a plant tissue can be found by immersing pieces of tissue in solutions of different water potentials and finding the solution that causes no mass or length change in the tissue. The solute potential can be found by observing tissue mounted in different solutions under a microscope, and finding the one which cells are at incipient plasmolysis.

- Substances can be moved across cell membranes against (up) their concentration gradient, through transporter proteins. This requires the input of energy from the cell, in the form of ATP, and is called active transport.

- Bulk liquids, small objects such as bacterial cells, or very large molecules, can be moved into or out of the cell by endocytosis or exocytosis respectively.

Questions

Multiple choice questions

1 The diagram on the right represents the fluid mosaic model of cell membrane structure.

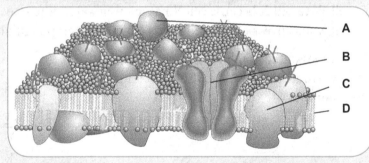

Which of the labelled structures refers to the 'fluid' part of the membrane structure?

2 What is the role of cholesterol in the cell membrane? To:
   A assist in diffusion of carbon dioxide molecules.
   B facilitate the movement of glucose into the cell.
   C regulate fluidity of the membrane.
   D increase the cell's permeability to small water-soluble molecules.

3 The components of the plasma membrane have specific functions. Which of the following structures is involved in cell recognition?
   A cholesterol   **B** glycoprotein   **C** extrinsic protein   **D** phospholipid

4 A cell produces enzymes which are secreted into the digestive tract. By which process does this occur?
   A pinocytosis   **B** exocytosis   **C** phagocytosis   **D** diffusion

continued ...

5   In the diagram to the right
    the structures labelled I to
    IV are different types of
    protein found
    in the membrane.

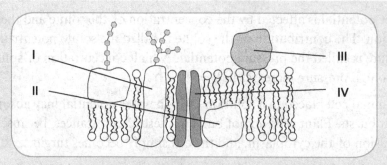

Which of the following correctly identifies the labelled structures?

	I	II	III	IV
A	intrinsic protein	glycoprotein	channel protein	extrinsic protein
B	extrinsic protein	channel protein	glycoprotein	intrinsic protein
C	glycoprotein	intrinsic protein	extrinsic protein	channel protein
D	extrinsic protein	intrinsic protein	channel protein	glycoprotein

6   The diagram below shows the water potentials (ψ) inside three mesophyll plant cells.

cell **W**	cell **X**	cell **Y**
−900 kPa	−800 kPa	−1300 kPa

Which one of the following explains the net movement of water among the cells?
Water moves from:
A cell **Y** to cell **X** because cell **Y** has a higher water potential.
B cell **X** to cells **W** and **Y** because cell **X** has a higher water potential.
C cell **W** to cell **X** because cell **X** has a lower water potential.
D cell **X** to cell **W** because cell **X** has a higher water potential.

7   The curves in the graph to the right show the
    transport rate of substance X by different
    mechanisms.

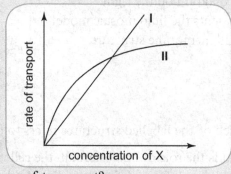

Which of the following correctly describes the mechanisms of transport?

	I	II
A	facilitated diffusion	diffusion
B	active transport	diffusion
C	diffusion	facilitated diffusion
D	active transport	facilitated diffusion

8   A cell uses ATP to transport sodium ions through its plasma membrane against
    a concentration gradient. Which method of transport is the cell using to transport
    the sodium ions?
    A facilitated diffusion   B osmosis   C endocytosis   D active transport

continued...

Structured questions

9 A CAPE® Biology class studied the fluid mosaic model of the cell membrane.
   They were then asked to carry out an investigation on the effect of temperature on
   the membrane's permeability. In the investigation, beetroot was used because it
   contains a red, soluble pigment, betalain, in its vacuoles. This pigment does not
   usually leave the vacuoles under normal conditions.

   The investigation consisted of three basic parts:
   A Making a series of standard solutions (colour standards) using 10% betalain solution.
   B Preparation of the beetroot. The steps were:
      • cutting the beetroot cylinders into 4 mm thickness discs;
      • placing 10 discs into water ranging in temperatures from 30 to 80 °C;
      • leaving them for 5 minutes in the particular temperature;
      • retaining the solution at each temperature and discarding the discs.
   C Comparing each solution with the standard solutions.

   a Students were given 10% betalain solution. They were asked to make standard
      solutions through dilutions. All volumes of the standard solutions were 10.0 cm³.
      Copy and complete the table below, which shows the volumes of 10% betalain and
      water needed for the standard solutions.

Volume of 10% betalain added / cm³	Volume of water added / cm³	Final concentration of standard solution / %
10.0	0.0	
8.0	2.0	
5.0	5.0	
4.0	6.0	
		2.0
		1.0
		0.0

[3 marks]

   b Students were asked to rinse the beetroot discs twice before placing them into the water
      of the required temperature. Give **one** reason why they were asked to do this.   [1 mark]
   c The results obtained for the colour standards and some of the temperatures are
      shown below.

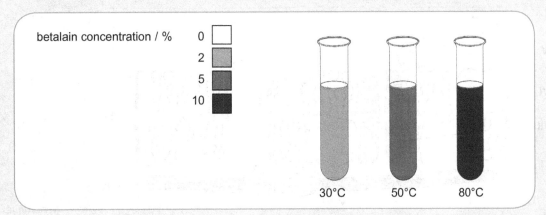

continued ...

    i Before comparing both the colour standards and the tubes at various temperatures, what should the students do to ensure valid results? [1 mark]

    ii Describe the results obtained above. [2 marks]

    iii Explain the results obtained. [4 marks]

    iv What conclusion can you draw from this investigation? [1 mark]

  d Give **one** limitation and **two** sources of error of the procedure used in this investigation. [3 marks]

10 An investigation was carried out to determine the solute potential of epidermal cells of purple onion. The method used was determination of the molar concentration of a sucrose solution which brought about plasmolysis of 50% of cells observed.

  a What is meant by the term 'solute potential'? [1 mark]

The investigation was carried out as follows:

- 5 mm squares of the onion epidermis were placed into test tubes containing sucrose solutions ranging in concentrations from 0.2 M to 0.80 M for 20 minutes.
- Slides of the onion squares were made and viewed using the light microscope under low power.
- 50 cells were observed and the plasmolysed cells were counted.

A student conducted the experiment described above and recorded the following in her notebook:

concentration of sucrose solution / M	0.2	0.4	0.6	0.8
total number of cells observed	50	50	50	50
number of cells plasmolysed	8	18	38	48
percentage plasmolysis				

  b Why was purple onion used in this investigation? [1 mark]

  c Make a labelled drawing of a fully plasmolysed cell. [2 marks]

  d Copy and complete the table shown above. [2 marks]

  e Explain the results obtained by the student. [3 marks]

  f Plot a graph to show the relationship between percentage plasmolysis and molarity of sucrose. [3 marks]

  g Determine the molarity of sucrose which corresponds to 50% plasmolysis. [1 mark]

  h Using the table on page 69, estimate the corresponding solute potential in kPa. [1 mark]

  i State one major source of error in this activity. [1 mark]

11 The diagram below is of a plasma (cell surface) membrane.

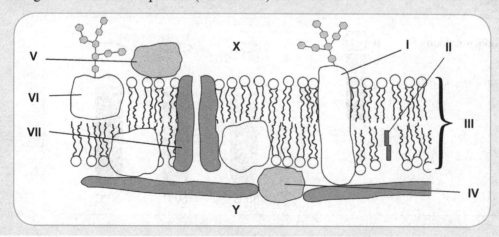

continued ...

a   The model of the cell membrane is described as 'fluid mosaic'. Explain what is
    meant by this term.   [2 marks]
b   Name the structures labelled **I** to **VII**.   [3 marks]
c   i   Which of the surfaces **X** and **Y** identify the inside environment of the
        cell membrane?   [1 mark]
    ii Give a reason for your answer.   [1 mark]
d   There are two basic types of protein in the cell membrane: extrinsic and intrinsic
    proteins. How do these types of protein differ in terms of their chemical properties?   [2 marks]
e   State **two** functions of proteins in the cell membrane.   [2 marks]
f   How are proteins held in membranes?   [2 marks]
g   Describe the chemical properties of the structure labelled **III**.   [2 marks]

Essay questions

12 a   i   By means of an annotated diagram only, describe the structure of the plasma
           (cell surface) membrane.
       ii State the functions of each of the components.   [7 marks]
   b   The internal environment of a cell is dynamic. The cell takes in water, oxygen and
       glucose among other materials. It expels excess water, gets rid of waste and secretes
       substances needed by other cells in the body.
       Describe how the cell transports the following substances across the membrane:
       i   oxygen and carbon dioxide
       ii   sodium and potassium ions
       iii water
       iv   glucose
       v   enzymes to be used by other cells of the body.   [8 marks]

13 a   Diffusion is one way by which substances can cross the plasma (cell surface) membrane.
       Define the term 'diffusion'.   [1 mark]
   b   Explain how the following factors would affect the rate of diffusion across
       the membrane:
       i   molecular size
       ii   temperature
       iii solubility in lipids
       iv   concentration of substance to be transported.   [8 marks]
   c   Active transport and facilitated diffusion are two other ways by which substances
       can cross the plasma (cell surface) membrane. Discuss the similarities and differences
       of both these processes.   [6 marks]

14 a   With the aid of diagrams, describe and explain what happens to red blood cells
       and to plant epidermal cells when placed in:
       i   an isotonic solution
       ii   a hypotonic solution
       iii a hypertonic solution.   [9 marks)]
   b   A nerve cell needs to secrete the neurotransmitter acetylcholine at a synapse,
       while it needs to take up the fat-soluble vitamin A and the water-soluble
       vitamin C. Describe the different routes taken by these three substances to cross
       the cell membrane.   [6 marks]

Chapter 4
Enzymes

By the end of this chapter you should be able to:

a explain that enzymes are globular proteins that catalyse metabolic reactions;

b define the terms metabolism, anabolism and catabolism;

c explain the mode of action of enzymes in terms of an active site, enzyme–substrate complex, lowering of activation energy and enzyme specificity;

d explain the effects of pH, temperature, enzyme concentration and substrate concentration on enzyme action;

e explain the effects of competitive and non-competitive inhibitors on enzyme activity;

f know how to investigate the effects of temperature and substrate concentration on enzyme-catalysed reactions and explain these effects.

Enzymes are protein molecules that can be defined as **biological catalysts**. A catalyst is a substance that speeds up a chemical reaction, but remains unchanged at the end of the reaction.

We can define **metabolism** as all the chemical reactions that take place within an organism's body. Some metabolic reactions are **anabolic** reactions – they involve the joining together of small molecules to produce large ones. An example of an anabolic reaction is the synthesis of protein molecules on ribosomes, by the formation of peptide bonds between amino acids. Anabolic reactions require input of energy. Other metabolic reactions are **catabolic** reactions – they involve breaking large molecules down to produce smaller molecules. An example of a catabolic reaction is respiration, in which glucose is broken down to carbon dioxide and water in a series of many small steps. Catabolic reactions release energy. Both anabolic and catabolic reactions are catalysed by enzymes.

How enzymes work

Some enzymes act inside cells, and are known as **intracellular** enzymes. Examples include **hydrolases** found inside lysosomes, which hydrolyse (break down) substances that a cell has taken in by phagocytosis. **ATPases** are also intracellular enzymes found, for example, inside mitochondria, where they are involved in the synthesis of ATP during aerobic respiration.

Some enzymes act outside cells, and are known as **extracellular** enzymes. These include the digestive enzymes in the alimentary canal, such as amylase, which hydrolyses starch to maltose.

Enzymes are globular proteins. Like all globular proteins, enzyme molecules are coiled into a precise three-dimensional shape – their tertiary structure – with hydrophilic R groups (side chains) on the outside of the molecules, making them soluble in water. Enzyme molecules also have a special feature in that they possess an **active site** (Figure 4.1). The active site of an enzyme is a region, usually a cleft or depression, to which another particular molecule can bind. This molecule is the **substrate** of the enzyme. The shape of the active site allows the substrate to fit perfectly, and to be held in place by temporary bonds that form between the substrate and some of the R groups of the enzyme's amino acids. This combined structure is called the **enzyme–substrate complex** (Figure 4.2). Each type of enzyme will usually act on only one type of substrate molecule. This is because the shape of the

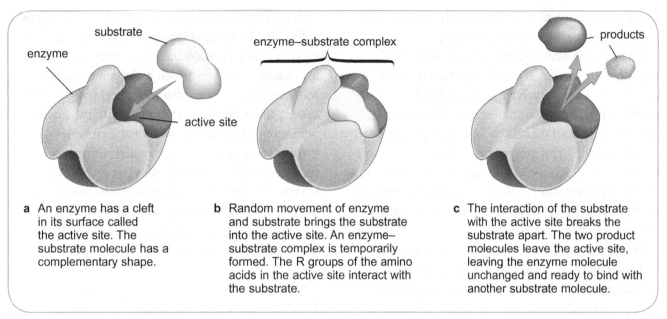

Figure 4.1 How an enzyme catalyses the breakdown of a substrate molecule into two product molecules.

active site will allow only one shape of molecule to fit, like a key fitting into a lock. In most enzymes, when the substrate fits into the active site, the shape of the whole enzyme changes slightly so that it can accommodate and hold the substrate in exactly the right position for the reaction to occur. This is called **induced fit** – the arrival of the substrate molecule causes a change in the shape of the enzyme. Whether it works by a simple lock-and-key mechanism, or by induced fit, the enzyme and

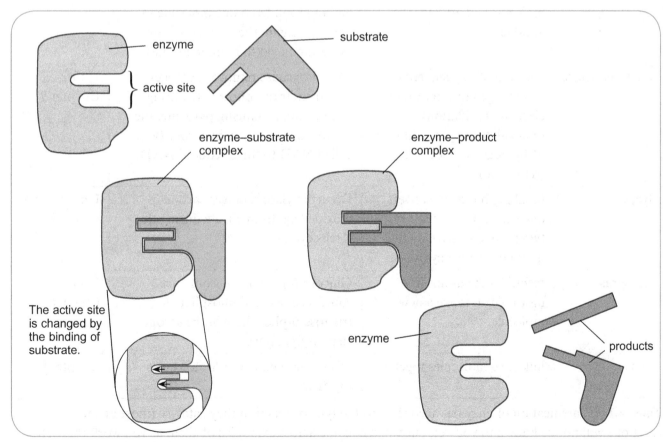

Figure 4.2 A simplified diagram of enzyme function.

its substrate must be a perfect match. The enzyme is said to be **specific** for this substrate.

The enzyme may catalyse a reaction in which the substrate molecule is split into two or more molecules. Alternatively, it may catalyse the joining together of two molecules – for example, when linking amino acids to form polypeptides during protein synthesis on ribosomes. Interaction between the R groups of the enzyme and the atoms of the substrate can break, or encourage formation of, bonds in the substrate molecule. As a result, one, two or more **products** are formed.

When the reaction is complete, the product or products leave the active site. The enzyme is unchanged by the process, so it is now available to receive another substrate molecule. The rate at which substrate molecules can bind to the enzyme's active site, be formed into products and leave can be very rapid. A molecule of the enzyme catalase, for example, can bind with hydrogen peroxide molecules, split them into water and oxygen and release these products at a rate of 10^7 molecules per second. In contrast, one of the enzymes involved in photosynthesis, called rubisco, can only deal with three molecules per second. The maximum number of substrate molecules that can be converted to product per minute is known as the enzyme's **turnover number**. A classification of enzymes is shown in Table 4.1.

SAQ

1 What is the turnover number for catalase?

Type of enzyme	Type of reaction it catalyses	Example	Read more
hydrolase	breaking down polymers by hydrolysis	Carbohydrases hydrolyse glycosidic bonds (e.g. amylase in the alimentary canal).	Unit 2 Chapter 8
transferase	removal of part of one molecule and adding it to another	Amino acyl transferase transfers an amino group from one molecule to another (e.g. in the liver, converting one amino acid to a different one).	
oxidoreductase	addition of oxygen, removal of hydrogen or removal of electrons (oxidation), or removal of oxygen, addition of hydrogen or electrons (reduction)	Dehydrogenase removes hydrogen (e.g. from triose phosphate during respiration, producing pyruvate; the hydrogen is taken up by a coenzyme called NAD to form reduced NAD).	Unit 2 Chapter 2
lyase	breaking bonds to convert one substrate molecule to two products, in a way that does not involve hydrolysis	Decarboxylase removes carbon dioxide (e.g. from citrate during the Krebs cycle).	Unit 2 Chapter 2
isomerase	reshuffles the atoms in a molecule, to make a new molecule	Glucose-6-phosphate isomerase converts glucose-6-phosphate to fructose-6-phosphate in the second step of glycolysis.	Unit 2 Chapter 2
ligase	links two molecules together	DNA ligase links DNA molecules together.	Chapter 8

Table 4.1 Classification of enzymes according to the type of reaction they catalyse (most of these reactions will not be known to you yet, but you will come across examples of them as you work through your biology course).

Activation energy

We have seen that enzymes increase the rates at which chemical reactions occur. Most of the reactions that occur in living cells would occur so slowly without enzymes that they would effectively not happen at all.

In many reactions, the substrate will not be converted to a product unless it is temporarily given some extra energy. This energy is called **activation energy** (Figure 4.3). One way of increasing the rates of many chemical reactions is to increase the energy of the reactants by heating them. You have probably done this on many occasions by heating substances that you want to react together. In the Benedict's test for reducing sugar, for example, you need to heat the Benedict's reagent and sugar solution together before they will react.

Mammals also use this method of speeding up their metabolic reactions. Our body temperature is maintained at 37 °C, which is usually considerably warmer than the temperature of the air around us. But even raising the temperature of cells to 37 °C is not enough to give most substrates the activation energy that they need to change into products. We cannot raise body temperature much more than this, because temperatures above about 40 °C begin to cause irreversible damage to many of the molecules from which we are made, especially protein molecules. Enzymes are a solution to this problem because they decrease the activation energy of the reaction that they catalyse (Figure 4.3b). They do this by holding the substrate or substrates in such a way that their molecules can react more easily. Reactions catalysed by enzymes take place rapidly at a much lower temperature than they would without them.

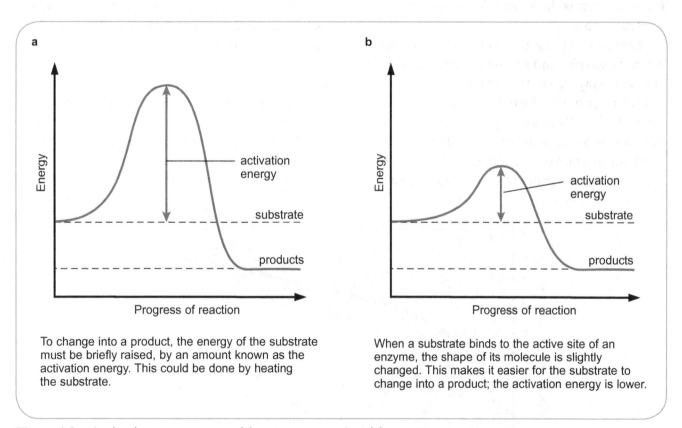

a

To change into a product, the energy of the substrate must be briefly raised, by an amount known as the activation energy. This could be done by heating the substrate.

b

When a substrate binds to the active site of an enzyme, the shape of its molecule is slightly changed. This makes it easier for the substrate to change into a product; the activation energy is lower.

Figure 4.3 Activation energy: **a** without enzyme; **b** with enzyme.

The course of a reaction

You may be able to carry out an investigation into the rate at which substrate is converted into product during an enzyme-controlled reaction. Figure 4.4 shows the results of such an investigation, using the enzyme catalase. This enzyme is found in the tissues of most living things and catalyses the breakdown of hydrogen peroxide into water and oxygen. (Hydrogen peroxide is a very toxic product of several different metabolic reactions.) It is an easy reaction to follow because the oxygen that is released can be collected and measured.

The reaction begins very swiftly. As soon as the enzyme and substrate are mixed, bubbles of oxygen are released. A large volume of oxygen can be collected in the first minute of the reaction. As the reaction continues, however, the rate at which oxygen is released gradually slows down. The reaction gets slower and slower, until it eventually stops.

The explanation for this is quite straightforward. When the enzyme and substrate are first mixed, there are many substrate molecules. Enzyme and substrate bind at incredible speed so, at any moment, virtually every enzyme molecule has a substrate molecule in its active site. The rate at which the reaction occurs will depend only on how many enzyme molecules there are, and the speed at which each enzyme molecule can bind with another substrate molecule.

However, as more and more substrate is converted into product, there are fewer and fewer substrate molecules to bind with enzymes. Enzyme molecules may be 'waiting' for a substrate molecule to move in their direction and, by chance, hit their active site. As fewer substrate molecules are left, the reaction gets slower and slower, until it eventually stops.

The curve in Figure 4.4 is therefore steepest at the beginning of the reaction: the rate of an enzyme-controlled reaction is always fastest at the beginning. This rate is called the **initial rate of reaction**. You can measure the initial rate of the reaction by calculating the slope of a tangent to the curve, as close to time 0 as possible. An easier way of doing this is simply to read off the graph the amount of oxygen given off in the first 30 seconds. In this case, the initial rate of oxygen production is 2.7 cm³ of oxygen per 30 seconds, or 5.4 cm³ per minute.

SAQ

2 Why is it better to calculate the initial rate of reaction from a curve such as the one in Figure 4.4, rather than simply measuring how much oxygen is given off in the first 30 seconds?

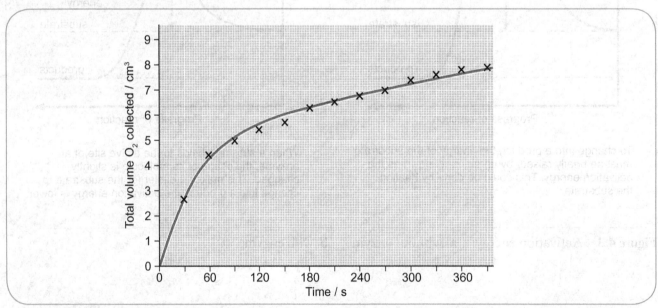

Figure 4.4 The course of an enzyme-catalysed reaction. Catalase was added to hydrogen peroxide at time 0. The gas released was collected in a gas syringe, and the volume read at 30 s intervals.

The effect of enzyme concentration

Figure 4.5 shows the results of an investigation in which different concentrations of catalase were added to the same volume of a hydrogen peroxide solution. You can see that the shapes of all five curves in Figure 4.5a are similar. In each case, the reaction begins very quickly (steep curve) and then gradually slows down (curve levels off). The amounts of hydrogen peroxide are the same in all five reactions, so the total amount of oxygen eventually produced will be the same. Eventually, all the curves will meet.

To compare the rates of these five reactions, in order to look at the effect of enzyme concentration on reaction rate, it is fairest to look at the rate *right at the beginning* of the reaction. This is because, once the reaction is under way, the amount of substrate in each reaction begins to vary, as substrate is converted to product at different rates in each of the five reactions. It is only at the very beginning of the reaction that we can be sure that differences in reaction rate are caused only by differences in enzyme concentration.

To work out this initial rate for each enzyme concentration, we can calculate the slope of the curve 30 seconds after the beginning of the reaction. (Ideally, we should do this for an even earlier stage of the reaction – as close to time 0 as possible, as explained earlier – but in practice this is impossible.) We can then draw a second graph (Figure 4.5b) showing this initial rate of reaction plotted against enzyme concentration.

This graph shows that the initial rate of reaction increases linearly as enzyme concentration increases. In these conditions, reaction rate is directly proportional to the enzyme concentration. This is just what common sense says should happen. If you double the number of enzyme molecules present, then twice as many active sites will be available for the substrate to slot into. As long as there is plenty of substrate available, the initial rate of a reaction increases linearly with enzyme concentration.

SAQ
3 Sketch the shape that Figure 4.5b would have if excess hydrogen peroxide was not available.

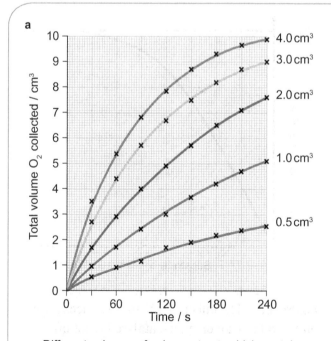

Different volumes of celery extract, which contains catalase, were added to the same volume of hydrogen peroxide. Water was added to make the total volume of the mixture the same in each case.

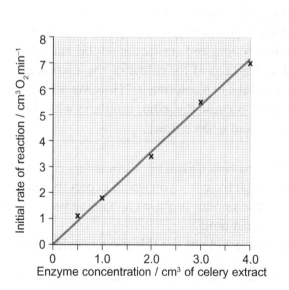

The rate of reaction in the first 30 s was calculated for each enzyme concentration.

Figure 4.5 The effect of enzyme concentration on the rate of an enzyme-catalysed reaction.

Measuring reaction rate

It is easy to measure the rate of the catalase–hydrogen peroxide reaction, because one of the products is a gas, which is released and can be collected. Unfortunately, it is not always so easy to measure the rate of a reaction. If, for example, you wanted to investigate the rate at which amylase breaks down starch to maltose, it would be very difficult to observe the course of the reaction because the substrate (starch) and the product (maltose) remain as colourless substances in the reaction mixture.

The easiest way to measure the rate of this reaction is to measure the rate at which starch disappears from the reaction mixture. This can be done by taking samples from the mixture at known times, and adding each sample to some iodine in potassium iodide solution. Starch forms a blue-black colour with this solution, but maltose does not. Using a colorimeter, you can measure the intensity of the blue-black colour obtained, and use this as a measure of the amount of starch still remaining. If you do this over a period of time, you can plot a curve of 'amount of starch remaining' against 'time'. You can then calculate the initial reaction rate in the same way as for the catalase–hydrogen peroxide reaction.

It is even easier to observe the course of this reaction if you mix starch, iodine in potassium iodide solution, and amylase in a tube, and take regular readings of the colour of the mixture in this one tube in a colorimeter. However, this is not ideal, because the iodine interferes with the reaction and slows it down.

SAQ

4 a Sketch the curve you would expect to obtain if the amount of starch remaining was plotted against time.

 b How could you use this curve to calculate the initial reaction rate?

The effect of substrate concentration

Figure 4.6 shows the results of an investigation in which the concentration of catalase was kept constant, and the concentration of hydrogen peroxide was varied. Once again, curves of 'oxygen released' against 'time' were plotted for each reaction, and the initial rate of reaction calculated for the first 30 seconds. These initial rates of reaction were then plotted against substrate concentration.

As substrate concentration increases, the initial rate of reaction also increases. Again, this is just what we would expect – the more substrate molecules there are around, the more often an enzyme's active site can bind with one. However, if we go on increasing substrate concentration, keeping the enzyme concentration constant, there comes a point where every enzyme active site is working continuously. If more substrate is added, the enzyme simply cannot work faster: substrate molecules are effectively 'queuing up' for an active site to become vacant. The enzyme is working at its maximum possible rate, known as V_{max}.

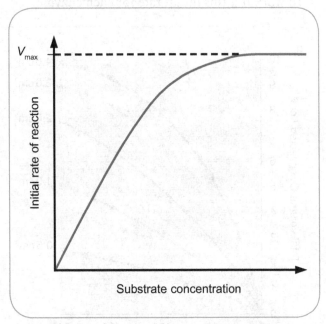

Figure 4.6   The effect of substrate concentration on the rate of an enzyme-catalysed reaction.

Temperature and enzyme activity

Figure 4.7 shows how the rate of a typical enzyme-catalysed reaction varies with temperature. At low temperatures, the reaction takes place only very slowly. This is because molecules are moving relatively slowly. Substrate molecules will not often collide with the active site of an enzyme molecule, and so binding between substrate and enzyme is a rare event. As temperature rises, the enzyme and substrate molecules move faster. Collisions happen more frequently, so that substrate molecules enter the active sites more often. Moreover, when they do collide, they do so with more energy, so more of them will have sufficient activation energy to react. It is easier for bonds to be broken so that the reaction can occur.

As temperature continues to increase, the speed of movement of the substrate and enzyme molecules also continues to increase. However, above a certain temperature the structure of the enzyme molecules vibrates so energetically that some of the bonds holding the enzyme molecule in its precise shape begin to break. This is especially true of hydrogen bonds. The enzyme molecule begins to lose its shape and activity and is said to be **denatured**. This is often irreversible.

In the reaction illustrated in Figure 4.7, at temperatures just above 40 °C, the substrate molecule fits less well into the active site of the enzyme, so the rate of the reaction slows down slightly compared with its rate just below 40 °C. At higher temperatures, the substrate no longer fits at all, or can no longer be held in the correct position for the reaction to occur.

The temperature at which an enzyme catalyses a reaction at the maximum rate is called the **optimum temperature**. Most human enzymes have an optimum temperature of around 40 °C. By keeping our body temperatures at about 37 °C, we ensure that enzyme-catalysed reactions occur at close to their maximum rate. It would be dangerous to maintain a body temperature of 40 °C, because even a slight rise above this would begin to denature enzymes.

Enzymes from other organisms may have different optimum temperatures. Some enzymes, such as those found in bacteria that live in hot springs, have much higher optimum temperatures (Figure 4.8). Some plant enzymes have lower optimum temperatures, depending on their habitat.

Figure 4.8 Bacteria living in hot springs such as this one in Yellowstone National Park, USA, are able to tolerate very high temperatures. Enzymes from such organisms are proving useful in various industrial applications.

SAQ

5 How could you carry out an experiment to determine the effect of temperature on the rate of breakdown of hydrogen peroxide by catalase?

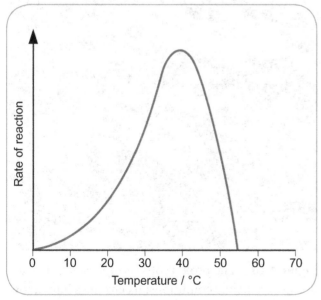

Figure 4.7 The effect of temperature on the rate of an enzyme-controlled reaction.

Q₁₀ temperature coefficient

The Q_{10} **temperature coefficient** is the increase in the rate of a reaction that takes place when the temperature is raised by 10 °C. For many reactions in living organisms, Q_{10} is 2. This means that the rate doubles for each increase of 10 °C. The rate increases because the reactants, or the enzyme and substrate, have more kinetic energy at higher temperatures.

However, this only applies below the optimum temperature. Above the optimum temperature denaturation of the enzyme significantly slows down the reaction.

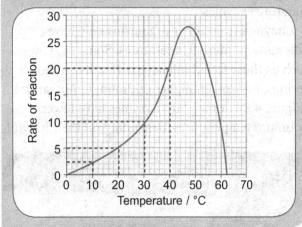

pH and enzyme activity

Figure 4.9 shows how the activity of an enzyme is affected by pH. Most enzymes work fastest at a pH somewhere around 7 – that is, in fairly neutral conditions. Some, however, such as the protease pepsin which is found in the acidic conditions of the stomach, or enzymes from certain bacteria, have a different optimum pH (Figure 4.10).

pH is a measure of the concentration of hydrogen ions in a solution; the lower the pH, the higher the hydrogen ion concentration. Hydrogen ions can interact with the R groups of amino acids – this influences the way in which they bond with each other and therefore affects tertiary structure. A pH that is very different from the optimum pH can cause denaturation of an enzyme.

Even very small changes in pH can have a large effect on the shape, and therefore the activity, of enzyme molecules. Many of the R groups of any protein molecule, including enzymes, contain

$-NH_2$ or $-COOH$ groups. As you saw on page 19, these can ionise to become $-NH_3^+$ or $-COO^-$, and the strong attraction between these groups on different parts of the molecule helps to maintain the tertiary structure (page 18).

However, if pH is low (acidic conditions), the concentration of H^+ in the solution is high, and this causes the $-COOH$ groups to remain unionised. If pH is high, the concentration of H^+ in the solution is low, and $-NH_3$ groups do not ionise. Either of these conditions can therefore cause ionic bonds between R groups in the enzyme molecule to break, thus denaturing the enzyme.

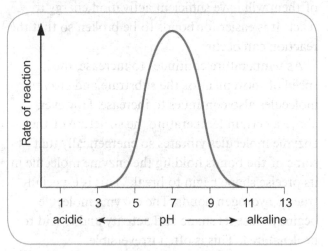

Figure 4.9 The effect of pH on the rate of an enzyme-controlled reaction.

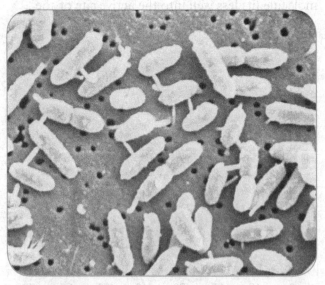

Figure 4.10 The prokaryote *Ferroplasma* is able to live in very acidic conditions; its enzymes have a low optimum pH. It may prove useful in breaking down acidic wastes from mining operations (× 7500).

Enzyme inhibitors

A substance that slows down or stops an enzyme-controlled reaction is said to be an **inhibitor**.

As we have seen, the active site of an enzyme fits, or adjusts to fit, one particular substrate perfectly. It is possible, however, for some *other* molecule to bind to an enzyme's active site if it is very similar to the enzyme's substrate. This would inhibit the enzyme's function.

If the inhibitor molecule binds only briefly to the site then there is competition between it and the substrate for the site. If there is much more of the substrate than the inhibitor present, substrate molecules can easily bind to the active site in the usual way and so the enzyme's function is hardly affected. However, if the concentration of the inhibitor rises or that of the substrate falls, it becomes more and more likely that the inhibitor will collide with an empty site and bind with the enzyme, rather than the substrate. This is known as **competitive inhibition** (Figure 4.11). It is said to be **reversible** because it can be reversed by increasing the concentration of substrate.

An example of competitive inhibition involves the enzyme succinic dehydrogenase. This enzyme is important in the Krebs cycle, where it converts succinate to fumarate. Its substrate is therefore succinate. However, another compound, malonate, has a similar shape to succinate and can fit into the active site of succinic dehydrogenase. Malonate is therefore a competitive inhibitor of succinic dehydrogenase (Figure 4.12).

Sometimes, the inhibitor can remain permanently bonded with the active site and cause a permanent block to the substrate. This kind of inhibition is **irreversible**. Even if more substrate is added, it cannot displace the inhibitor from the active site. The antibiotic penicillin works like this. It permanently occupies the active site of an enzyme that is essential for the synthesis of bacterial cell walls.

A different kind of inhibition takes place if a molecule can bind to another part of the enzyme, called an **allosteric site**. This can seriously disrupt the normal arrangement of bonds holding the enzyme in shape. The resulting distortion ripples across the molecule to the active site, making it unsuitable for the substrate. The enzyme's function is blocked no matter how much substrate is present so this is **non-competitive inhibition** (Figure 4.13). It can be **reversible inhibition** or **irreversible inhibition**, depending on whether the inhibitor bonds briefly or permanently with the enzyme. Digitalis – a substance extracted from foxglove plants – is an example of a non-competitive inhibitor. It binds with the enzyme ATPase, resulting in an increase in the contraction of heart muscle.

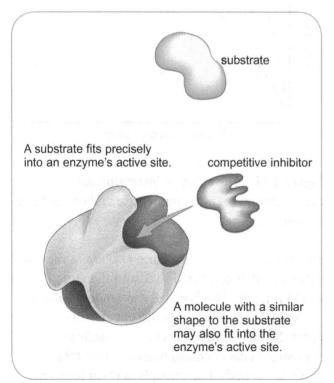

A substrate fits precisely into an enzyme's active site.

substrate

competitive inhibitor

A molecule with a similar shape to the substrate may also fit into the enzyme's active site.

Figure 4.11   Competitive inhibition.

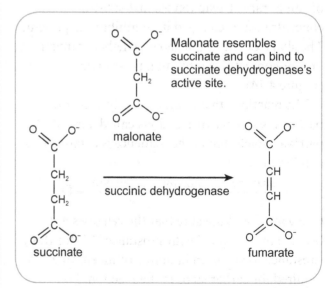

Malonate resembles succinate and can bind to succinate dehydrogenase's active site.

malonate

succinate

succinic dehydrogenase

fumarate

Figure 4.12   Competitive inhibition by malonate.

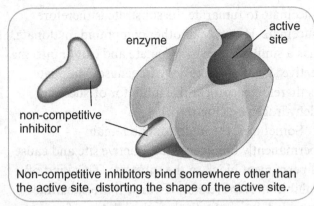

Non-competitive inhibitors bind somewhere other than the active site, distorting the shape of the active site.

Figure 4.13   Non-competitive inhibition.

Heavy metals, such as lead and mercury, may affect enzyme activity in several different ways. One of these is by binding permanently with −SH groups within enzyme molecules, either within or outside the active site. This breaks the disulphide bridges that help to maintain the tertiary structure of the enzyme, altering the shape of the active site and therefore preventing the enzyme from binding with its substrate.

Another example of non-competitive inhibition is the action of organophosphates on acetylcholinesterase, an enzyme found at synapses where it breaks down the neurotransmitter acetylcholine. Organophosphates, such as malathion, are used as insecticides or as nerve gases in warfare. By preventing acetylcholinesterase functioning, they stop synapses working and therefore interfere with normal neuro-muscular activity.

In an enzyme-catalysed reaction, the graph of initial rate of reaction against substrate concentration is changed if an inhibitor is present. The shape of the graph shows whether competitive or non-competitive inhibition is occurring (Figure 4.14).

The reactions that enzymes catalyse are often part of a sequence of reactions called a metabolic pathway. Each step in the sequence is catalysed by a different enzyme.

$$A \xrightarrow[1]{enzyme} B \xrightarrow[2]{enzyme} C \xrightarrow[3]{enzyme} D$$

One way of making sure that the cell does not become oversupplied with substance D is for this substance to act as an inhibitor of an enzyme required for earlier steps in the reaction. For example, substance D might inhibit enzyme 1. This

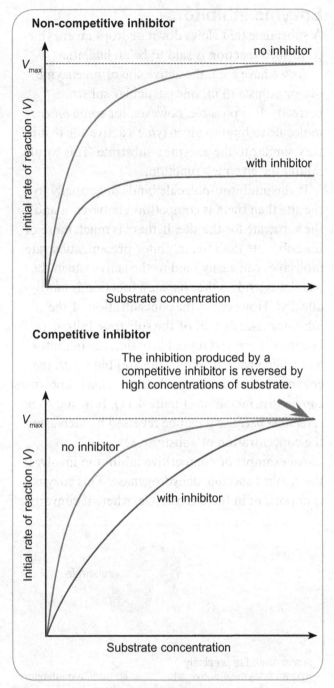

Figure 4.14   The effect of increasing substrate concentration on non-competitive and competitive inhibition.

means that the more substance D there is, the less substance A is converted to B, and therefore the less D is made. This is called **end-product inhibition**.

Inhibitors that seriously disrupt enzyme-controlled reactions can act as **metabolic poisons**, preventing vital chemical reactions that take place in the body. For example, a toxin found in the death cap mushroom, called alpha-amanitin, inhibits enzymes that catalyse the production of RNA from DNA. When this happens, cells are no longer able to synthesise proteins. This mushroom is one of the deadliest fungi known.

Summary

- Enzymes are globular proteins that act as catalysts by lowering activation energy. They control virtually all metabolic reactions, including both anabolic and catabolic reactions, both intracellularly and extracellularly.

- Each enzyme has an active site whose shape and chemical structure allow it to bind specifically with one substrate, forming an enzyme–substrate complex. Usually, the presence of the substrate causes changes in the shape of the active site so that it binds perfectly, a process called induced fit.

- Anything that affects the shape of the active site, such as temperatures above the enzyme's optimum temperature, high or low pH values, or the presence of a non-competitive inhibitor, reduce the ability of the substrate to bind with the enzyme and therefore reduce its activity.

- Competitive inhibitors have a shape that is similar to the enzyme's normal substrate, and can bind with the active site. Their presence therefore reduces the likelihood of the real substrate binding, and therefore reduces the rate of reaction.

- Increasing substrate concentration in relation to the concentration of a competitive inhibitor lessens the degree of inhibition. If substrate concentration is high, then the enzyme's maximum rate of activity, V_{max}, remains unchanged. However, increasing the substrate concentration does not reduce the effect of non-competitive inhibitors, and V_{max} remains lower than normal.

- Malonate is a competitive inhibitor of succinic dehydrogenase, whose normal substrate is succinate. Organophosphates are non-competitive inhibitors of acetylcholinesterase, which normally breaks down acetylcholine at synapses.

Questions

Multiple choice questions

1 The diagram below represents an enzyme and four other molecules that could combine with it.

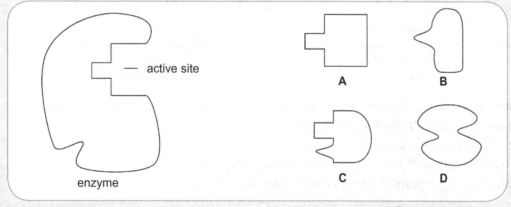

Which molecule is the substrate for the enzyme?

continued ...

2 How do enzymes increase the rate of a reaction? They:
   A cause random collisions more frequently.
   B bring the reacting molecules into precise orientation.
   C increase the activation energy of the reaction.
   D shift the point of equilibrium of the reaction.

3 In non-competitive inhibition, the inhibitor attaches to:
   A the substrate, preventing it from attaching to the active site.
   B the active site, preventing the substrate from attaching to it.
   C an allosteric site, thereby changing the shape of the enzyme.
   D both substrate and active site, preventing the formation of enzyme–substrate complexes.

4 In an experiment to determine the effect of temperature on the activity of the enzyme
   amylase, the times taken for the starch to be hydrolysed at various temperatures were
   recorded. Which graph shows the results of this experiment?

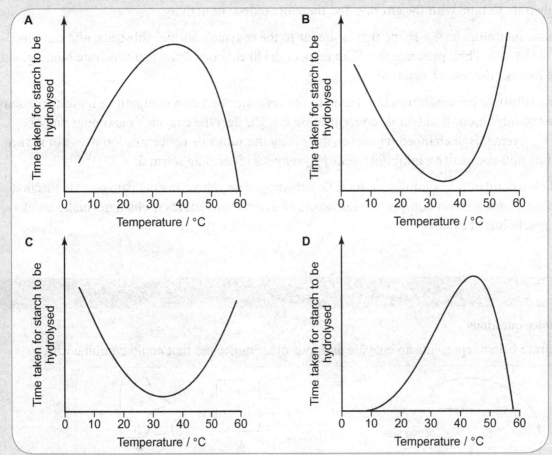

5 Enzymes are denatured at high temperatures. What happens when an enzyme is denatured?
   A The turnover number of the enzyme increases.
   B The activation energy is lowered.
   C The tertiary structure of the enzyme molecule is disrupted.
   D The optimum temperature of the enzyme is increased.

continued ...

6 The graph below shows the effect of substrate concentration on the rate of an enzyme-controlled reaction. The concentration of the enzyme remains constant.

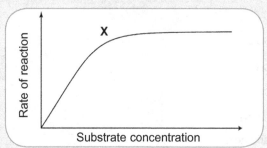

The reason that the curve reaches a plateau at X and does not increase any further at high substrate concentration is that:

A the number of enzyme molecules is in excess.

B the active sites of the enzyme molecules are saturated.

C there are no more substrate molecules present.

D there is a competitive inhibitor present.

7 The diagram below shows a reaction with and without an enzyme.

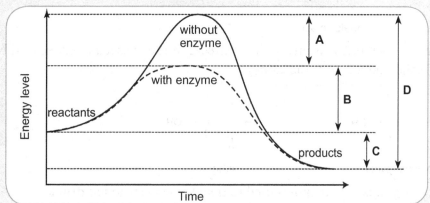

Which of the labelled sections of the curves represents the activation energy of the reaction with enzyme?

8 A metabolic pathway is shown below.

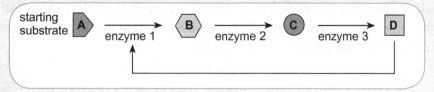

Which of the following correctly identifies how this pathway is controlled?

	Where inhibitor binds to enzyme	How long inhibitor binds to enzyme	Type of inhibitor
A	active site	temporarily	competitive
B	active site	permanently	irreversible non-competitive
C	allosteric site	permanently	reversible non-competitive
D	allosteric site	temporarily	reversible non-competitive

continued ...

Structured questions

9 The reaction below occurs during aerobic respiration. The reaction is catalysed by succinic dehydrogenase.

COOH   COOH
| |
CH₂   succinic   CH
| dehydrogenase   ‖
CH₂   ────────►   CH
| |
COOH   COOH

succinic acid   fumaric acid

    a Name the substrate in the above reaction.   [1 mark]

    b The molecule malonic acid, which is shown below, inhibits this reaction.
It does not bind permanently to the enzyme. Using diagrams, describe how
malonic acid inhibits the enzyme succinic dehydrogenase.   [3 marks]

COOH
|
CH₂
|
COOH
malonic acid

    c Ethylene glycol, shown below, is present in antifreeze. It is converted to oxalic acid by
an enzyme. Oxalic acid damages the kidney.

        H   H   H   H
        | | | |
HO — C — C — OH   H — C — C — OH
        | | | |
        H   H   H   H

        ethylene glycol   ethanol

Explain why a large dose of ethanol is used to treat someone who has drunk
antifreeze accidently.   [3 marks]

    d Heavy metals such as silver and mercury bind permanently to −SH groups of amino
acids present in enzymes. These groups could be in the active site or elsewhere in
the enzyme. Using diagrams, explain how heavy metals inhibit enzyme activity.   [4 marks]

    e The graph below represents the relationship between an enzyme and the concentration
of its substrate under optimal conditions. Sketch **two** curves that would represent
the progress of the reactions taking place in **b** and **d**.   [4 marks]

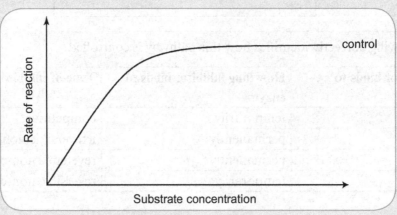

continued ...

10 a Enzymes catalyse metabolic reactions.
What is meant by the term 'metabolic'? [2 marks]

b Describe the ways in which an enzyme interacts with its substrate. [3 marks]

c Many factors affect the activity of an enzyme. Temperature and pH are two of them.

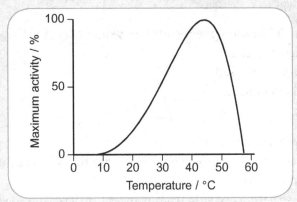

i With reference to the graph above, describe and explain the effect of temperature on the rate of enzyme reaction. [5 marks]

ii Copy and complete the diagram below to show the effect of pH changes on the interacting groups in the active site of an enzyme which has an optimum pH of 7.

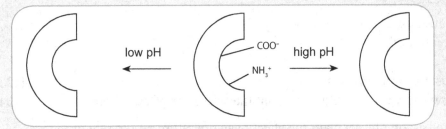

State the expected activity of the enzyme at the high and low pH values. [5 marks]

11 The reaction mechanism of an enzyme is shown below.

enzyme (E) + substrate (S) $\rightleftharpoons$ ES $\longrightarrow$ product (P)

Investigations into enzyme activity are common in CAPE® Biology. There are two ways by which enzyme activity is measured.

a State the **two** ways of measuring the rate of enzyme activity. [2 marks]

An experiment was conducted to investigate the effect of pH on the rate of the enzyme, catalase. The reaction is shown below:

$$2H_2O_2 \rightleftharpoons H_2O + O_2$$

The activity of the enzyme was measured at different values of pH using buffered solutions. The temperature was kept constant at 37 °C. All other variables were kept constant.

The results of the experiment are shown below.

pH	Volume of oxygen collected / cm³ per minute
3	0.5
5	1.0
6	3.5
7	8.5
9	7.2

continued ...

b i Plot a graph to show the relationship between pH and volume of
oxygen collected. [4 marks]

ii With reference to the data, describe the effect of pH on activity of catalase. [2 marks]

iii Suggest an explanation for the activity of the enzyme at pH 3, pH 7
and pH 9. [3 marks]

c Draw a curve on the graph in **b i** to show the results you would expect if the
same investigation was carried out at 22 °C. [2 marks]

d Explain why buffered solutions were used. [1 mark]

e List **two** variables that were kept constant. [1 mark]

Essay questions

12 a i Explain what is meant by 'activation energy'. [2 marks]

ii Draw and label a graph to show the energy changes during the course of a
reaction with and without an enzyme. [2 marks]

b With reference to molecular structure, explain what is meant by 'enzyme specificity'. [4 marks]

c With reference to molecular structure, explain the mode of action of an enzyme. [7 marks]

13 a Enzyme activity is affected by many factors. Describe and explain the effect of
the following factors on enzyme activity:

i pH

ii enzyme concentration

iii substrate concentration

iv inhibitors

Use graphs to illustrate your answers. [12 marks]

b Briefly explain how a substrate is held in place at the active site of an enzyme. [3 marks]

14 a Using an annotated graph only, explain the effect of temperature on the rate of
an enzyme-catalysed reaction. [4 marks]

Bioactive detergents are now on the market. These detergents contain enzymes
which break down various biological stains such as blood, food and grass.

b i What types of enzymes do you expect to be present in the detergent? [2 marks]

ii Suggest, with reasons, what would be the best conditions of use to obtain
maximum cleaning power from the bioactive detergents. [5 marks]

c A number of drugs affect enzyme activity in the body. Aspirin and
penicillin (an antibiotic) are two such drugs.

i Pain is cause by release of chemicals known as prostaglandins. One of the enzymes
in the metabolic pathway to produce prostaglandins is cyclo-oxidase (COX).
Aspirin attaches to a side group of an amino acid close to the active site of COX.
Suggest how aspirin affects the action of the enzyme COX. [2 marks]

ii Penicillin stops bacterial infection by preventing a reaction involved in the
formation of cell walls. The reaction is catalysed by the enzyme transpeptidase.
Penicillin resembles the substrate that binds to transpeptidase.
It forms a permanent bond with the enzyme. Suggest how penicillin affects the
action of transpeptidase, thereby killing bacteria. [2 marks]

Chapter 5
Structure and role of nucleic acids

By the end of this chapter you should be able to:

a illustrate the structure of RNA and DNA using simple labelled diagrams;

b recognise the structural formulae of nucleotides, ribose, deoxyribose, pyrimidines and purines;

c explain the importance of hydrogen bonds and base pairing in DNA replication;

d recognise the significance of semiconservative replication;

e explain the relationship between the sequence of nucleotides in DNA and the amino acid sequence in a polypeptide, and recognise the significance of the genetic code;

f describe the roles of DNA and the different types of RNA in protein synthesis;

g recognise the significance of 5' and 3', initiation, transcription, translation and termination;

h explain the relationship between the structure of DNA, protein structure and the phenotype of an organism;

i describe the relationship between DNA, chromatin and chromosomes.

Polynucleotides

It is amazing to realise that until the middle of the 20th century we did not even know that DNA is the genetic material. Our DNA carries the genetic code – a set of instructions telling the cell the sequence in which to link together amino acids when proteins are being synthesised. Slight differences in the structure of these proteins may result in slight differences in our metabolic reactions. Partly for this reason, we are all slightly different from one another.

You probably know that DNA is a 'double helix'. A DNA molecule is made of two long chains of **nucleotide** molecules, linked together to form a twisted ladder. Each chain is called a **polynucleotide**.

The structure of DNA

DNA stands for **deoxyribonucleic acid**. When it was discovered, it was given the name 'nucleic acid' because it was mostly found in the nuclei of cells and is slightly acidic.

Each nucleotide in a DNA molecule contains:
- a phosphate group;
- the five-carbon sugar **deoxyribose**;
- an organic base.

Figure 5.1 shows the components of a nucleotide in DNA. There are four nucleotides, determined by the base present, which can be **adenine**, **guanine**, **thymine** or **cytosine**. The bases are usually abbreviated to A, G, T and C. Adenine and guanine each contain two rings in their structure. They are known as **purine bases**. Thymine and cytosine have only one ring. They are known as **pyrimidine bases**.

Figure 5.2 and Figure 5.3 show how these components are linked in nucleotides and how the nucleotides link together to form long chains called polynucleotides.

You can see that the base in each nucleotide sticks out sideways from the chain. In DNA, two chains of nucleotides lie side by side, one chain running one way and the other in the opposite direction (Figure 5.4, page 97). They are said to be

anti-parallel. The bases of one chain link up with the bases of the other by means of **hydrogen bonds** (Figure 5.5). The whole molecule twists to produce the double helix shape.

The key to the ability of DNA to hold and pass on the code for making proteins in the cell is the way in which these bases link up. There is just the right amount of space for one large base – a purine – to link with one smaller base – a pyrimidine.

And the linking is even more particular than that: A can only link with T, and C can only link with G. This is called **complementary base pairing**.

Complementary base pairing ensures that the code carried on a molecule of DNA can be copied perfectly over and over again, so it can be passed down from cell to cell and from generation to generation. It also enables the code on the DNA to be used to instruct the protein-making machinery in a cell to construct exactly the right proteins.

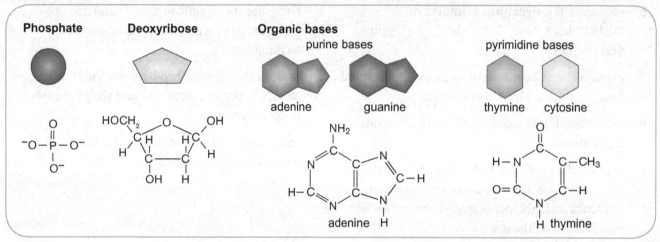

Figure 5.1   The components of a nucleotide in DNA.

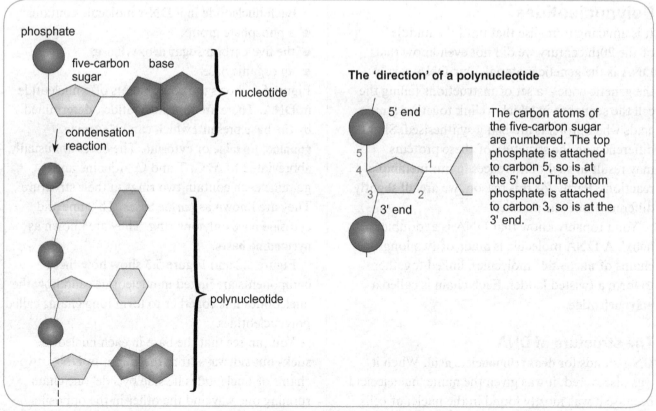

Figure 5.2   How nucleotides join to form a polynucleotide.

free nucleotide
growing polynucleotide
H_2O
phosphodiester bond
A condensation reaction forms a phosphodiester bond.

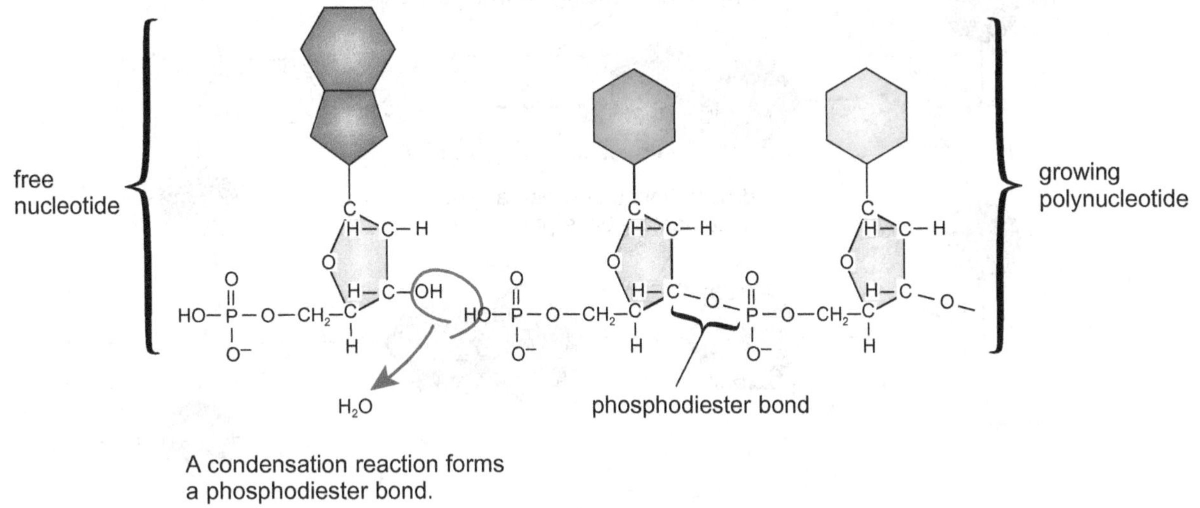

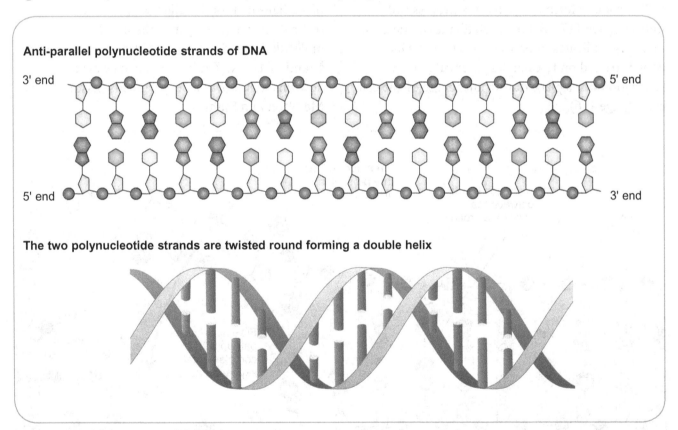

Figure 5.3 Polymerisation of DNA by formation of phosphodiester bonds.

Anti-parallel polynucleotide strands of DNA
3' end
5' end
5' end
3' end
The two polynucleotide strands are twisted round forming a double helix

Figure 5.4 The structure of DNA.

SAQ

1 One end of a DNA strand is called the 5' ('five prime') end, and the other is the 3' end. Look at Figure 5.2 and explain why they are given these names.

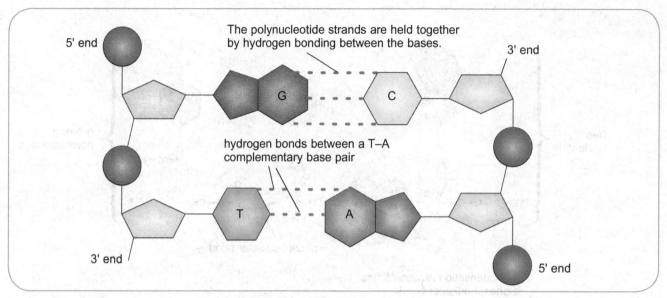

Figure 5.5   Hydrogen bonding joining the bases in DNA.

DNA replication

Cells divide to form new cells, in a process called mitosis (page 117). Before a cell divides by mitosis, its DNA replicates to produce two copies. One copy is passed on to each daughter cell. DNA replication takes place during interphase of the cell cycle (page 116).

Figure 5.6 shows how DNA replication takes place. This method is called **semiconservative replication**, because each of the new DNA molecules is made of one old strand and one new strand of DNA. The box on page 99 explains how this was shown by experiments performed by Meselson and Stahl.

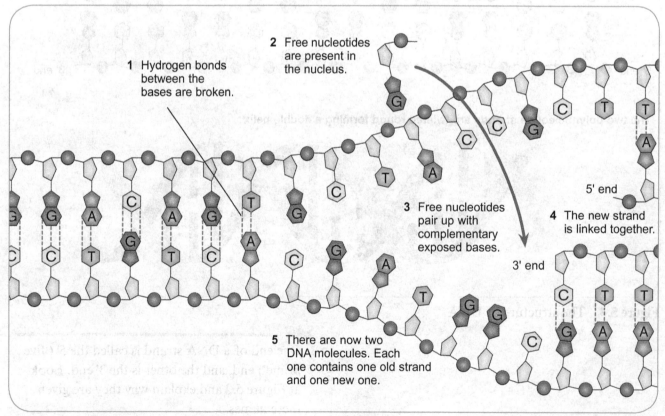

Figure 5.6   DNA replication.

Meselson and Stahl's proof of semiconservative replication

Semiconservative replication of DNA was suggested by Watson and Crick in 1953 when they discovered the structure of DNA. The experiments to determine whether this was correct were carried out by Meselson and Stahl and represent an early example of the use of chemical isotopes in biological research. Meselson and Stahl had to show that two other theories that had been proposed to explain how DNA was replicated were not correct.

One theory was that a completely new DNA molecule, with two new strands, might be made from the original one. This was called conservative replication. Another possibility, called dispersive replication, involved parts, but not all, of each strand of the old DNA molecule becoming part of the new ones, with new bits scattered in between.

Meselson and Stahl used the bacterium *Escherichia coli*, a common and usually harmless rod-shaped bacterium found in the human gut. They grew *E. coli* in a medium (food source) whose only nitrogen-containing substance was ammonium chloride (NH_4Cl). All the nitrogen in the ammonium chloride was the heavy isotope, ^{15}N. The bacteria had to use this nitrogen to make new DNA when their cells divided. The *E. coli* were grown in this medium for so many generations that it was certain that all the bases in the DNA molecules must contain virtually 100% ^{15}N. Their DNA was therefore heavy, or dense, because of the quantity of the heavy N atoms it contained. The mass of the DNA affects where the extracted DNA comes to rest in a tube using CsCl density gradient centrifugation.

---

newly synthesised parts of DNA shown dark blue

parent DNA molecule

The two new DNA molecules produced by each method would be like this:

semiconservative replication

one strand all old, the other all new

one strand all old, the other all new

conservative replication

both strands all new

both strands all old

dispersive replication

both strands partly old and new

both strands partly old and new

Density gradient centrifugation

less dense

more dense

As a result of high-speed centrifugation, a density gradient is set up. In this gradient, the density of CsCl gets greater nearer the base of the tube.

DNA is extracted from many bacteria and added to a CsCl density gradient tube.

tube centrifuged at high speed again

DNA rests where its density matches the density of the CsCl around it. The lower the resting point, the more dense (heavier) the DNA. The broader the band, the more DNA is present.

continued ...

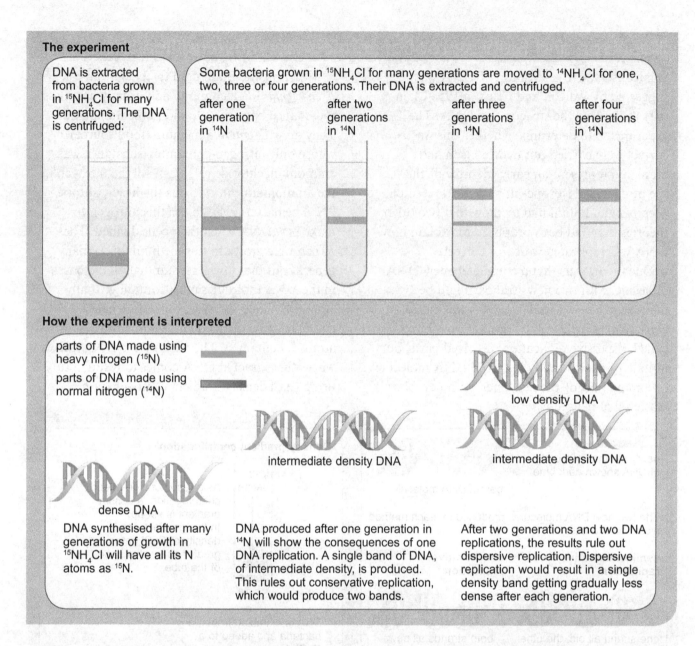

The experiment

DNA is extracted from bacteria grown in $^{15}NH_4Cl$ for many generations. The DNA is centrifuged:

Some bacteria grown in $^{15}NH_4Cl$ for many generations are moved to $^{14}NH_4Cl$ for one, two, three or four generations. Their DNA is extracted and centrifuged.

after one generation in ^{14}N

after two generations in ^{14}N

after three generations in ^{14}N

after four generations in ^{14}N

How the experiment is interpreted

parts of DNA made using heavy nitrogen (^{15}N)

parts of DNA made using normal nitrogen (^{14}N)

low density DNA

intermediate density DNA

intermediate density DNA

dense DNA

DNA synthesised after many generations of growth in $^{15}NH_4Cl$ will have all its N atoms as ^{15}N.

DNA produced after one generation in ^{14}N will show the consequences of one DNA replication. A single band of DNA, of intermediate density, is produced. This rules out conservative replication, which would produce two bands.

After two generations and two DNA replications, the results rule out dispersive replication. Dispersive replication would result in a single density band getting gradually less dense after each generation.

The new strands are both synthesised from 5' to 3' ends. But the two new strands do not run in the same direction. This requires the synthesis of one strand to be completed in sections (the lagging strand), whereas in the other strand, synthesis is continuous (Figure 5.7).

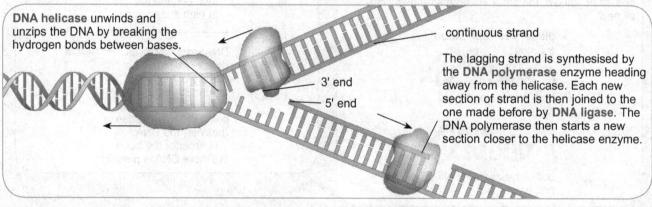

DNA helicase unwinds and unzips the DNA by breaking the hydrogen bonds between bases.

continuous strand

3' end

5' end

The lagging strand is synthesised by the **DNA polymerase** enzyme heading away from the helicase. Each new section of strand is then joined to the one made before by **DNA ligase**. The DNA polymerase then starts a new section closer to the helicase enzyme.

Figure 5.7 Enzymes involved in DNA replication.

The role of DNA

DNA carries a code that is used by the cell when making proteins. The sequence of bases in the DNA molecules determines the sequence of amino acids that are strung together when a protein molecule is made on the ribosomes.

A length of DNA that codes for making one polypeptide is called a **gene**. It is thought that there are around 30 000 genes in our cells.

The code is read in groups of three 'letters' – that is, triplets of bases. As we have seen, there are four bases in a DNA molecule: A, T, C and G. A sequence of three bases in a DNA molecule codes for one amino acid.

The structure of RNA

DNA is not the only polynucleotide in a cell. There are also polynucleotides which contain the sugar **ribose** rather than deoxyribose. They are therefore called **ribonucleic acids**, or **RNA** for short. Figure 5.8 shows the structure of RNA. RNA is generally single stranded, while DNA is generally double stranded. Another difference between them is that RNA always contains the base **uracil** (U) instead of thymine.

While DNA stores the genetic information in the nucleus of a cell, RNA is involved with using that information to make proteins. As we will see, there are different types of RNA and they have different roles in this process.

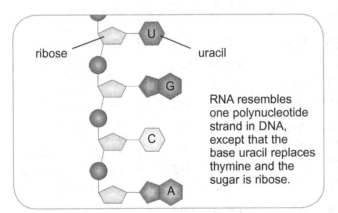

RNA resembles one polynucleotide strand in DNA, except that the base uracil replaces thymine and the sugar is ribose.

Figure 5.8   The structure of RNA.

SAQ ⎯⎯⎯⎯⎯⎯⎯⎯⎯⎯

2 Use a table, or a list of bullet points, to summarise the differences between DNA and RNA.

The sequence of bases on part of a DNA molecule is used to build an RNA molecule with the complementary base sequence. This RNA molecule then travels out into the cytoplasm and attaches to a ribosome. Working with other RNA molecules, the base sequence is used to determine the sequence of amino acids that are strung together to make a protein molecule. The base sequence on the DNA therefore determines the primary structure of the protein that is made. We will now look at this process in detail.

The genetic code

A length of DNA that codes for one polypeptide is called a **gene**. Some of the proteins for which DNA codes are structural ones, such as keratin or collagen. Others have physiological roles, such as haemoglobin or the hundreds of different enzymes that control our metabolic reactions. As you will see later in this chapter, even a small change in a DNA molecule that codes for a protein can have a very large effect on the appearance or body chemistry of an organism, and may even mean that it cannot survive at all. If, for example, the DNA coding for a particular enzyme is faulty, then the enzyme may not be able to catalyse its reaction and a whole bundle of other metabolic reactions that depend on that one could also be affected.

How the genetic code works

The four different bases in a DNA molecule can be put together in any order along one of the polynucleotide chains that make up the DNA. Normally, only one of the strands is used as the code for making proteins. We will refer to this strand as the **coding strand** (Figure 5.9).

The code is a three-letter code. A sequence of three bases, known as a base **triplet**, on the coding strand of part of a DNA molecule codes for one amino acid.

The genetic code is almost universal – the same DNA triplets code for the same amino acids in almost every kind of organism. This indicates that it evolved very, very early on in the evolution of life on Earth. The triplets of bases on the DNA coding strand that code for each amino acid are shown in Table 5.1. If you look at Table 5.1, you will see that most amino acids have more than one base triplet coding for them. The code is therefore said to be **degenerate**.

Another feature of the genetic code is that it is only read in one direction along the DNA strand.

It is always read from the 5' end to the 3' end. Moreover, each base on the code is part of only one triplet, so the code is said to be **non-overlapping**.

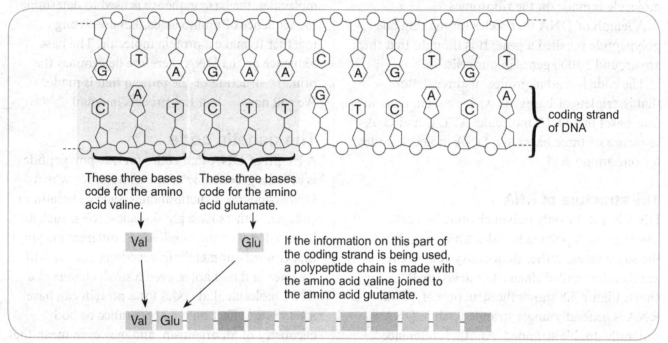

These three bases code for the amino acid valine.

These three bases code for the amino acid glutamate.

Val

Glu

If the information on this part of the coding strand is being used, a polypeptide chain is made with the amino acid valine joined to the amino acid glutamate.

Val – Glu

coding strand of DNA

Figure 5.9 How DNA codes for amino acid sequences in proteins.

		Second base							
		A		**G**		**T**		**C**	
First base	**A**	AAA	Phe	AGA	Ser	ATA	Tyr	ACA	Cys
		AAG	Phe	AGG	Ser	ATG	Tyr	ACG	Cys
		AAT	Leu	AGT	Ser	ATT	*stop*	ACT	*stop*
		AAC	Leu	AGC	Ser	ATC	*stop*	ACC	Trp
	G	GAA	Leu	GGA	Pro	GTA	His	GCA	Arg
		GAG	Leu	GGG	Pro	GTG	His	GCG	Arg
		GAT	Leu	GGT	Pro	GTT	Gln	GCT	Arg
		GAC	Leu	GGC	Pro	GTC	Gln	GCC	Arg
	T	TAA	Ile	TGA	Thr	TTA	Asn	TCA	Ser
		TAG	Ile	TGG	Thr	TTG	Asn	TCG	Ser
		TAT	Ile	TGT	Thr	TTT	Lys	TCT	Arg
		TAC	Met (*start*)	TGC	Thr	TTC	Lys	TCC	Arg
	C	CAA	Val	CGA	Ala	CTA	Asp	CCA	Gly
		CAG	Val	CGG	Ala	CTG	Asp	CCG	Gly
		CAT	Val	CGT	Ala	CTT	Glu	CCT	Gly
		CAC	Val	CGC	Ala	CTC	Glu	CCC	Gly

Ala	alanine
Arg	arginine
Asn	asparagine
Asp	aspartate
Cys	cysteine
Gln	glutamine
Glu	glutamate
Gly	glycine
His	histidine
Ile	isoleucine
Leu	leucine
Lys	lysine
Met	methionine
Phe	phenylalanine
Pro	proline
Ser	serine
Thr	threonine
Trp	tryptophan
Tyr	tyrosine
Val	valine

Table 5.1 A DNA dictionary, showing the base triplets on the DNA coding strand that code for each amino acid. Some triplets do not code for amino acids but have other functions (explained on page 104).

SAQ

3 a There are 20 different naturally occurring amino acids. There are four different bases. Remembering that the sequence of bases is always read in the same direction, work out how many different base triplets there can be.

b There are many more possible base triplets than there are amino acids. Using the information in Table 5.1, explain how these 'spare' triplets are used.

c Explain why a two-letter code, rather than a three-letter code, would not work.

Protein synthesis

The DNA is in the nucleus. It is good to keep the DNA safely shut away from the rest of the cell, because it makes it much less likely that the DNA will be affected by any of the metabolic reactions taking place in the cytoplasm.

But proteins are made in the cytoplasm, on the ribosomes. So there needs to be a messenger to take the instructions from the DNA to the ribosomes. This is done by a substance called **messenger RNA**, or **mRNA** for short.

The process of using the DNA code to make a polypeptide or protein takes place in two stages.

- First, the instructions on part of a DNA molecule are transferred to an mRNA molecule. This is called **transcription**.
- Next, the mRNA takes the instructions to a ribosome, and they are used to build a polypeptide. This is called **translation**.

Transcription

Transcription is the production of an mRNA molecule that has a complementary base sequence to one strand of a length of DNA. Usually, only a small part of a DNA molecule is transcribed at one time – a section that contains the code for making one polypeptide, called a gene.

The process begins as this part of the DNA molecule 'unzips'. An enzyme called **DNA helicase** breaks the hydrogen bonds between the bases, and the helix unwinds (Figure 5.10).

Next, free RNA nucleotides slot into place against one of the exposed DNA strands. The nucleotides pair exactly. C and G always pair together, just as they did in the DNA molecule. However, there are no RNA nucleotides that

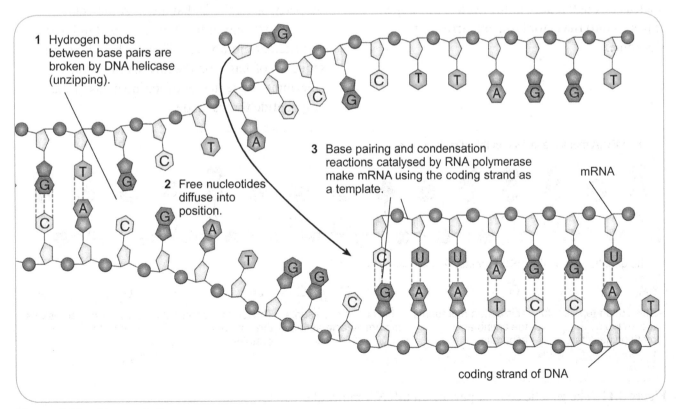

1 Hydrogen bonds between base pairs are broken by DNA helicase (unzipping).

2 Free nucleotides diffuse into position.

3 Base pairing and condensation reactions catalysed by RNA polymerase make mRNA using the coding strand as a template.

mRNA

coding strand of DNA

Figure 5.10 Transcription.

contain the base T – they have a base called **uracil**, U, instead. So the base A on the DNA links up with the base U on an RNA nucleotide, and the base T on the DNA links up with the base A.

As the RNA nucleotides slot into place and form hydrogen bonds with their complementary bases on the DNA strand, condensation reactions take place to form phosphodiester bonds between the adjacent RNA nucleotides. These reactions are catalysed by **RNA polymerase**. This enzyme also checks that the bases have paired up correctly – it will not link the RNA nucleotides if they don't have the correct base pairing with the exposed DNA strand.

Working steadily along the DNA strand, a complementary RNA strand is built up. This is a messenger RNA molecule. The elongation of the mRNA continues until the end of the gene is reached, when the complete mRNA molecule breaks away. The end is signalled by a particular triplet of bases on the DNA (Table 5.1) that, instead of coding for an amino acid, signifies 'stop here'. The DNA molecule may stay unzipped for more transcription, or it may zip back up again.

The mRNA molecule is now guided out of the nucleus through a pore in the nuclear envelope. It passes into the cytoplasm and arrives at a ribosome.

Translation

Translation is the process by which the code for making the protein – now carried by the mRNA molecule – is used to line up amino acids in a particular sequence and link them together to make a polypeptide.

Each group of three bases on the mRNA molecule is called a **codon**. Each codon stands for a particular amino acid (Figure 5.11). Some codons also act as 'stop' and 'start' codons. The sequence AUG, which denotes methionine, can indicate that this is where the amino acid chain should be begun. AUG is a 'start' codon.

Transfer RNA

In translation, yet another type of nucleotide comes into play. This is a different kind of RNA, known as **transfer RNA**, or **tRNA** for short.

Each tRNA molecule has a group of three exposed (unpaired) bases at one end (Figure 5.12). This is called an **anticodon**. An anticodon can undergo complementary base pairing with a codon on an mRNA molecule.

At the other end of the tRNA molecule there is a site where an amino acid can bind. The crucial property of tRNA is that a tRNA molecule with a particular anticodon can only bind with a particular amino acid. This is what allows the sequence of bases on the mRNA molecule to determine the sequence of amino acids in the polypeptide that is made.

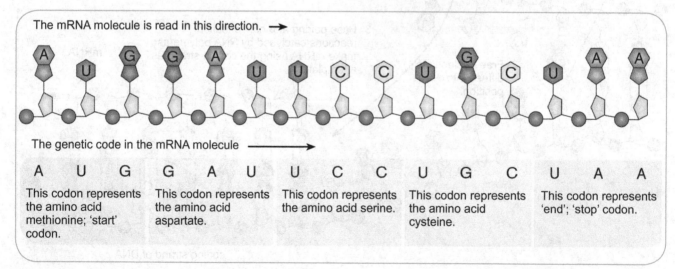

The mRNA molecule is read in this direction. →

The genetic code in the mRNA molecule →

A U G	G A U	U C C	U G C	U A A
This codon represents the amino acid methionine; 'start' codon.	This codon represents the amino acid aspartate.	This codon represents the amino acid serine.	This codon represents the amino acid cysteine.	This codon represents 'end'; 'stop' codon.

Figure 5.11 The genetic code in part of an mRNA molecule.

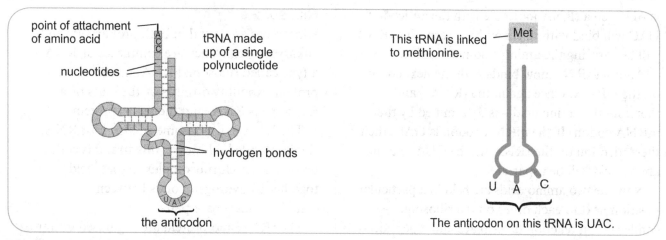

Figure 5.12 Transfer RNA.

In the cytoplasm, specific enzymes load specific amino acids onto specific tRNA molecules. These enzymes are called **tRNA transferases**, and there is a different kind for each type of tRNA.

For example, a tRNA with the anticodon UAC will have the amino acid methionine loaded onto its amino acid binding site. You can imagine thousands of tRNA molecules in the cytoplasm, each loaded with its particular amino acid, waiting for the opportunity to offload it at the polypeptide-making production line on a ribosome.

Building the polypeptide

The mRNA molecule, carrying the code copied from part of a DNA molecule, is held in a cleft in the ribosome, so that just six of its bases are exposed. This is two codons.

A tRNA with an anticodon that is complementary to the first ('start') mRNA codon then binds with it (Figure 5.13). This is called **initiation** of the translation process. Complementary base pairing makes sure that only the 'correct' tRNA can bind. The mRNA codon

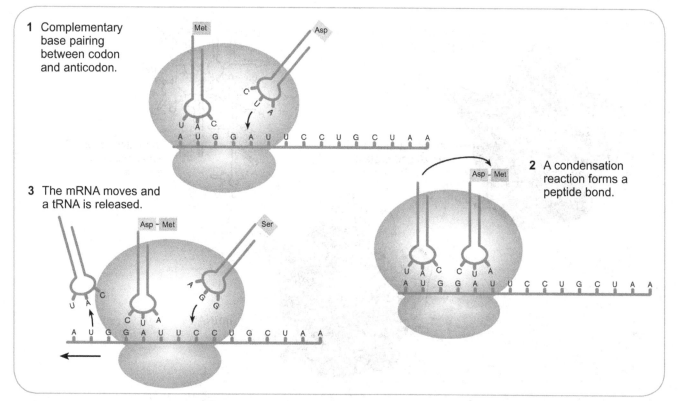

Figure 5.13 Translation on a ribosome.

is AUG, so a tRNA molecule with the anticodon UAC will bind with it. As we have seen, this tRNA will be carrying the amino acid methionine.

Another tRNA now binds with the next codon on the mRNA. Once again, the tRNA – and therefore the amino acid – is determined by the mRNA codon. If the mRNA codon is GAU, then the anticodon on the tRNA will be CUA and the amino acid will be aspartate.

Now the two amino acids are held in a particular position next to each other on the ribosome. A condensation reaction takes place and a peptide bond is formed between the two amino acids resulting in elongation of the polypeptide.

The mRNA moves on through the cleft in the ribosome, bringing a third codon into place. A third tRNA binds with it, and a third amino acid is added to the chain. Meanwhile, the first tRNA (the one that brought methionine) has completed its role. It breaks away, leaving the methionine behind. This released tRNA is now available to be reloaded with another methionine molecule. In this way, the whole polypeptide chain is gradually built up. It is released when a 'stop' codon is reached on the mRNA. This is called **termination**.

Ribosomes

Ribosomes are found in both prokaryotic and eukaryotic cells. They are complexes of RNA – a type called ribosomal RNA, or rRNA – and protein. About two-thirds of the mass of a ribosome is RNA, and the rest is protein.

The RNA in a ribosome is, like most RNA, single stranded. However, this strand is coiled up on itself to form complex shapes, held together by hydrogen bonds between complementary base pairs.

The rRNA and protein are organised into two subunits, known as the large and small subunits (Figure 5.14 and Figure 5.15). The large subunit helps to hold the ribosome, mRNA and tRNA firmly together during translation, while the small subunit is more flexible and allows movement of the various components involved in translation to occur. When they are not active, the two subunits of a ribosome are separated. They come together to form a complete ribosome when translation is about to begin.

Amino acids are linked by peptide bonds, in a cavity and groove in between the two subunits. The process is catalysed by rRNA (in the large subunit)

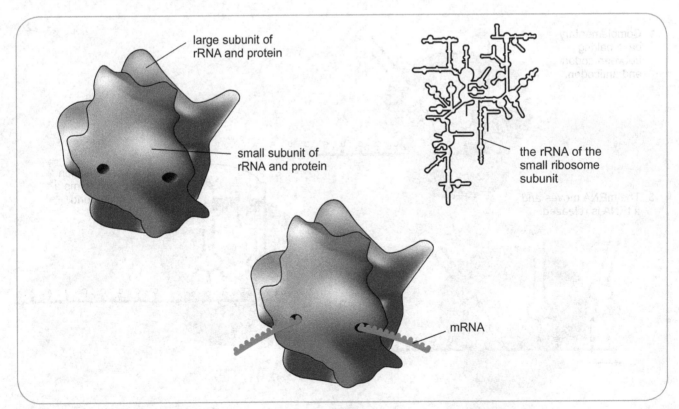

Figure 5.14 A ribosome, some of its rRNA, and a ribosome attached to mRNA.

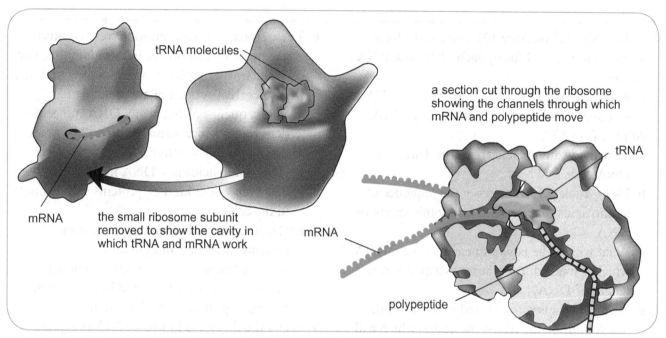

Figure 5.15 Messenger RNA moves through a channel to the cavity between large and small ribosome subunits where translation takes place. The polypeptide leaves through a channel in the large subunit.

and not protein. This means that the rRNA is acting as an enzyme.

Many ribosomes can work on the same mRNA at the same time if the mRNA is large enough (Figure 5.16).

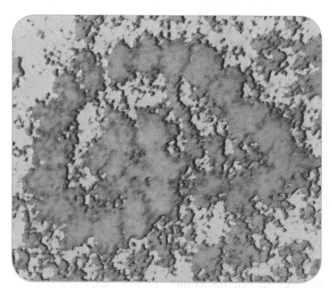

Figure 5.16 Each large blob in the spiral is a ribosome. They are all working on the same strand of mRNA, which is too thin to be seen. A group of ribosomes like this is sometimes called a polyribosome or polysome. This electron micrograph is of a tiny part of the cytoplasm of a brain cell ($\times 225\,000$).

DNA, chromatin and chromosomes

When you look at a cell through a microscope, you cannot usually see chromosomes. Instead, lighter and darker areas can be seen in the nucleus. These are called **euchromatin** and **heterochromatin** respectively (Figure 5.17).

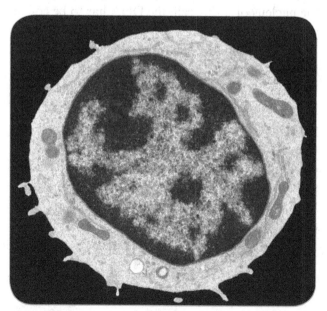

Figure 5.17 This is a transmission electron micrograph of a lymphocyte (a type of white blood cell). The darkly staining areas in the nucleus are heterochromatin, and the lighter areas are euchromatin ($\times 13\,300$).

SAQ

4 Using Table 5.1 on page 102, work out which amino acid is coded for by each of these mRNA codons.

a AAA **b** ACG **c** GUG **d** CGC **e** UAG

5 A length of DNA has the base sequence ATA AGA TTG CCC.

a How many amino acids does this length of DNA code for?

b Using Table 5.1, write down the sequence of amino acids that is coded for by this length of DNA.

c What will be the base sequence on the mRNA which is made during the transcription of this length of DNA?

d Using your answers to **b** and **c**, work out the anticodons of the tRNA molecules which will carry these amino acids to this mRNA:

i tyrosine

ii asparagine.

6 These statements contain some very common errors made by candidates in examinations. For each statement, explain why it is wrong and then write a correct version of the statement.

a 'The sequence of bases in a DNA molecule determines which amino acids will be made during protein synthesis.'

b 'The amino acids in a DNA molecule determine what kind of proteins will be made in the cell.'

c 'The four bases in DNA are adenosine, cysteine, thiamine and guanine.'

d 'During transcription, a complementary mRNA molecule comes and lies against an unzipped part of a DNA molecule.'

7 Using the diagrams in Figure 5.13 as a guide, make annotated drawings of the next stage in the synthesis of the polypeptide shown in the figure. The codon UGC codes for cysteine.

The individual chromosomes are not visible because they are so thin. In a human cell, the total length of all the DNA molecules is about 2 m, but they are only 2 nm wide. To fit this huge length into the nucleus of a tiny cell, the DNA has to be coiled and supercoiled.

The DNA is associated with proteins called **histones** (Figure 5.18). These help to hold the DNA in coils, and they also have important roles in determining which genes will be expressed at particular times and in particular cells. DNA that is not being used for transcription is mostly coiled up compactly, and this produces the darkly staining heterochromatin. DNA that is being used for transcription is not so tightly coiled, and this is present in euchromatin.

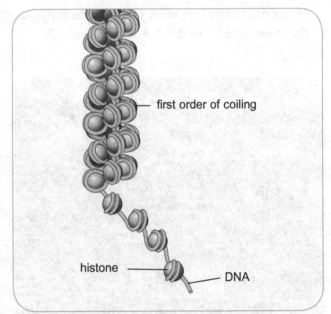

Figure 5.18 The DNA in a chromosome is coiled around histone molecules. This is then coiled again, as shown here. It can have two more orders of coiling (not shown) to produce very compact chromatin, as in heterochromatin.

Control of gene expression in eukaryotes

'Gene expression' is the production of a protein for which a gene codes. In any one cell, only a tiny number of its genes will be expressed at any one time. There are many different ways in which the expression of genes is controlled, and a great deal of research is being carried out into it. A better understanding could help the development of treatments and cures for many different diseases, including cancers (where genes that control cell division do not work correctly). There are control mechanisms at every step of the pathway from the original gene through to the construction of the final protein.

The diagram shows some of the areas of a DNA molecule that are involved in determining whether or not a particular gene is expressed, and also in the final form of the protein that is made.

A crucial area is the **promoter**. This is a short length of DNA to which the enzyme RNA polymerase must bind before transcription can begin. It always contains a sequence of the bases TATAAAA, known as the 'TATA box'. The binding of RNA polymerase to the promoter is called **initiation**, as it starts off the entire process of transcription.

Many genes are regulated by substances called transcription factors binding with their promoter region. For example, some hormones act as **transcription factors**. They bind with the promoter region of a particular gene, allowing RNA polymerase to bind and therefore initiating transcription, leading to expression of the gene.

The part of the gene from which RNA is transcribed is made up of **introns** and **exons**. After the RNA is made, the sections copied from the introns are cut out and discarded. The exons are spliced together to make the mRNA that will be translated.

In many cases, there are several different ways in which the exons can be spliced together. This means you can get several different mRNAs, and therefore several different polypeptides, produced from the instructions on a single gene.

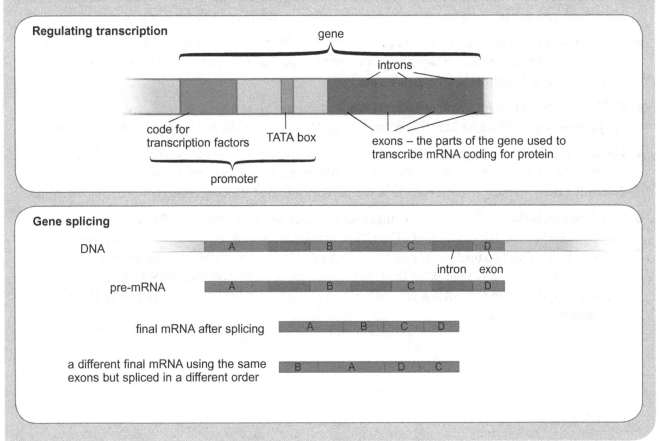

Regulating transcription

gene

introns

code for transcription factors

TATA box

exons – the parts of the gene used to transcribe mRNA coding for protein

promoter

Gene splicing

DNA — A B C D

intron exon

pre-mRNA — A B C D

final mRNA after splicing — A B C D

a different final mRNA using the same exons but spliced in a different order — B A D C

Summary

- DNA is deoxyribonucleic acid. It is a double-stranded molecule made up of two strands of nucleotides.

- A DNA nucleotide is made up of a phosphate group, a five-carbon sugar called deoxyribose, and a base. There are four bases in DNA – adenine, guanine, cytosine and thymine. They are usually abbreviated to A, G, C and T.

- Adenine and guanine are purine bases. Cytosine and thymine are pyrimidine bases.

- The nucleotides in a strand of DNA are linked to each other by strong covalent bonds between the phosphate groups and deoxyribose. The phosphate groups bond to carbon 5 and to carbon 3 of the deoxyribose ring. The end of the molecule where the phosphate is bonded to carbon 5 is called the 5' end, while the other is the 3' end.

- The two strands of a DNA molecule are linked to each other by weak hydrogen bonds between the bases. A always bonds with T, and C always bonds with G. A and T are linked by two hydrogen bonds. C and G are linked by three hydrogen bonds.

- The two strands of a DNA molecule run in opposite directions. They are said to be anti-parallel. They twist around each other to form a double helix.

- The DNA molecules in a cell nucleus are replicated before cell division takes place. First, the two strands of the molecule are untwisted and unzipped. Free DNA nucleotides pair up with the exposed bases on both strands. They are then linked together by the formation of bonds between their deoxyribose and phosphate groups. This is catalysed by the enzyme DNA polymerase. Two new molecules are therefore formed, each identical to the original one. Each new molecule contains one old strand and one new strand, so the process is called semiconservative replication.

- The sequence of bases in a DNA molecule codes for the sequence of amino acids in a protein to be made on the ribosomes. Three bases code for one amino acid. A sequence of DNA nucleotides that codes for one polypeptide is known as a gene.

- RNA is ribonucleic acid. There are several kinds of RNA. Most are single stranded. They contain the pentose sugar ribose, rather than deoxyribose. They contain the base uracil instead of thymine.

- During protein synthesis, an mRNA molecule is built up against one of the DNA strands in a gene. The mRNA then travels out of the nucleus to a ribosome, where its sequence of bases is used to determine the sequence of amino acids in the polypeptide that is being constructed in the ribosome.

- Transcription is the production of a complementary mRNA molecule by building it against the coding strand of the DNA in a gene. The DNA strands are separated by DNA helicase, and free mRNA nucleotides line up against one of the exposed strands, with complementary bases pairing with each other. C and G pair. A on the DNA strand pairs with U on the mRNA. T on the DNA strand pairs with A on the mRNA. The mRNA nucleotides are then linked together by RNA polymerase.

continued ...

- Translation is the production of a polypeptide following the base sequence on the mRNA. It takes place on a ribosome, where two codons (that is, two sets of three bases) on the mRNA are exposed at one time. The amino acids are brought to the ribosome by tRNA molecules, each of which has an anticodon that binds with the mRNA codon by complementary base pairing. The anticodon on the tRNA determines which specific amino acid it brings. As successive amino acids are brought to the ribosome, they are linked by peptide bonds.

- DNA is stored in the nucleus associated with proteins called histones. Each chromosome contains one very long molecule of DNA. DNA that is not being transcribed is usually tightly packed and coiled, forming heterochromatin. DNA that is being transcribed is less tightly packed, and forms euchromatin.

Questions

Multiple choice questions

1 Which of the following describes a nucleotide?
   A a phosphate group, a hexose sugar and a nitrogenous base
   B a phosphate group, a pentose sugar and a nitrogenous base
   C one strand of the DNA molecule
   D phosphate group, amino acid, nitrogenous base

2 The diagram shows part of a DNA molecule.

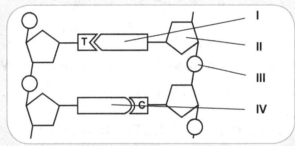

Which of the following correctly identifies the component parts of a DNA molecule?

	I	II	III	IV
A	deoxyribose sugar	phosphate	adenine	guanine
B	phosphate	adenine	guanine	deoxyribose sugar
C	adenine	deoxyribose sugar	phosphate	guanine
D	guanine	adenine	deoxyribose sugar	phosphate

3 A strand of messenger RNA is transcribed from a DNA strand with bases
   —A—G—C—C—T—T— . What is the base sequence on the mRNA strand produced?
   A —T—C—G—G—A—A—
   B —U—C—G—G—A—A—
   C —T—G—C—C—A—A—
   D —U—C—G—C—T—A—

4 Biochemical analysis of a sample of DNA shows that thymine forms 15% of the nitrogenous bases. What percentage of the bases are cytosine?
   A 15%   B 30%   C 35%   D 70%

continued ...

5 What is the role of ribosomes in protein synthesis? They:

    A transcribe DNA into mRNA.

    B provide amino acids to form a polypeptide.

    C are the site where codons bond to anticodons in protein synthesis.

    D produce rRNA.

6 The following are events occurring during protein synthesis.

    I   An anticodon on a tRNA molecule attaches to an mRNA codon.

    II The sense (coding) strand of DNA acts as the template for mRNA production.

    III The ribosome moves down one codon.

    IV The mRNA leaves the nucleus through a nuclear pore.

    V Amino acids join together by a peptide bond.

    Which of the following shows the correct sequence of protein synthesis?

    A I → II → III → IV → V

    B II → III → V → IV → I

    C II → IV → I → V → III

    D III → IV → I → V → II

7 The following molecules are components of nucleotides.

HOCH₂ O OH	NH₂	OH	O	HOCH₂ O OH
OH H		O=P−OH		OH OH
I deoxyribose	**II** purine	**III** phosphate	**IV** pyrimidine	**V** ribose

    Which of the following would be the components of the nucleotide that contains the base thymine?

    A I, II, III **B** II, III, IV **C** I, III, IV **D** III, IV, V

8 The diagram below shows a tRNA molecule for the amino acid leucine.

    What is the DNA sequence which codes for leucine?

    A CUC **B** GAG **C** CTC **D** GTG

9 DNA and RNA are both nucleic acids. Which of the following is **not** correct when comparing DNA and RNA?

	DNA	RNA
A	contains a pentose sugar	contains a pentose sugar
B	made up of nucleotides	made up of nucleotides
C	double stranded	single stranded
D	contains nitrogenous base, uracil	contains nitrogenous base, thymine

continued ...

Structured questions

10 The diagram below shows a short section of a DNA molecule.

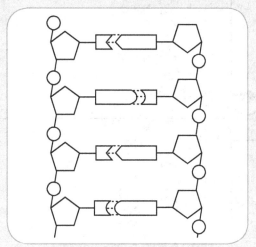

   a Identify the purines and pyrimidines present in DNA. [2 marks]

   b **i** Copy the diagram above and label the following:

       P (phosphate), S (sugar), A (adenine), T (thymine), C (cytosine), G (guanine) [3 marks]

     ii What do the dotted lines represent? [1 mark]

    iii The basic unit of DNA is a nucleotide. Draw a circle around a nucleotide

       on your diagram. [2 marks]

   c DNA is made up of two strands.

     i Describe the polarity of the two strands of DNA. [2 marks]

    ii Illustrate your answer to **c i** on your diagram. [1 mark]

    iii How are the two strands of DNA held together? [1 mark]

   d DNA in the nucleus of a cell is described as chromatin or chromosome depending

     on the stage of the cell cycle. Differentiate between the terms 'chromosome'

     and 'chromatin'. [3 marks]

11 **a** List the **two** basic functions of DNA. [2 marks]

   b The diagram below shows part of a DNA molecule.

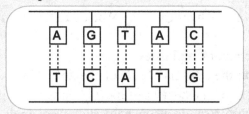

     Show by means of a diagram what happens to this part of DNA during replication. [3 marks]

   c It was proposed by Watson and Crick that, during replication, the two strands of

     the DNA molecule would separate from each other and each would serve as a

     template for a new strand, giving an exact replication.

     Suggest why it is necessary to have exact replication of DNA. [2 marks]

   d In 1957, experiments were done to determine the method of DNA replication.

     Bacteria was grown for many generations in ^{15}N (heavy nitrogen), then transferred

     to a medium with light nitrogen, ^{14}N. Samples were taken from each generation.

     DNA of light, intermediate and heavy densities were separated by centrifugation.

continued ...

The results are shown below:

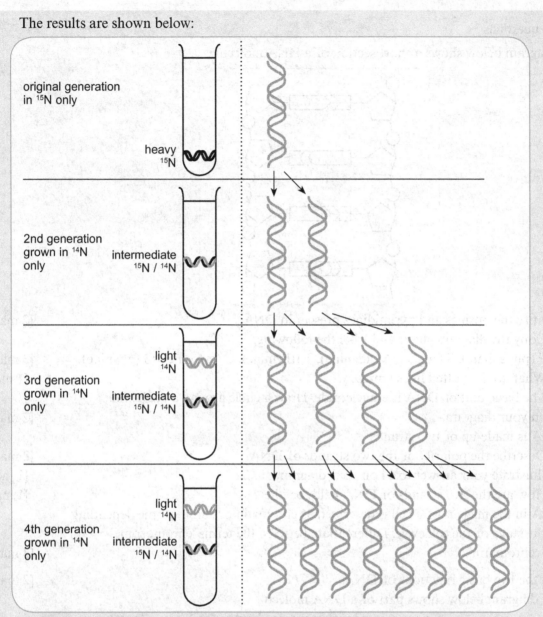

original generation
in ^{15}N only

heavy
^{15}N

2nd generation
grown in ^{14}N
only

intermediate
^{15}N / ^{14}N

3rd generation
grown in ^{14}N
only

light
^{14}N

intermediate
^{15}N / ^{14}N

4th generation
grown in ^{14}N
only

light
^{14}N

intermediate
^{15}N / ^{14}N

i Do the results above support the hypothesis by Watson and Crick in **c**?
Explain your answer. [3 marks]

ii Name the method by which DNA replicates. [1 mark]

iii Copy the diagram above on the right hand side of the dotted line. On the diagram
indicate the labelling pattern of the heavy isotope, ^{15}N, in the four generations. [2 marks]

e In another experiment, the DNA strands of the bacterium were analysed.
The percentage of each nitrogenous base present was determined.
Some of the results are shown below.

DNA sample	Percentage of base present			
	Cytosine	Adenine	Guanine	Thymine
Strand 1				12
Strand 2	26	12	24	

Copy and complete the table to show the percentage of bases. [2 marks]

continued ...

Essay questions

12 a Give **two** ways in which the structure of a molecule of tRNA differs from the structure of a molecule of mRNA. [2 marks]

   b Discuss briefly the roles of the following in protein synthesis:

     i RNA polymerase;

     ii tRNA;

     iii the ribosome. [7 marks]

   c Explain how the structure of DNA results in accurate replication. [3 marks]

   d Cancer cells are uncontrollable and copy themselves more quickly than healthy cells, resulting in a tumour. Drugs have been developed to treat cancer. This is known as chemotherapy. Suggest how these drugs may work to stop the uncontrollable replicating of the tumour cells. [3 marks]

13 a List the structural differences between DNA and RNA. [4 marks]

   b Describe the processes of transcription and translation that occur during the expression of a nucleotide sequence of a DNA molecule. [8 marks]

   c Describe **three** ways in which transcription differs from translation in protein synthesis. [3 marks]

14 a Discuss briefly the roles of the following during the replication of DNA:

     i DNA helicase;

     ii DNA polymerase;

     iii DNA ligase. [6 marks]

   b Explain why base pairing is important during semiconservative replication. [2 marks]

   c During DNA replication, the replicated strands are produced at different rates. This leads to the production of the leading and lagging strands. Explain why this occurs during replication. [2 marks]

   d List **two** ways in which transcription differs from replication. [2 marks]

   e Telomeres are bits of DNA found at the ends of the chromosomes. The telomeres allow cells to divide without losing genes, but they themselves become shorter. The enzyme telomerase repairs the telomeres. It has been suggested that telomeres play a role in aging and cancer. Suggest how telomeres may be involved in aging and tumour formation. [3 marks]

Chapter 6
Mitotic and meiotic cell division

By the end of this chapter you should be able to:

a describe, with the aid of diagrams, the processes involved in mitotic cell division and interphase;

b know how to make drawings from prepared slides and/or a freshly prepared root tip squash to show the stages of mitosis;

c explain the importance of DNA replication for maintaining genetic stability;

d discuss the role and importance of mitosis in growth, repair and asexual reproduction;

e explain what is meant by homologous pairs of chromosomes, and the terms 'haploid' and 'diploid';

f describe, with the aid of diagrams, the processes involved in meiotic cell division, including crossing over, alignment of chromosomes at metaphase and random segregation at anaphase;

g know how to construct models to demonstrate chromosome behaviour in meiosis;

h describe how meiosis contributes to heritable variation.

The mitotic cell cycle

Like most animals, you began your life as a single cell. This cell was a **zygote** – a cell that forms when two gametes fuse. The zygote contained a set of chromosomes from your father and a set of chromosomes from your mother.

All the cells in your body have developed from this single original cell. Soon after it was formed, the zygote divided to form two cells, which then each divided to form a total of four cells. This division went on and on, eventually forming your body containing many millions of cells. Some cells continue to divide even in an adult.

The repetitive process of growing and dividing, growing and dividing is called the **cell cycle** (Figure 6.1). The cell cycle is made up of two main phases: **interphase** and **mitosis**.

Interphase

In a cell in a human embryo, one complete cell cycle lasts about 24 hours. About 95% of this time is spent in **interphase**. During interphase, the cell is carrying out all the normal cell activities, such as

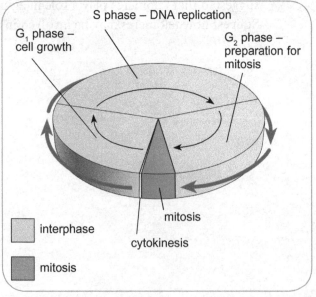

Figure 6.1 The cell cycle.

respiration and protein synthesis. The DNA that makes up its chromosomes is replicated – a perfect copy is made (page 98), so that the DNA can be divided up equally into the two new cells that will be made when the cell divides.

In a human cell, there are 46 **chromosomes**, each of which is made up of one enormously long molecule of DNA. Some time before the cell divides, each DNA molecule is copied. The pair of identical DNA molecules that are now contained in each chromosome remain attached to each other, at a point called the **centromere**. The two identical strands of DNA are called **chromatids** (Figure 6.2 and Figure 6.3).

It is very important that the new DNA molecules that are made are the same as the old ones. Even a small error – a **mutation** – could have harmful effects on the cell. Cells therefore run a 'checking' process on the new DNA. Special proteins work along the DNA molecules, checking for any errors and, where possible, correcting them.

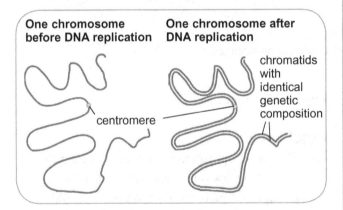

Figure 6.2   Chromosomes in interphase.

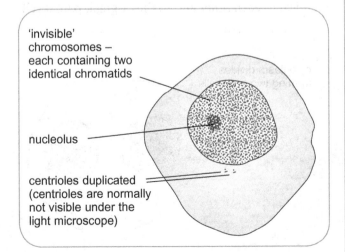

Figure 6.3   Late interphase.

Mitosis

The cell then moves into the next stage of the cell cycle, called **mitosis**. This is the stage during which the nucleus of the cell divides into two nuclei. During mitosis, the two chromatids which make up each chromosome break apart. One of them goes into one new nucleus and one into the other. In this way, the new cells will be genetically identical to each other and to the original parent cell.

Mitosis is made up of four stages: **prophase**, **metaphase**, **anaphase** and **telophase**. The four stages run into one another, without breaks between them.

Prophase

During prophase (Figure 6.4), the chromosomes become visible. Up to now, they have been lying in the nucleus as extremely long and thin threads, so thin that they cannot be seen at all with a light microscope. As prophase begins to get under way, the DNA molecules coil and supercoil, shortening and getting thicker until they eventually form threads that are thick enough to be visible if they are stained.

When the chromosomes appear, they can sometimes be seen to be made of two threads – the chromatids. The chromatids are held together at the centromere. The two chromatids of each chromosome contain identical molecules of DNA, formed in DNA replication during interphase.

As prophase proceeds, the nucleolus disappears. It is also at this stage that the **spindle** begins to form. The **centrioles** move away from each other to opposite ends of the cell. The centrioles organise

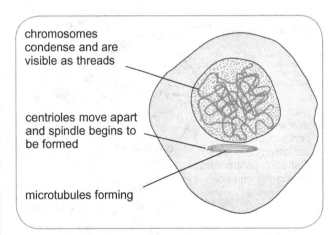

Figure 6.4   Prophase.

the formation of the microtubules – long, thin tubes of protein (Figure 6.4).

Metaphase

Now the nuclear membrane breaks down. Its loss means that the whole of the space in the cell is available for manoeuvring the chromosomes. By the time the nuclear membrane breaks down, many of the microtubules have attached themselves to the centromeres of the chromosomes. Each centromere is grabbed by one microtubule on either side. The microtubules pull in opposite directions on the centromeres, bringing the chromosomes to lie at the **equator** of the cell (Figure 6.5).

Anaphase

Now the centromeres split. The microtubules are still pulling on them, so the centromeres and the chromatids are pulled apart and moved to either end, or **pole**, of the cell (Figure 6.6).

Telophase

The two groups of chromatids have now arrived at the poles. Each group contains a complete set of chromatids, which we can now call chromosomes again. The microtubules making up the spindle fibres break down, so the spindle disappears. New nuclear envelopes form around each group of chromosomes. The chromosomes slowly uncoil and become thinner again, so they effectively disappear (Figure 6.7).

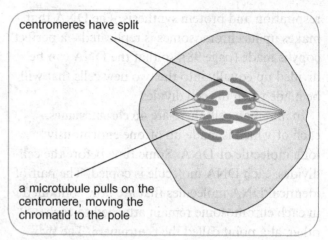

Figure 6.6   Anaphase.

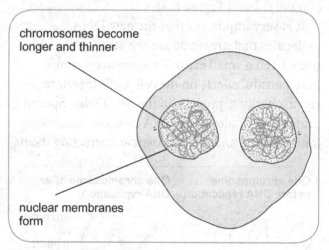

Figure 6.7   Telophase.

Cytokinesis

Usually, the cytoplasm now divides (Figure 6.8). This forms two new cells, each with a nucleus containing a complete set of chromosomes, and each with a centriole. The new cells are genetically identical to each other and to the original, parent cell.

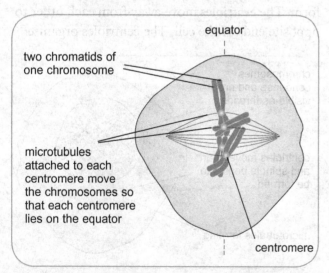

Figure 6.5   Metaphase.

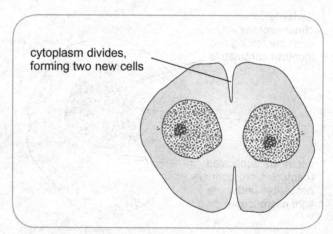

Figure 6.8   Cytokinesis.

Centrioles and spindle fibres

Every animal and plant cell contains many tiny structures called **microtubules**. These are part of the cytoskeleton, which helps to hold the cell in shape and gives it mechanical strength. The particular role of microtubules is to help with the movement of organelles within the cell, and this includes the movement of chromosomes during cell division.

Microtubules are made of a soluble, globular protein called **tubulin**. Tubulin monomers polymerise to produce very long, thin, hollow cylinders. The cylinders are approximately 25 nm in diameter, and can be anything up to 25 µm in length. They grow or shrink by the addition or removal of tubulin subunits at one end, known as the plus end.

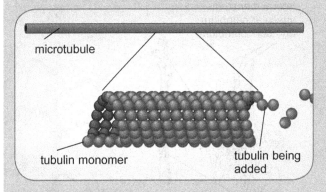

During mitosis in an animal cell, microtubules grow from the two pairs of centrioles, which are microtubule organising centres (MTOCs). These add tubulin units to the plus ends of numerous microtubules, which grow outwards from the MTOCs forming a series of radiating spokes known as **asters**. It is these microtubules that form the spindle.

Plant cells do not have centrioles, and it is thought that the nuclear envelope somehow functions as an MTOC during the formation of the spindle in a dividing plant cell, but currently very little is known about this.

Some of the spindle microtubules from each MTOC contact and then push against microtubules from the other MTOC, and this pushes the two MTOCs apart from each other so that they end up at opposite poles of the cell.

At this stage of mitosis, each chromosome in the cell is made up of two chromatids, joined at a centromere. During prophase, a protein structure forms on each centromere, called a kinetochore. Some of the growing microtubules from each MTOC attach to the kinetochores. As the microtubules grow or shrink, they first move the chromosomes to the equator of the cell and then pull them apart, drawing the separated chromatids of each chromosome to opposite poles of the cell.

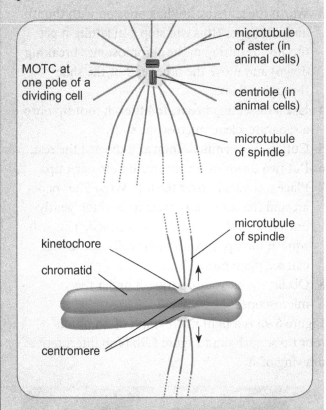

The formation of the spindle fibres is inhibited by the presence of certain chemicals, one of which is the toxin colchicine. Colchicine is sometimes used in research into mitosis, because it prevents the chromatids being separated into the two daughter cells. Failure of the correct separation of chromatids is called non-disjunction. In some instances, this might affect just one or two of the chromosomes, but in some cases one daughter cell might get all the chromatids while the other gets none.

Observing mitosis

A good place to find cells undergoing mitosis is just behind the tip of a growing root. The root tip can be stained with a stain such as acetic orcein, which is absorbed by the chromosomes but not other parts of the cell. The root tip is gently squashed to spread out the cells into a single layer, so that they can be seen clearly using a light microscope. An outline method for this is:

1 Cut approximately 5 mm off the ends of two intact garlic or onion roots.
2 Put these two tips into a watch glass containing 10 drops of acetic orcein stain and one drop of $1\,mol\,dm^{-3}$ hydrochloric acid.
3 Warm the glass by holding one end with thumb and forefinger (this will stop you letting it get too hot, resulting in the chromosomes breaking down) and move the other end of the slide through a flame.
4 Use a dissecting needle to lift each root tip onto a separate, clean microscope slide.
5 Cut off the terminal 2 mm and discard the rest.
6 Put two drops of acetic orcein onto each tip.
7 Place a coverslip over the top. Wrap filter paper around the slide and coverslip, and tap gently onto it with the blunt end of a pencil. This will squash the tip, spreading the cells out so you can see them more clearly.
8 Observe your root tip squash under the microscope.

Figure 6.9a is a light micrograph of an onion root tip squash, and Figure 6.9b is an interpretive drawing of it.

The roles of mitosis

Mitosis is the normal way in which cells of plants and animals divide. It produces two daughter cells that are genetically identical to the parent cell. Mitosis has several different roles in living organisms.

● Mitosis is involved in the **growth** of an organism. For example, a human begins life as a single cell, a zygote, which divides repeatedly by mitosis to form all of the millions of cells that make up the human body. Each of these cells has the same number of chromosomes (46), and each of these chromosomes is identical to those in all the other cells in the body. In animals, growth can normally occur in all parts of the body. In plants, it can only take place at particular areas called **meristems** (Figure 6.10).

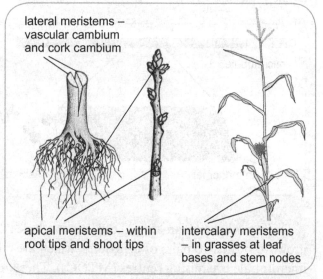

lateral meristems – vascular cambium and cork cambium

apical meristems – within root tips and shoot tips

intercalary meristems – in grasses at leaf bases and stem nodes

Figure 6.10 Some plant meristems.

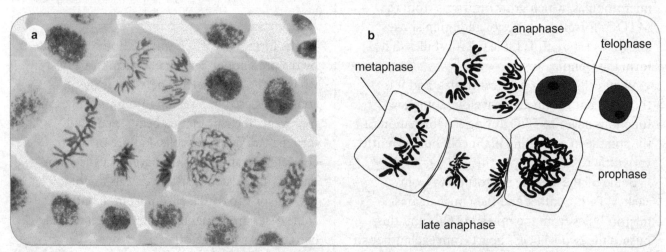

a

b

metaphase

anaphase

telophase

late anaphase

prophase

Figure 6.9 **a** Stained onion root tip squash (×250); **b** interpretive diagram of **a**.

- Mitosis is involved in the **repair of tissues** in an organism. For example, if you cut your skin, some of the cells at the edges of the cut will divide by mitosis to produce new, genetically identical cells that can cover the wound and join the cut edges together.
- Mitosis is involved in the **immune response**. When a lymphocyte (a type of white blood cell) comes into contact with a bacterium carrying antigens that bind with the particular receptors in the cell's plasma membrane, the lymphocyte is stimulated to divide repeatedly by mitosis and produce a large population of genetically identical plasma cells. These all secrete antibody that binds with the antigen, leading to the destruction of the bacteria.
- Mitosis is involved in **asexual reproduction**. Many plants and some non-vertebrate animals are able to reproduce by growing a new organism directly from the body of a single adult (Figure 6.11). The cells of the new organism are produced by mitosis, so they are all genetically identical to the parent's cells. You can read more about asexual reproduction in plants, and its advantages and disadvantages with respect to sexual reproduction, in Chapter 10.

Meiosis

We have seen that some organisms can reproduce asexually. Animals, however, and also plants for much of the time, generally use **sexual reproduction**. This involves the production of specialised sex cells called **gametes**. The nuclei of two gametes (usually, but not necessarily, from two different parents) fuse together in a process called **fertilisation**. The new cell that is formed is called a **zygote**. The zygote then divides repeatedly by mitosis to form a new organism that is genetically different from its parents and its siblings.

During mitosis, chromosomes are divided equally between the two daughter cells. Perfect copies of each chromosome are made before the division of the cell begins, so that each cell gets a complete set of chromosomes, containing an exact replica of all the DNA that was present in the parent cell. This is how most eukaryotic cells divide most of the time.

In sexual reproduction, however, another type of cell division is needed. This is **meiosis**. This type of division *halves* the chromosome number. A **diploid** cell, with two sets of chromosomes, divides to produce four **haploid** cells, each with one set of chromosomes. These become gametes. When the nuclei of two gametes fuse, the two single sets of

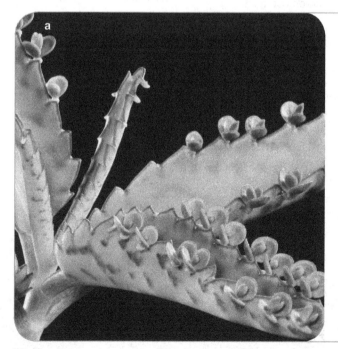
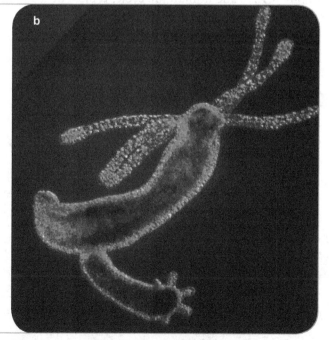

Figure 6.11 Asexual reproduction produces genetically identical new organisms by mitosis: **a** vegetative reproduction in *Kalanchoe*; **b** asexual reproduction in *Hydra* (×50).

chromosomes are brought together to produce a zygote with two sets (Figure 6.12, Figure 6.13 and Figure 6.14).

In a diploid cell, there are therefore two copies of each chromosome. Any two 'matching' chromosomes are known as **homologous** chromosomes. Homologous chromosomes carry the same genes at the same positions, or loci (singular: locus).

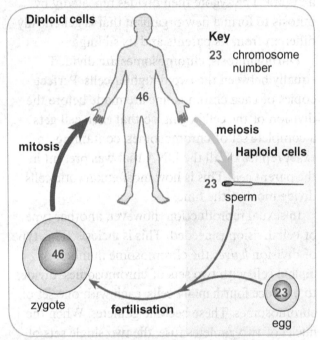

Figure 6.12 The human life cycle, typical of a sexually reproducing animal. Mitosis produces genetically identical diploid body cells, and meiosis produces genetically varying haploid gametes.

SAQ

1 The fruit fly *Drosophila melanogaster* has eight chromosomes in its body cells. How many chromosomes will there be in a *Drosophila* sperm cell?

2 The symbol *n* is used to indicate the number of chromosomes in one set – the haploid number of chromosomes. For example, in humans $n = 23$. In a horse, $n = 32$.

   a How many chromosomes are there in a gamete of a horse?

   b What is the diploid number of chromosomes, $2n$, of a horse?

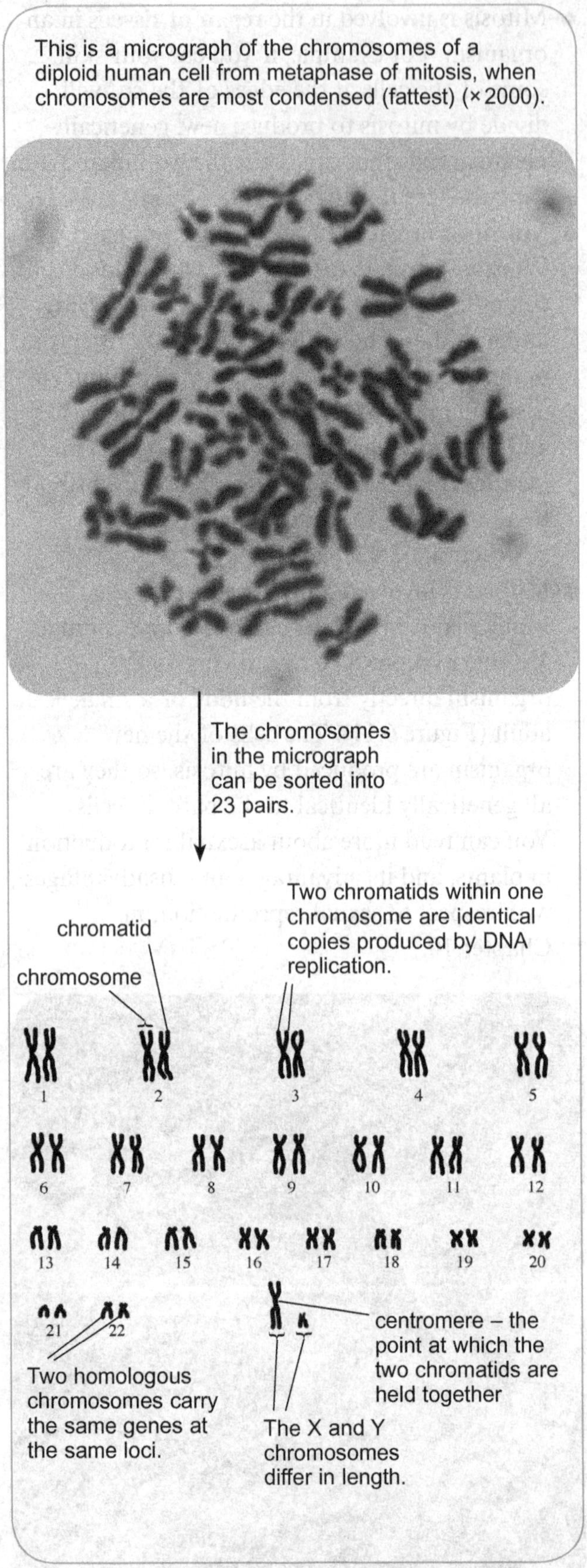

This is a micrograph of the chromosomes of a diploid human cell from metaphase of mitosis, when chromosomes are most condensed (fattest) (× 2000).

The chromosomes in the micrograph can be sorted into 23 pairs.

chromatid

chromosome

Two chromatids within one chromosome are identical copies produced by DNA replication.

Two homologous chromosomes carry the same genes at the same loci.

The X and Y chromosomes differ in length.

centromere – the point at which the two chromatids are held together

Figure 6.13 Chromosomes of a diploid cell.

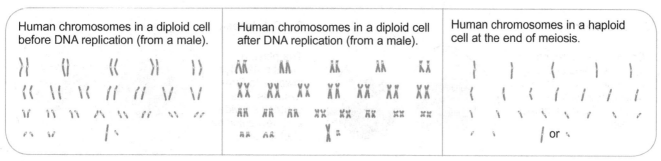

| Human chromosomes in a diploid cell before DNA replication (from a male). | Human chromosomes in a diploid cell after DNA replication (from a male). | Human chromosomes in a haploid cell at the end of meiosis. |

Figure 6.14   Chromosomes of a diploid and haploid cell.

Stages of meiosis

The main events that take place during meiosis are shown in Figure 6.15. The diagrams show a cell with four chromosomes – that is, a haploid number of 2.

You will see that there are actually two divisions: meiosis I and meiosis II. The second division, meiosis II (page 124), is very like mitosis.

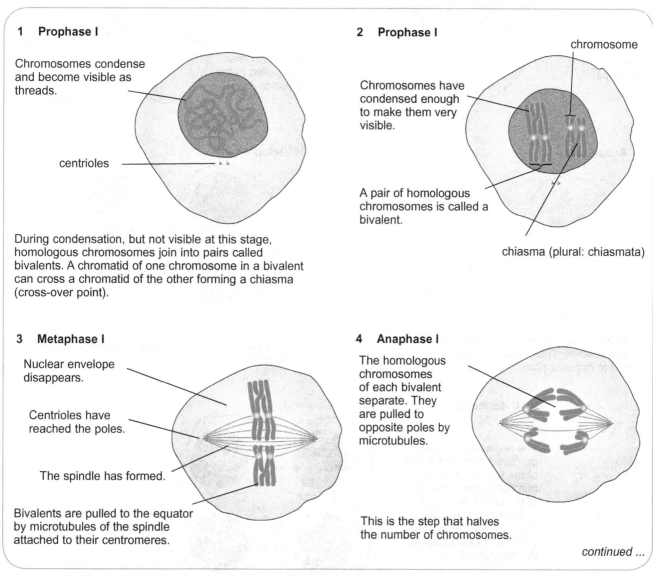

1   Prophase I

Chromosomes condense and become visible as threads.

centrioles

During condensation, but not visible at this stage, homologous chromosomes join into pairs called bivalents. A chromatid of one chromosome in a bivalent can cross a chromatid of the other forming a chiasma (cross-over point).

2   Prophase I

chromosome

Chromosomes have condensed enough to make them very visible.

A pair of homologous chromosomes is called a bivalent.

chiasma (plural: chiasmata)

3   Metaphase I

Nuclear envelope disappears.

Centrioles have reached the poles.

The spindle has formed.

Bivalents are pulled to the equator by microtubules of the spindle attached to their centromeres.

4   Anaphase I

The homologous chromosomes of each bivalent separate. They are pulled to opposite poles by microtubules.

This is the step that halves the number of chromosomes.

continued ...

Figure 6.15   Meiosis in an animal cell (continued overleaf).

5   Telophase I

Chromosomes reach opposite poles and may decondense and form two nuclei, now with half the number of chromosomes in each.

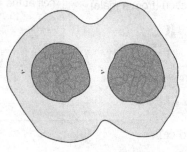

6   Cytokinesis I

The plasma membrane folds inwards to form two cells. The centrioles divide and new spindles start to form.

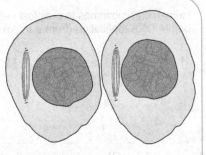

7   Prophase II

Chromosomes condense.

Spindle develops.

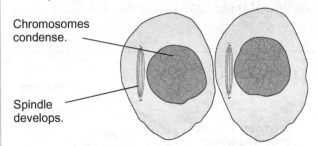

8   Metaphase II

Nuclear envelope disappears.

Chromosomes are pulled to the equator.

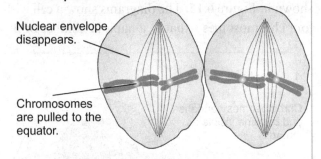

9   Anaphase II

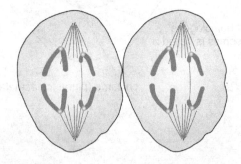

Centromeres divide so each chromatid is now a chromosome. These chromosomes are pulled to opposite poles.

10 Telophase II

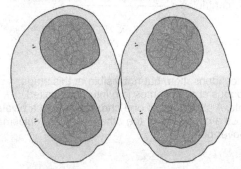

Chromosomes reach the poles, decondense and nuclei form.

11 Cytokinesis II

The original cell has now produced four cells. Each of the four has half the number of chromosomes of the parent cell. Each cell has one chromatid (now chromosome) from each homologous pair in the original cell.

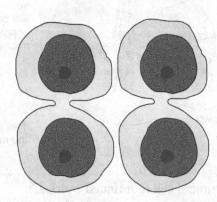

Figure 6.15   Meiosis in an animal cell (continued).

3 During which division, meiosis I or meiosis II, is the chromosome number halved?

4 Which of these divisions would be possible? Explain your answers.

a a diploid cell dividing by mitosis to form diploid cells

b a diploid cell dividing by meiosis to form haploid cells

c a haploid cell dividing by mitosis to form haploid cells

d a haploid cell dividing by meiosis to form haploid cells

5 Name the stage of meiosis at which each of these events occurs. Remember to state whether the stage is in meiosis I or meiosis II.

a Homologous chromosomes pair to form bivalents.

b Chiasmata form between chromatids of homologous chromosomes.

c Homologous chromosomes separate.

d Centromeres divide and chromatids separate.

e Haploid nuclei are first formed.

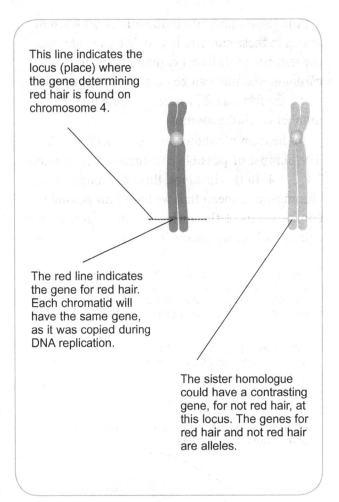

This line indicates the locus (place) where the gene determining red hair is found on chromosome 4.

The red line indicates the gene for red hair. Each chromatid will have the same gene, as it was copied during DNA replication.

The sister homologue could have a contrasting gene, for not red hair, at this locus. The genes for red hair and not red hair are alleles.

Figure 6.16   Different alleles for a gene can exist on homologous chromosomes.

How meiosis causes variation

We have already seen one way in which the new cells formed by meiosis are different from their parent cell. The new cells are haploid whereas the parent cell was diploid. But meiosis also produces variation among the **genes** that these cells contain.

Consider a human cell, with two sets of 23 chromosomes, 46 in all. There are two chromosome 1s, two chromosome 2s and so on. One of each pair came from the father, and one from the mother. Both of the chromosomes of a homologous pair carry genes for the same feature at the same place, called a **locus**. For example, both chromosome 4s carry a gene that determines whether red hair will be produced.

Most genes exist in different versions, called **alleles**. The alleles have slight differences in the base sequences in their DNA. As a human cell has two copies of each chromosome, they have two copies of each gene. The cell could therefore contain two different alleles of that gene (Figure 6.16).

Independent assortment

During meiosis, as pairs of chromosomes line up on the equator, each pair behaves independently of every other pair. Figure 6.17 shows this for two pairs of chromosomes. One pair carries the gene for red hair, and the other pair carries the gene for colour blindness to blue. In Figure 6.17, chromosomes from the father are shown in blue, and chromosomes from the mother in grey.

This is called **independent assortment**. It mixes up alleles that originally came from an organism's father and its mother, so that the gametes it produces contain a mixture of alleles from both of the organism's parents. Each sperm or egg that you produce contains a mixture of alleles from your father and your mother.

And the number of combinations of different alleles in these gametes is vast. We can calculate the number of different combinations of *chromosomes* that can be present in the gametes using the formula 2^n, where n is the haploid number of chromosomes.

In the example shown in Figure 6.17, $n = 2$. The number of possible combinations is therefore $2 \times 2 = 4$. In this instance, these combinations of chromosomes mean that we have four possible combinations of the alleles that they carry for hair colour and colour vision. They are:

- red hair / blue colour blindness
- red hair / normal blue vision
- not red hair / blue colour blindness
- not red hair / normal blue vision

But in a human cell, the haploid number is 23. The number of different combinations of chromosomes is therefore 2^{23}. Try working this out (you have to multiply 2 by itself 23 times). No wonder we never look exactly like either of our parents, or our brothers or sisters. The only exception is identical twins, who each inherit exactly the same combination of genes.

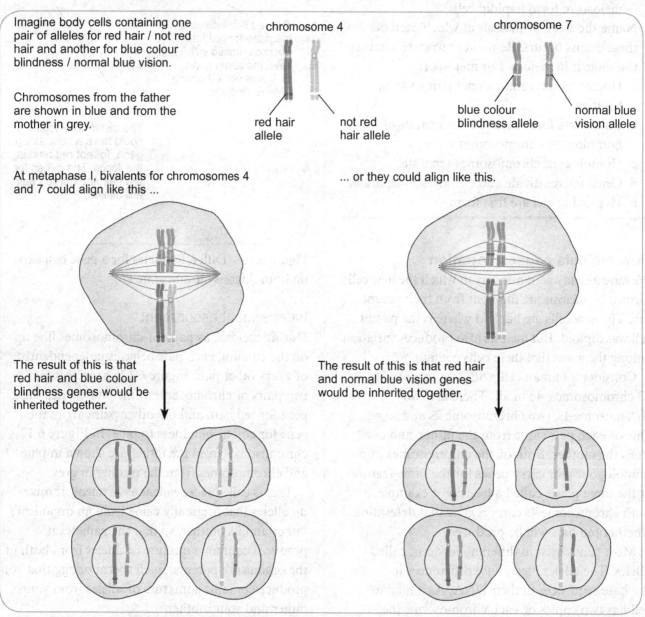

Imagine body cells containing one pair of alleles for red hair / not red hair and another for blue colour blindness / normal blue vision.

Chromosomes from the father are shown in blue and from the mother in grey.

chromosome 4

red hair allele

not red hair allele

chromosome 7

blue colour blindness allele

normal blue vision allele

At metaphase I, bivalents for chromosomes 4 and 7 could align like this ...

... or they could align like this.

The result of this is that red hair and blue colour blindness genes would be inherited together.

The result of this is that red hair and normal blue vision genes would be inherited together.

Figure 6.17 How independent assortment produces variation. As a result of the randomness of alignment of the bivalents during metaphase I, either of a pair of alleles of one gene may end up in the same cell as either of a pair of alleles of another gene on a different chromosome.

Crossing over

Crossing over happens during prophase I. It is a result of the chromatids within a bivalent (pair of homologous chromosomes) getting tangled up with one another. They form **chiasmata** (singular: chiasma). The chromatids break and rejoin at each chiasma, producing a different arrangement of alleles on each one (Figure 6.18).

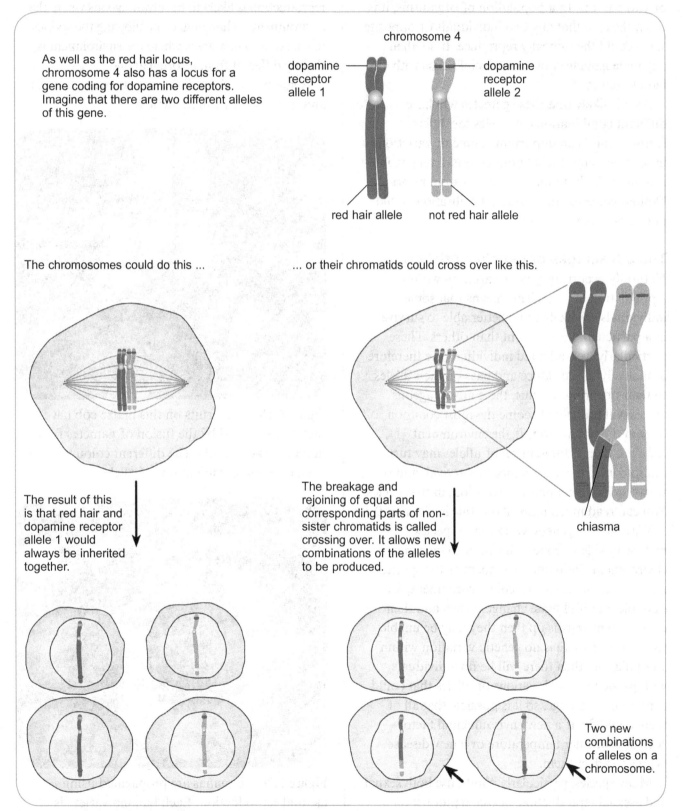

As well as the red hair locus, chromosome 4 also has a locus for a gene coding for dopamine receptors. Imagine that there are two different alleles of this gene.

chromosome 4

dopamine receptor allele 1

dopamine receptor allele 2

red hair allele

not red hair allele

The chromosomes could do this ...

... or their chromatids could cross over like this.

chiasma

The result of this is that red hair and dopamine receptor allele 1 would always be inherited together.

The breakage and rejoining of equal and corresponding parts of non-sister chromatids is called crossing over. It allows new combinations of the alleles to be produced.

Two new combinations of alleles on a chromosome.

Figure 6.18   How crossing over produces variation.

Random fertilisation

A central event in sexual reproduction is the fusion of the nuclei of two gametes, in the process of fertilisation. This brings together the chromosomes of two parents. In a population of organisms, it is often the case that any two individuals of opposite sexes could theoretically reproduce. If so, then any male gamete could, in principle, fuse with any female gamete.

As it is likely that these gametes will have different combinations of alleles for different features, this is an important source of variation in sexually reproducing populations. Every zygote is unique. Without meiosis, there would be no haploid gametes, and random fertilisation would not be possible.

The advantages of genetic variation

Naturally occurring genetic variation within a population of organisms means that some individuals are likely to be better able to survive in a particular environment than others. These particularly well-adapted individuals are therefore more likely to reproduce and pass on their alleles to their offspring. In time, these advantageous alleles will probably become the most common ones in the population. If the environment changes, then a different set of alleles may turn out to confer the best chance of survival, and so we would see a change, or evolution, in the species. You can read much more about this in Chapter 9.

Without any genetic variation, a species may be unable to cope well with a change in its environment. Individual members of the species may be well adapted to a cold environment, for example, but if climate change occurs and their environment warms up then they may be unable to survive. If there is no genetic variation within a population, then there will be no individuals with particular combinations of alleles that could confer an advantage, so it is possible that all of them could die if a new environmental factor – such as a warmer temperature or a new disease – begins to affect them.

Many species, particularly plants, use both sexual and asexual reproduction. Sexual reproduction provides genetic variation (Figure 6.19), while asexual reproduction provides genetic stability (Figure 6.20). The latter can be useful when the environment is stable and the plant is well adapted and is spreading within one area. Sexual reproduction is likely to be advantageous where the environment is changing, or in allowing the species to spread into new areas where the environment is slightly different from its original one.

A summary of the differences between mitosis and meiosis is shown in Table 6.1.

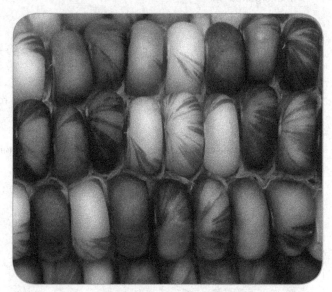

Figure 6.19   The fruits on this maize cob have each been formed by the fusion of gametes from the same two parents. The different colours result from different combinations of alleles.

Figure 6.20   Bananas are propagated using asexual reproduction. Each banana variety is therefore genetically uniform.

Feature	Mitosis	Meiosis
number of divisions	one	two
homologous chromosome behaviour	homologous chromosomes behave independently of one another	during meiosis I, homologous chromosomes associate to form bivalents
separation at anaphase	chromatids are pulled apart during anaphase	during anaphase I, homologous chromosomes are pulled apart but chromatids remain attached at the centromere; the chromatids are separated at anaphase II
crossing over	does not occur in mitosis	crossing over results in exchange of alleles between the chromatids of homologous chromosomes during prophase I
number of daughter cells produced	two daughter cells are produced from one parent cell	four daughter cells are produced from one parent cell
genes in daughter cells	daughter cells are genetically identical to the parent cell	daughter cells are genetically different from the parent cell, as a result of crossing over at prophase I and independent assortment at metaphase I followed by random segregation at anaphase I
chromosome number of daughter cells	daughter cells have the same number of chromosomes as the parent cell	daughter cells have half the number of chromosomes of the parent cell

Table 6.1 A comparison of mitosis and meiosis.

Summary

- The cell cycle consists of interphase, mitosis and cytokinesis. Mitosis occupies only about 5% of the time, while the rest is used for replication and checking of DNA.

- In interphase, DNA replicates, so that each chromosome is made up of two identical chromatids.

- During mitosis, the nuclear membrane breaks down, spindle fibres form, attach themselves to the condensed chromosomes and manoeuvre them to the equator of the cell. The centromeres then break, and the spindle fibres pull the separated chromatids to opposite ends of the cell. New nuclear membranes form around each group of chromatids. The phases of mitosis are prophase, metaphase, anaphase and telophase.

- During cytokinesis, the cytoplasm splits and two daughter cells are formed.

- Mitosis produces daughter cells that are genetically identical and is used for growth, repair of tissues and asexual reproduction.

- A cell with one complete set of chromosomes is said to be haploid, while a cell with two complete sets of chromosomes is diploid. The two similar chromosomes from the two sets are said to be homologous.

continued ...

- Cell division by meiosis produces four haploid cells from one diploid cell. The daughter cells are genetically different.

- In the first division of meiosis, homologous chromosomes pair up to form bivalents, and chiasmata form between the chromatids of the two chromosomes in the pair. The homologous chromosomes are then pulled to opposite ends of the cell, reducing the chromosome number by half. Each cell then undergoes a second division similar to that which takes place in mitosis.

- Meiosis is used to produce gametes for sexual reproduction. It is a very important source of genetic variation. This occurs because of independent assortment of homologous chromosomes, crossing over, and the random fusion of gametes.

Questions

Multiple choice questions

1 During mitosis, a double-stranded chromosome is attached to a spindle fibre at the:
   A centriole.   **B** centrosome.   **C** centromere.   **D** chromatid.

2 What happens to chromosomes in prophase of mitosis?
   A They are formed by replication of DNA.
   B They attach to the spindle fibres.
   C They divide to form chromatids.
   D They shorten and become visible.

3 The diagram below shows one stage in cell division in a locust.

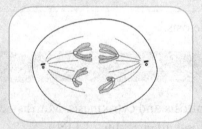

   Which of the following correctly identifies the type of division shown in the diagram, the stage and where it may occur?

	Type of division	Stage	Where it occurs
A	mitosis	anaphase	testis
B	mitosis	anaphase	wing
C	meiosis	anaphase I	wing
D	meiosis	anaphase I	testis

4 Which statement correctly describes homologous chromosomes?
   A They are held together by centromeres.
   B They are present in meiotic but not mitotic cells.
   C They carry the same alleles.
   D They carry the same gene loci.

continued ...

130

5   The following events occur during either mitosis or meiosis or in both.
    Which only occurs in meiosis?
    A   nuclear envelope breaks down
    B   condensation of DNA
    C   pairing up of homologous chromosomes
    D   centromeres separate

6   The micrographs below show the stages of mitosis in *Allium* root tip.

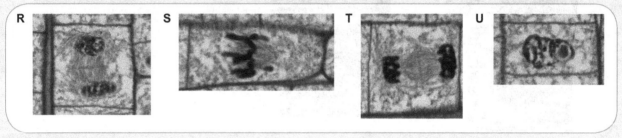

    In which order do the stages occur?

	first			last
A	R	S	T	U
B	S	R	T	U
C	U	S	R	T
D	T	R	S	U

7   Which of the following is the stage at which sister chromatids separate in meiosis?
    A prophase I
    B metaphase I
    C anaphase I
    D anaphase II

8   The diagram on the right represents a pair of homologous
    chromosomes at the beginning of meiosis I. The loci of
    three genes are shown. All three genes have two alleles.
    Four gametes are formed at the end of meiosis.

    How many genetically different gametes would be produced
    at the end of meiosis?
    A 1   B 2   C 3   D 4

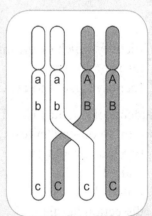

9   Which of the following occurs during mitosis but not meiosis I?
    A Sister chromatids separate.
    B The nuclear membrane breaks up into vesicles.
    C The chromosome number is reduced.
    D The homologous chromosomes pair up.

10 If a cell with 16 chromosomes divides by meiosis, how many chromosomes will each nucleus contain
    at telophase I?
    A 32   B 24   C 16   D 8

continued ...

Structured questions

11 The micrographs below show cells undergoing mitosis in the root tip of the onion, *Allium*.

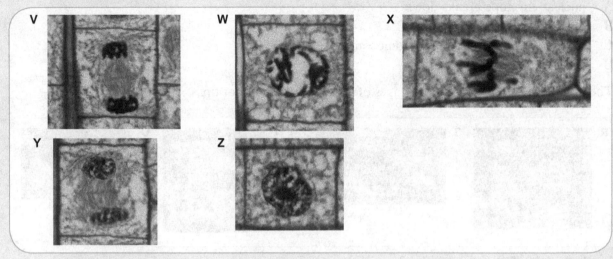

a Name the stages **V** to **Z**. [3 marks]

b Make drawings to show the alignment of the chromosomes in the stages labelled **W** and **X** above. Drawings should be magnified five times. [4 marks]

c The micrographs below show stages in meiosis.

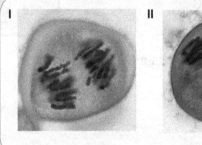  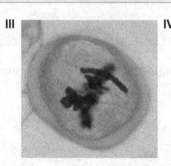 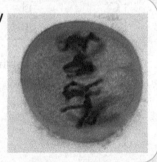

  i Identify which of the micrographs represent stages in meiosis II. Give a reason for your answer. [3 marks]

 ii Identify stages **I** to **IV**. Give a reason for each one. [5 marks]

12 a Students of a Unit 1 CAPE® Biology class were asked to prepare a slide of a garlic root tip to observe mitosis. They were provided with the following materials:

- two intact root tips
- scalpel
- watch glass
- dissection needle
- coverslip, slide
- acetic orcein
- 1 mol dm⁻³ hydrochloric acid.

Briefly describe how the students should prepare the slide. [3 marks]

b Another set of students used a prepared slide of *Allium* to observe and count cells in different stages of the cell cycle. Each student observed 50 cells in each slide. The class data were pooled and then analysed.

continued ...

The table below shows the data obtained.

	Total number of cells in each stage of mitosis				
	interphase	prophase	metaphase	anaphase	telophase
Aneela	14	13	8	8	7
Faeryal	17	10	9	10	4
Alex	15	12	7	9	7
Damian	15	8	9	9	9
Leon	13	11	9	9	8
class total	74				
class average	14.8				
% of cells	29.6				

    i Copy and complete the table above. (The first column has been done for you.)   [4 marks]

    ii Using your own graph paper, plot a bar graph to show the percentage of cells in each stage of the cell cycle.   [4 marks]

c There was a difference between Faeryal's data and the other students for the telophase stage. Suggest a possible reason for this.   [1 mark]

d Make a conclusion from the graph you have drawn.   [1 mark]

e During the interphase stage of the cell cycle, the DNA of the cell is replicated and sister chromatids are formed. If a parent plant cell has 16 chromosomes, what are the number of chromosomes and chromatids present in prophase I, metaphase I, anaphase I, telophase I, prophase II, metaphase II, anaphase II and telophase II?   [2 marks]

Essay questions

13 a With the aid of diagrams, give an account of the process of meiosis.   [7 marks]

   b Explain the significance of meiosis in sexually reproducing organisms.   [6 marks]

   c Name **two** places in a plant where meiosis takes place.   [2 marks]

14 a During mitosis, the chromosomes condense and spiral while the nuclear envelope breaks down. With the aid of diagrams, describe what happens after this until the nuclear envelope reforms.   [5 marks]

   b List **four** places in a plant where mitosis occurs.   [2 marks]

   c Explain the role of mitosis.   [4 marks]

   d Cancer is a result of uncontrolled cell division. As a result, anti-cancer drugs have been developed to interrupt different parts of the cell cycle. One such drug is Oncovin. It is specifically made to disrupt the mitotic phase of the cell cycle. Suggest how it may work.   [4 marks]

15 a Explain what is meant by 'homologous chromosomes' and how two homologous chromosomes may differ from each other.   [3 marks]

   b Explain why the DNA in the sister chromatids in early prophase I of meiosis is identical, while in metaphase I the sister chromatids are no longer identical.   [3 marks]

   c **i** Explain what is meant by the term 'variation'.   [2 marks]

      ii Explain how chiasma formation and independent assortment might result in several different phenotypes in the offspring of two parents.   [7 marks]

Chapter 7
Patterns of inheritance

By the end of this chapter you should be able to:

a explain the terms gene, allele, dominant, recessive, codominant, homozygous and heterozygous;

b use genetic diagrams to solve problems involving monohybrid and dihybrid crosses, including sex linkage, codominance, multiple alleles and epistasis;

c analyse the results of a genetic cross by applying the chi-square test;

d determine whether the difference between the observed and expected ratio is significant, using the results of the chi-square test, including the use of 0.05 confidence limits and a null hypothesis.

Genetics and inheritance

The study of the inheritance of genes is called **genetics**. We will begin by looking at some characteristics that are affected by just one gene **locus**, and then consider some patterns of inheritance that may be seen when alleles found at two different gene loci interact with one another.

Single gene inheritance

We will use the inheritance of **cystic fibrosis** as an example. This is a genetic disease in which abnormally thick mucus is produced in the lungs and other parts of the body. A person with cystic fibrosis is very prone to bacterial infections in the lungs because it is difficult for the mucus to be removed, allowing bacteria to breed in it.

Cystic fibrosis is caused by a faulty version of a gene that codes for the production of a protein called **CFTR**. The protein normally sits in the plasma membranes of cells in the lungs and other organs, where each protein molecule forms a channel that allows chloride ions to pass from inside the cell to the outside. The gene for CFTR is found on chromosome 7. It consists of about 250 000 bases. Mutations in this gene have produced several different alleles. The commonest of these alleles results from the deletion of three bases. The CFTR protein made using this code is therefore missing one amino acid. The machinery in the cell recognises that this is not the right protein, and it does not place it in the plasma membrane.

This faulty allele is a **recessive allele**. The normal allele is a **dominant allele**. A recessive allele only has an effect on an organism when the dominant allele is not present. A dominant allele has an effect whether or not the recessive allele is present.

We can use symbols to represent these two alleles. Because they are alleles of the same gene, we should use the same letter to represent both of them. By convention, a capital letter is used to represent the dominant allele, and a small letter to represent the recessive allele. It is a good idea to choose letters where the capital and small letter look different, so that neither you nor an examiner is in any doubt about what you have written.

In this case, we will use the letter F for the allele coding for the normal CFTR protein, and the letter f for the allele coding for the faulty version.

Because human cells are diploid, we have two copies of each gene. There are three possible gene combinations – called **genotypes** – that may be present in any one person's cells. They affect the person's **phenotype** – their observable characteristics.

The three possible genotypes are:

Genotype	Phenotype
FF	unaffected
Ff	unaffected
ff	cystic fibrosis

A genotype in which both alleles of a gene are the same is said to be **homozygous**. A genotype in which the alleles of a gene are different is **heterozygous**. FF and ff are homozygous, and Ff is heterozygous.

Inheritance of the *CFTR* gene

When gametes are made by meiosis, the haploid daughter cells get only one copy of each pair of chromosomes. So they only contain one copy of each gene. A sperm or an egg can therefore contain only one allele of the *CFTR* gene.

Genotype of parent	Possible genotypes of their gametes
FF	all F
Ff	50% F and 50% f
ff	all f

At fertilisation, any gamete from the father can fertilise any gamete from the mother. We can show all of this by drawing a **genetic diagram**. This is a conventional way of showing the chances of a child of a certain genotype or phenotype being born to parents having a particular genotype or phenotype. The genetic diagram at the top of the next column shows the chances of a heterozygous man and a heterozygous woman having a child with cystic fibrosis.

The genetic diagram shows that the phenotype ratio among the offspring is 3 unaffected : 1 affected. This means that every time the couple have a child, there is a 25% chance that the child will inherit the genotype FF and a 50% chance that it will inherit the genotype Ff. Thus, there is a 75% chance that the child will not have cystic fibrosis. The chance of the child inheriting the genotype ff and having cystic fibrosis is 25%.

phenotypes of parents:	male not affected	×	female not affected
genotypes of parents:	Ff		Ff
genotypes of gametes:	Ⓕ and Ⓕ		Ⓕ and Ⓕ

genotypes and phenotypes of offspring:

		gametes from father	
		Ⓕ	Ⓕ
gametes from mother	Ⓕ	FF unaffected	Ff unaffected (carrier)
	Ⓕ	Ff unaffected (carrier)	ff cystic fibrosis

Expected offspring phenotype ratio is 3 unaffected : 1 cystic fibrosis.

Another way of expressing this is to say that the probability of the child *not* having cystic fibrosis is 0.75, while the probability of it having the disease is 0.25. This can also be stated as a probability of 1 in 4 that a child born to these parents will have this disease.

Yet another way of expressing this is to say that the expected ratio of children without cystic fibrosis to those with cystic fibrosis is 3 : 1.

Test crosses

If an organism shows a characteristic that is determined by a recessive allele in its phenotype, we know that its genotype must be homozygous for that recessive allele. However, if an organism shows a characteristic that is determined by a dominant allele in its phenotype, we cannot tell if it is heterozygous or homozygous.

To find out, we can cross the organism showing the dominant characteristic with one showing the recessive characteristic. If we get any offspring that show the recessive characteristic, then the genotype of the parent with the dominant characteristic must be heterozygous. If all the offspring show the dominant characteristic, then it is probable (but not absolutely certain) that the parent in question is homozygous for the dominant allele.

You may like to draw two genetic diagrams to work this out for yourself.

1 Explain what is wrong with each of these statements.

   a 'A couple who are both carriers for cystic fibrosis will have four children, one with cystic fibrosis and three without.'

   b 'If a couple's first child has cystic fibrosis, their second child will not have it.'

2 Copy and complete the genetic diagram to determine the chance of a heterozygous man and a woman with the genotype FF having a child with cystic fibrosis.

F is the normal allele; f is the cystic fibrosis allele

phenotypes of parents: male not affected × female not affected

genotypes of parents: Ff   FF

genotypes of gametes: (F) and (f)   all (F)

genotypes and phenotypes of offspring: gametes from father

gametes from mother (F)

Chance of child with cystic fibrosis is …

3 Explain why, in the genetic diagram you have drawn for SAQ 2, it is not necessary to show two gametes from the female parent.

Codominance

So far, we have looked at examples where one allele of a gene is recessive and another is dominant. The alleles controlling the ABO blood group phenotypes, and those responsible for sickle cell anaemia (Chapter 9), behave differently.

ABO blood group inheritance

Red blood cells contain a glycoprotein in their plasma membranes that determines the ABO blood group. There are two forms of this protein, known as antigens A and B. The gene that encodes this protein is on chromosome 9. It has three alleles, coding for antigen A, antigen B or no antigen at all.

The symbols for these alleles are written differently from those for *CFTR*.

Each symbol includes the letter I to represent the gene locus. A superscript represents one particular allele.

I^A allele for antigen A

I^B allele for antigen B

I^O allele for no antigen

They are written like this because alleles I^A and I^B show **codominance**. They each have an effect when they are together. However, both I^A and I^B are dominant with respect to allele I^O, which is recessive. There are four possible phenotypes:

Genotype	Phenotype
$I^A I^A$	Group A
$I^A I^B$	Group AB
$I^A I^O$	Group A
$I^B I^B$	Group B
$I^B I^O$	Group B
$I^O I^O$	Group O

4 Using the correct symbols, draw a complete and fully labelled genetic diagram to find the chance of a child with blood group O being born to a heterozygous man with blood group B and a heterozygous woman with blood group A.

Sex linkage

Genes whose loci are on the X or Y chromosomes (**sex chromosomes**) have different inheritance patterns from genes on all the other chromosomes (**autosomes**). Women have two X chromosomes, while men have one X and one Y.

The X chromosome is much larger than the Y chromosome. It has many genes that are not present on the Y. Most of these two chromosomes are therefore not homologous (Figure 7.1). These genes are said to be **sex-linked**, because their inheritance is affected by whether a person is male or female. If one of these genes has a recessive allele that causes a particular condition, then this condition is much more common in males than in females and, indeed, may not ever occur in females at all (Figure 7.2).

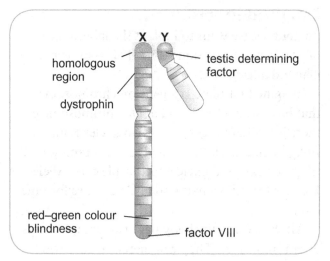

Figure 7.1 X and Y chromosomes showing the positions of some genes.

One such gene determines the production of a factor that is needed to enable blood to clot, a protein called **factor VIII**. There is a recessive allele of this gene that codes for a faulty version of factor VIII. With this faulty version, blood does not clot properly, a condition called **haemophilia**. Bleeding occurs into joints and other parts of the body, which can be very painful and eventually disabling. Haemophilia can nowadays be treated by giving the person factor VIII throughout their life.

When writing symbols of genes carried on the X chromosome, they are written as superscripts. The symbol X^H can be used to stand for the normal allele, and X^h for the haemophilia allele.

In a woman, there are two X chromosomes, so a woman always has two factor VIII genes. Her possible genotypes and phenotypes are:

Genotype	Phenotype
$X^H X^H$	normal blood clotting
$X^H X^h$	normal blood clotting (but she is a carrier)
$X^h X^h$	lethal

A foetus with the genotype $X^h X^h$ does not develop, so no babies are born with this genotype.

In a man, however, there is only one X chromosome present, so he can only have one allele of this gene. His possible genotypes and phenotypes are:

Genotype	Phenotype
$X^H Y$	normal blood clotting
$X^h Y$	haemophilia

The genetic diagram at the top of the next page shows how a woman who is a carrier for haemophilia, and a man who has normal blood clotting, can have a son with haemophilia.

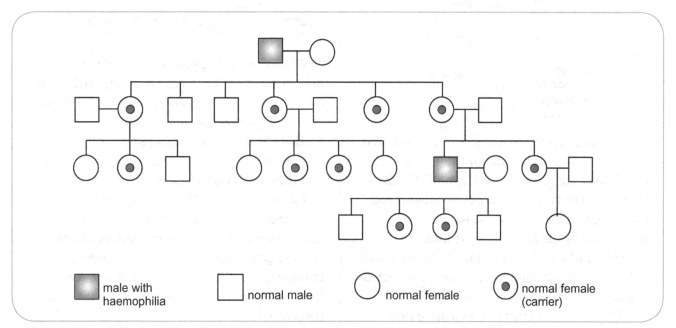

male with haemophilia   normal male   normal female   normal female (carrier)

Figure 7.2 Pedigree for a sex-linked recessive disease, such as haemophilia.

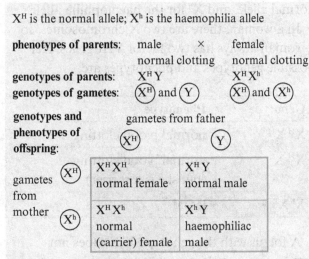

X^H is the normal allele; X^h is the haemophilia allele

phenotypes of parents:	male ×	female
	normal clotting	normal clotting
genotypes of parents:	X^H Y	X^H X^h
genotypes of gametes:	(X^H) and (Y)	(X^H) and (X^h)

genotypes and phenotypes of offspring:

gametes from father

	(X^H)	(Y)
gametes from mother (X^H)	X^H X^H normal female	X^H Y normal male
(X^h)	X^H X^h normal (carrier) female	X^h Y haemophiliac male

Expected offspring phenotype ratio is
3 normal : 1 haemophilia.

SAQ

5 Explain why a man with haemophilia cannot pass it on to his son.

6 The family tree shows the occurrence of a genetic condition known as brachydactyly (short fingers). Use the tree to deduce:

  a whether the allele for this condition is dominant or recessive

  b if this condition is sex-linked.

  Explain your answers.

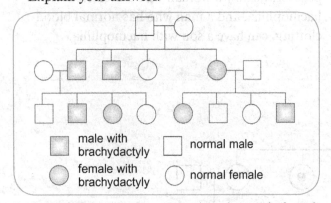

■ male with brachydactyly   □ normal male
● female with brachydactyly   ○ normal female

7 One of the genes for coat colour in cats is found on the X chromosome but not the Y. The allele C^O of this gene gives orange fur, while C^B gives black fur. The two alleles are codominant, and when both are present the cat has patches of orange and black, known as tortoiseshell.

  a Explain why male cats cannot be tortoiseshell.

  b Draw a genetic diagram to show the expected genotypes and phenotypes of the offspring from a cross between an orange male and a tortoiseshell female cat.

Dihybrid inheritance

Sometimes, we want to look at the inheritance of two genes at the same time. This is known as **dihybrid inheritance**.

Imagine that there is a gene on chromosome 4 that has two alleles, A and a. On chromosome 6 there is a different gene with two alleles B and b. Imagine that allele A, in the Rainbow family, codes for green ears and allele a for purple ears. Allele B codes for yellow hair and allele b codes for blue hair.

All the cells in the body have two complete sets of chromosomes. They will therefore have two chromosome 4s and two chromosome 6s, so they will have two copies of each gene. There are nine different genotypes that any one person could have, and four different phenotypes:

Genotype	Phenotype
AABB	green ears, yellow hair
AABb	green ears, yellow hair
AAbb	green ears, blue hair
AaBB	green ears, yellow hair
AaBb	green ears, yellow hair
Aabb	green ears, blue hair
aaBB	purple ears, yellow hair
aaBb	purple ears, yellow hair
aabb	purple ears, blue hair

When meiosis happens and gametes are made, only one copy of each gene goes into each gamete. So, if a man has the genotype AABB, all of his sperm will get one of the A alleles and one of the B alleles. If he has the genotype AaBB, half of his sperm will get allele A and the other half allele a, and they will all get allele B.

We saw on page 125 that independent assortment in meiosis I means that each pair of chromosomes behaves entirely independently. If these genes A/a and B/b are on different chromosomes, then either allele of one may find itself in a gamete with either allele of the other (Figure 7.3).

We can work out the results of a dihybrid cross in just the same way as for a monohybrid cross, but showing the alleles of *both* genes. Notice that we always write the two alleles for one gene next to each other.

In the example below, both parents are heterozygous at both gene loci. The 9 : 3 : 3 : 1 ratio of phenotypes resulting from this cross is typical of a dihybrid cross between two parents who are both heterozygous at both gene loci.

A is the green ear allele; a the purple ear allele; B the yellow hair allele; b the blue hair allele

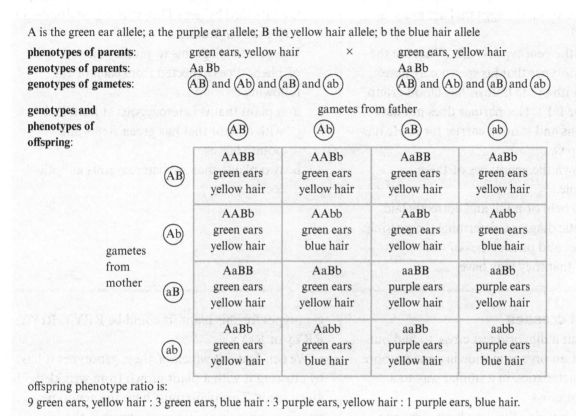

phenotypes of parents:	green ears, yellow hair		×	green ears, yellow hair	
genotypes of parents:	Aa Bb			Aa Bb	
genotypes of gametes:	(AB) and (Ab) and (aB) and (ab)			(AB) and (Ab) and (aB) and (ab)	

offspring phenotype ratio is:

9 green ears, yellow hair : 3 green ears, blue hair : 3 purple ears, yellow hair : 1 purple ears, blue hair.

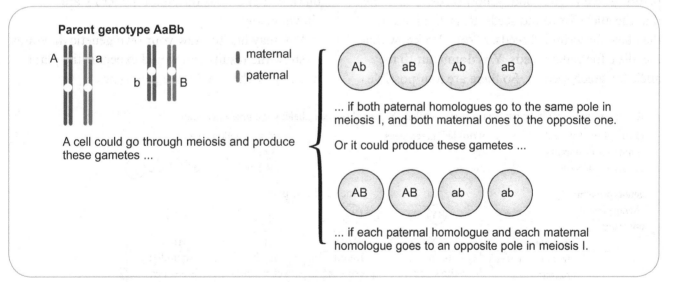

Figure 7.3 Independent assortment in dihybrid inheritance.

SAQ

8 Copy the two groups of four gametes in Figure 7.3 and draw the appropriate chromosomes inside each one.

9 A woman has the genotype AAbb. What is the genotype of all the eggs that are made in her ovaries?

10 A man has the genotype AABb. What are the possible genotypes that his sperm may have?

11 A woman with cystic fibrosis has blood group A (genotype $I^A I^o$). Her partner does not have cystic fibrosis and is not a carrier for it. He has blood group O.

   a Write down the genotypes of these two people.

   b With the help of a full and correctly laid out genetic diagram, determine the possible genotypes and phenotypes of any children that they may have.

12 Tomato plants can have purple or green stems, and potato (smooth) or cut (jagged) leaves. Stem colour is controlled by gene A/a, where A is dominant and gives purple stem. Leaf shape is controlled by gene D/d, where D is dominant and gives cut leaves.

   Use genetic diagrams to predict the ratios of phenotypes expected from each of the following crosses:

   a a plant that is heterozygous at both loci with a plant that has green stems and potato leaves

   b two plants that are heterozygous at both loci.

Dihybrid test crosses

We can carry out a dihybrid test cross, to find out the genotype of an organism showing one or more dominant characteristics, in a similar way to a monohybrid test cross.

For example, imagine that you have some pea plants grown from round, yellow seeds. You know that the allele for round seeds, R, is dominant to the allele for wrinkled seeds, r. You also know that the allele for yellow seeds, Y, is dominant to the allele for green seeds, y. So there are four possible genotypes for this plant. It could be RRYY, RrYY, RRYy or RrYy.

We can find out which of these genotypes it has by crossing it with a plant grown from wrinkled, green seeds. This plant must have the genotype rryy. The offspring you would expect if the unknown plant has the genotype RrYy are shown below.

You may like to draw your own genetic diagrams to show the results you would expect if the plant had any of the other three genotypes.

R is the round pea allele; r the wrinkled pea allele; Y the yellow pea allele; y the green pea allele

phenotypes of parents:	wrinkled, green peas	×	round, yellow peas
genotypes of parents:	rryy		RrYy
genotypes of gamete:	all (ry)		(R) and (r) and (Y) and (y)

genotypes and phenotypes of offspring:

gametes of one parent

	(RY)	(Ry)	(rY)	(ry)
gamete from other parent (ry)	RrYy round yellow peas	Rryy round green peas	rrYy wrinkled yellow peas	rryy wrinkled green peas

Expected offspring phenotype ratio is:
1 round, yellow peas : 1 round, green peas : 1 wrinkled, yellow peas : 1 wrinkled, green peas.

Epistasis

Quite frequently, two different genes both affect the same characteristic. This is often because the two genes code for two enzymes that help to control the same metabolic pathway.

For example, a particular plant might produce the pigments that colour its petals in a two-step pathway:

$$\text{colourless substance} \xrightarrow{\textit{enzyme 1}} \text{yellow pigment} \xrightarrow{\textit{enzyme 2}} \text{orange pigment}$$

The gene that codes for enzyme 1 could have two alleles. A is the dominant allele and does not produce a working enzyme. The recessive allele, a, codes for the enzyme. B codes for enzyme 2, while b does not produce any enzyme 2.

Before the plant can produce any colour at all, it must have a working version of enzyme 1. So it must have two a alleles. Only if it has the genotype aa can it produce yellow pigment. It does not matter what alleles of the B/b gene it has, if there is no yellow pigment for their products to work on. All the possible genotypes and phenotypes are shown below.

Genotype	Phenotype
AABB	white
AABb	white
AAbb	white
AaBB	white
AaBb	white
Aabb	white
aaBB	orange
aaBb	orange
aabb	yellow

You can see from this example that the genotype for one gene affects the expression of another quite separate gene. This situation is called **epistasis**. Allele A masks the effect of the alleles at the B locus. As A is dominant, this epistasis is **dominant epistasis**. The alleles at the B locus are said to be **hypostatic**.

Coat colour in animals is quite often determined by epistatic genes. Commonly, one gene determines whether there is any pigment produced at all, while another determines its pattern or precise colour. Obviously, the 'pattern' gene cannot have any effect unless there is some pigment there. In fact, the situation is often even more complicated than this, with many different genes all interacting to determine coat colour. You only have to look at all the different coat colours in cats and dogs to get an indication of this.

For example, the colours of 'wild type' and black mice are determined by a gene, A/a, which codes for the distribution of the pigment melanin in the hairs (Figure 7.4). The coat of a wild type mouse is made up of banded hairs, which produces a grey-brown colour called agouti. Allele A determines the presence of this banding. Allele a determines the uniform black colour of the hair of a black mouse.

A second gene, C/c, determines the production of melanin. The dominant allele C allows colour to develop, while a mouse with the genotype cc does not make melanin and so is albino. Allele c therefore masks the effect of the alleles at the A/a locus. This is **recessive epistasis**.

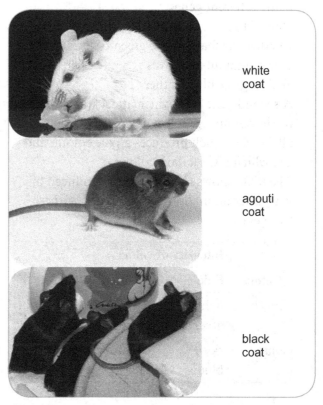

white coat

agouti coat

black coat

Figure 7.4 The coat colour of mice is controlled by epistatic genes.

SAQ

13 a List the three possible genotypes of each of the three mice in Figure 7.4.

  b What would be the expected results of a cross between two agouti mice with genotypes AaCc?

  c Suggest how you could do a breeding experiment to determine if an agouti mouse was heterozygous for the C/c alleles.

14 Cats with either black or white fur are common, but brown fur is rarer. The dominant allele of one gene, B, gives black fur and the recessive allele, b, gives brown fur.

  Many white cats carry a dominant allele, A, of a second gene which inhibits pigment production no matter which pigment-producing alleles are present in the genotype. The recessive allele, a, has no effect on fur colour.

  Draw a genetic diagram to show the expected genotypes and phenotypes of a cross between two white cats that are heterozygous at both loci.

15 Feather colour in budgerigars is affected by many different genes.

  One of these is the gene G/g, which determines whether the feathers are green or blue. Allele G is dominant and gives green feathers, while allele g gives blue feathers.

  A second gene is C, which affects the intensity of the colouring. It has two codominant alleles: C^P, which produces a pale colour, and C^D, which gives a dark colour.

  The table shows the six colours produced by various combinations of the alleles of these two genes.

	Intensity of colour		
Colour	**Pale**	**Medium**	**Dark**
green	light green	dark green	olive green
blue	sky blue	cobalt blue	mauve

  a What is the genotype of:
    i a dark green bird
    ii a sky blue bird?

  b Draw a genetic diagram to show the possible offspring produced from a cross between a dark green bird and a cobalt blue bird. Indicate the phenotype of each of the different genotypes produced in the cross.

16 The synthesis of carotenoid pigments in sweet pepper fruits is controlled by two unlinked genes. As a pepper fruit ripens, the dominant allele, A, of one gene results in the synthesis of red pigment, while the recessive allele, a, gives yellow pigment. The recessive allele, b, of the second gene reduces the quantity of carotene produced by the gene A/a so that potentially red peppers are orange coloured and potentially yellow peppers are a paler, lemon yellow. The dominant allele B has no effect on the gene A/a.

  Two pepper plants with the genotypes AABB and aabb were crossed and the resulting F_1 generation interbred to give an F_2 generation. Draw a genetic diagram of this cross to show:

  - the phenotypes of the parent plants
  - the genotypes of the gametes
  - the genotypes and phenotypes of the F_1 and F_2 generations
  - the ratios of the phenotypes expected in the F_2 generation.

The chi-square (χ^2) test

The results of the cross between two tomato plants which both have genotype AaDd (described in SAQ **12**, page 140), would be expected to show a 9:3:3:1 ratio of phenotypes in the offspring – 9 purple cut:3 purple potato:3 green cut:1 green potato. It is very important to remember that this ratio is just a *probability*. We would be rather surprised if we got precisely this ratio among the offspring, just as you would not necessarily expect to get exactly five heads and five tails if you tossed a coin ten times.

But just how much difference might we be happy with, before we began to worry that the situation was not quite what we had thought? For example, let us imagine that the two plants produced a total of 144 offspring. If the parents really were both heterozygous, and if the purple stem and cut leaf alleles really are dominant, and if the alleles really do assort independently (that is, they are on different chromosomes and not linked) then we would expect the following numbers of each phenotype to be present in the offspring:

$$\text{purple, cut} \quad = \frac{9}{16} \times 144 \quad = 81$$

$$\text{purple, potato} = \frac{3}{16} \times 144 \quad = 27$$

$$\text{green, cut} \quad = \frac{3}{16} \times 144 \quad = 27$$

$$\text{green, potato} \quad = \frac{1}{16} \times 144 \quad = 9$$

But imagine that, among these 144 offspring, the results we actually observed were as follows:

purple, cut	86
purple, potato	26
green, cut	24
green, potato	8

We might ask: are these results sufficiently close to the ones we expected that the differences between them have probably just arisen by chance? Or are they so different that something unexpected must be going on?

To answer these questions, we can use a statistical test called the χ^2 (**chi-square**) **test**. This test allows us to compare our observed results with the expected results, and decide whether or not there is a significant difference between them.

A statistical test usually begins by setting up a **null hypothesis**. We then use the statistical test to determine the probability of the null hypothesis being true.

In this case, our null hypothesis would be: *the observed results are not significantly different from the expected results.*

In biology, if our statistical test tells us that the chance of the null hypothesis being correct is equal to or greater than 0.05, then we can accept this hypothesis as being correct.

The first stage in carrying out this test is to work out the expected results, as we have already done. These, and the observed results, are then recorded in a table like the one below. We then calculate the difference between observed and expected for each set of results, and square each difference. (Squaring it gets rid of any minus signs – it is irrelevant whether the differences are negative or positive.) Then we divide each squared difference by the expected value and add up all of these answers.

$$\chi^2 = \Sigma \frac{(O - E)^2}{E}$$

where Σ = the sum of
O = the observed value
E = the expected value

	Purple stems, cut leaves	Purple stems, potato leaves	Green stems, cut leaves	Green stems, potato leaves
Observed number, O	86	26	24	8
Expected number, E	81	27	27	9
$O - E$	+5	−1	−3	−1
$(O - E)^2$	25	1	9	1
$\dfrac{(O - E)^2}{E}$	0.31	0.04	0.33	0.11
$\Sigma \dfrac{(O - E)^2}{E} = 0.79$				
$\chi^2 = 0.79$				

So now we have our value for χ^2. Next we have to work out what it means.

To do this, we look in a table that relates χ^2 values to probabilities (Table 7.1). The table tells us *the probability that the null hypothesis is correct.*

For example, a probability of 0.05 means that we would expect these differences to occur in 5 out of every 100 experiments, or 1 in 20, just by chance. A probability of 0.01 means that we would expect them to occur in 1 out of every 100 experiments. For biological data, we have seen that we take a probability of 0.05 as being the critical one. If our χ^2 value represents a probability of 0.05 or larger, then it is reasonable to assume that the differences between our observed and expected results may simply be due to chance – the differences between them are not **significant**. However, if our χ^2 value represents a probability smaller than this, then it is likely that the difference *is* significant, and we must reconsider our assumptions about what was going on in this cross.

There is one more aspect of our results to consider, before we can look up our value of χ^2 in Table 7.1. This is the number of **degrees of freedom** in our results. This takes into account the number of comparisons made. (Remember that to get our value of χ^2 we added up all our calculated values, so obviously the larger the number of observed and expected values we have, the larger χ^2 is likely to be. We need to compensate for this.) To work out the number of degrees of freedom, simply calculate: (number of classes of data – 1). Here we have four classes of data (the four possible phenotypes) so the number of degrees of freedom is 4 – 1 = 3.

Now, at last, we can look at Table 7.1 to determine whether our results show a significant deviation from what we expected. The numbers in the body of the table are χ^2 values. We look at the third row in the table, because that is the one for 3 degrees of freedom, and find the χ^2 value that represents a probability of 0.05. You can see that this is 7.82. Our calculated value of χ^2 was 0.79. So our value is much, much smaller than the one we have read from the table. In fact, we cannot find anything like this number in the table – it would be way off the left-hand side, representing

a probability of much more than 0.1 (1 in 10) that the difference in our results is just due to chance. So we can say that the difference between the observed and expected results could well be due to chance, and there is *no significant difference* between what we expected and what we actually got.

Degrees of freedom	Probability greater than			
	0.1	0.05	0.01	0.001
1	2.71	3.84	6.64	10.83
2	4.60	5.99	9.21	13.82
3	6.25	7.82	11.34	16.27
4	7.78	9.49	13.28	18.46

Table 7.1 Table of χ^2 values.

SAQ

17 The allele for grey fur in a species of mammal is dominant to white, and the allele for long tail is dominant to short.

a Using the symbols G and g for coat colour, and T and t for tail length, draw a genetic diagram to show the genotypes and phenotypes of the offspring you would expect from a cross between a pure-breeding (homozygous) grey animal with a long tail and a pure-breeding white animal with a short tail.

b If the first generation of offspring were bred together, what would be the expected phenotypes in the next generation, and in what ratios would you expect them to occur?

c In an actual cross between the animals in the first generation, the numbers of each phenotype obtained in the offspring were:

grey, long   54
grey, short   4
white, long   4
white, short 18

Use the χ^2 test to determine whether or not the difference between these observed results and the expected results is significant.

Summary

- A gene is a length of DNA that codes for the formation of a polypeptide or protein. A gene for a particular polypeptide or protein is always found at the same locus on a particular chromosome. It may have several different forms, called alleles.

- Diploid cells contain two copies of each gene. If the two copies are the same allele of the gene, the cell is homozygous for this gene. If they are different, it is heterozygous.

- Haploid cells, such as gametes, contain only one copy of each gene.

- An allele that has an effect even if a different allele of the same gene is present in the cell is said to be dominant. A recessive allele only has an effect if the dominant allele is not present. If both alleles have an equal effect when they are present together, they are said to be codominant.

- Many genes, such as those which determine the ABO blood groups, have several different alleles, and these are known as multiple alleles.

- Genetic diagrams show the different genotypes that can occur among the offspring of two parents, and the probabilities of each of these genotypes occurring.

- Genes that are found on the X chromosome, but not on the Y chromosome, are said to be sex-linked. Recessive alleles of a sex-linked gene are much more likely to have an effect in a male than in a female, because they will be the only allele of that gene in its cells. Haemophilia and red–green colour-blindness are examples of sex-linked conditions in humans.

- A cross involving two different genes is called a dihybrid cross. If one of the parents is homozygous for both of the genes, and the other parent is heterozygous for both of the genes, the expected ratio of phenotypes in the offspring is 1 : 1 : 1 : 1. If both parents are heterozygous for both genes, the expected ratio is 9 : 3 : 3 : 1.

- The effect of one gene may be determined by another. This is called epistasis.

- The chi-square test can be used to determine whether the differences between the observed and expected results in a genetic cross are significant. The null hypothesis is that there is no difference between the observed and expected results. If the results of the chi-square test tell us that the probability of this null hypothesis being correct is equal to or greater than 0.05, then we can accept that there is no significant difference between the observed and expected results.

Questions

Multiple choice questions

1 The allele for cystic fibrosis is recessive to the normal allele. Which of the statements correctly describes the term 'allele'?
   A the genetic constitution of the organism
   B an observable characteristic of an individual
   C position of a gene on a chromosome
   D alternative form of a gene determining contrasting characteristics

2 The diploid condition in which alleles at a given locus are identical is described as:
   A heterozygous   B codominant   C homozygous   D dominant.

continued ...

3 In the Australian shepherd dog, coat pattern is controlled by a multiple allele series. The allele for coat pattern no trim (T) is dominant to all others. The allele for sable trim (t^s) is dominant to the allele for copper (t^c). Which of the following genotypes correspond to the coat pattern, no trim?

A $t^c t^c$ B TT   C $t^s t^s$ D $t^s t^c$

4 The diagram below shows the inheritance of a form of breast cancer. It is associated with a gene *BRCA* 1.

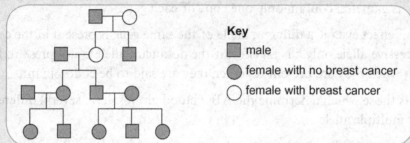

Key
■ male
● female with no breast cancer
○ female with breast cancer

Which statement is true about the inheritance of the gene, *BRCA* 1?
The condition is a:
A sex-linked recessive trait.
B recessive autosomal trait.
C sex-linked dominant trait.
D dominant autosomal trait.

5 Red–green colour blindness is inherited as an X-linked recessive trait. A woman whose father is colour-blind, but herself has normal vision, marries a colour-blind man. If they have a son, what is the probability that he will be colour-blind?
A 0   B 0.25   C 0.50   D 0.75

6 A snapdragon pure-breeding for red flowers was bred with one that was pure-breeding for white flowers. The F_1 generation had only pink flowers. A plant with red flowers was crossed with one with pink flowers. What will be the phenotypic ratio of the offspring?
A 1 red : 1 white   B 1 red : 1 pink   C 1 red : 2 pink : 1 white   D 1 red : 1 pink : 2 white

7 The critical value of a chi-square test is 5%. What does this indicate?
A Observed results would be expected to occur by chance in more than 1 in 20 experiments.
B Observed results would be expected to occur by chance in fewer than 1 in 20 experiments.
C If the calculated probability is less than 5%, the results are not significant.
D The value is harmful to the experiment.

8 A test cross was done on the genotype BbCc. Which of the following genotypes would you **not** expect among the offspring?
A bbcc   B BbCc   C Bbcc   D BbCC

9 A cross was made between a tall plant with round leaves and a dwarf plant with round leaves. The following results were obtained from the cross: 121 tall, round leaf; 124 dwarf, round leaf; 42 tall, oval leaf; 37 dwarf, oval leaf.

A chi-square test was performed on the results. To determine if the difference between the observed and expected results was significant, which row in the degrees of freedom column of the distribution table would be read?
A 1   B 2   C 3   D 4

continued...

10 Mouse hair colour is controlled by two autosomal genes. The gene at one locus, C/c, determines whether the pigment for hair colour is produced and deposited. Mice with the genotypes cc are albino. A gene at a second locus, A/a, determines the colour. The allele A is dominant and results in agouti hair and the recessive allele, a, results in the black hair.

A mouse with agouti coat is mated with an albino mouse of genotype aacc. The agouti mouse is heterozygous at both loci.

What will be the expected phenotypic ratio of the offspring of this cross?

A 1 agouti : 2 black : 1 albino
B 13 agouti : 3 black
C 9 agouti : 3 black : 4 albino
D 1 agouti : 1 black : 2 albino

Structured questions

11 Gregor Mendel is described as the father of genetics. He studied the inheritance patterns of pea plants differing in one or two traits. The following is one such experiment. He crossed pure-breeding plants for smooth, yellow seeds with pure breeding plants for green, wrinkled seeds to produce an F_1 generation. All seeds of the F_1 had smooth, yellow seeds. The F_1 plants were self-fertilised and produced the following results:

315 smooth yellow
108 smooth green
101 wrinkled yellow
32 wrinkled green

a What type of inheritance patterns was Mendel studying in the cross above ?
Give **one** reason for your answer. [1 mark]

b State the phenotypic ratio obtained in the F_1 cross. [1 mark]

c Mendel went through his data and examined each characteristic separately. He compared the total numbers of round versus wrinkled and yellow versus green seeds, as shown in the tables below. Copy and complete the tables.

	Smooth	Wrinkled
Number of seeds		
Phenotypic ratio		

	Yellow	Green
Number of seeds		
Phenotypic ratio		

[6 marks]

d Using the results of the phenotypic ratios obtained in **c**, state **one** conclusion which can be made. [2 marks]

e Do you expect all crosses which involve two traits to yield the same phenotypic ratio as in **b**? Explain your answer. [2 marks]

continued ...

12 a Explain briefly what is meant by the term 'dominant epistasis'. [2 marks]

b In squash fruits, the dominant allele, Y, of a gene for pigment production gives yellow fruits. The recessive allele, y, gives green fruits. The dominant allele, W, of another unlinked gene inhibits the effect of the gene, Y/y, producing white fruits, while the recessive allele, w, has no effect.
State the colour of the ripe fruit produced by squash plants with each of the following genotypes:
WwYy   Wwyy   wwYY   wwyy [2 marks]

c Two squash plants with the genotypes WWYY and wwyy were crossed and the resulting F_1 generation interbred to give the F_2 generation.
  i Draw a genetic diagram to show this cross. [4 marks]
  ii Give the ratio of the phenotypes expected in the F_2 generation. [1 mark]

d The inheritance of shape of squash fruits was also studied. Pure-breeding plants for pear-shaped fruits were crossed with pure-breeding plants for scallop fruit. All the fruits of this cross were scallop. The plants resulting from this cross were then self-fertilised (selfed). The following results were obtained:
  scallop fruits: 76, pear fruits: 34
A chi-square test was then conducted on the data obtained for the F_2 generation.
Copy and complete the table and calculate the value of chi-square.

	Scallop fruits	Pear fruits
Observed results (O)	76	34
Expected ratio		
Expected results (E)		
$O - E$		
$(O - E)^2$		
$\dfrac{(O - E)^2}{E}$		

$$\chi^2 = \Sigma \frac{(O - E)^2}{E}$$

[4 marks]

e A chi-square probability table as shown below was used to determine whether the value obtained in **d** indicates that there was no significant difference between the observed numbers and the expected numbers.

Degrees of freedom	Probability				
	0.9	**0.5**	**0.1**	**0.05**	**0.01**
1	0.02	0.46	2.71	3.84	6.64
2	0.21	1.39	4.61	5.99	9.21
3	0.58	2.37	6.25	7.82	11.35
4	1.06	3.36	7.78	9.49	13.28
5	1.61	4.35	9.24	11.07	15.09

Was the result significant? Give a reason for your answer. [2 marks]

continued ...

13 a Explain what is meant by:

    i sex-linkage

    ii recessive allele.   [3 marks]

  b A pedigree chart for colour-blindness is shown below.

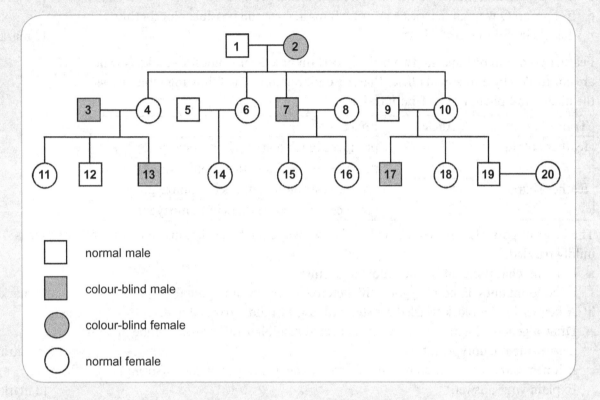

normal male

colour-blind male

colour-blind female

normal female

    Using the pedigree chart above, what conclusions can you draw about the pattern of inheritance of red–green colour-blindness?   [3 marks]

  c Name the relationship between the two alleles that control colour vision. Give **one** reason for your answer.   [2 marks]

  d Using the symbols X^N for normal vision and X^n for the allele for red–green colour-blindness, state the genotypes of the following individuals:

      2

      3

      7

      11

      13

      19   [3 marks]

  e If individuals **3** and **4** have another son, what is the probability that he would be colour-blind? Show how you arrive at your answer.   [4 marks]

continued ...

Essay questions

14 a Using examples, explain the types of crosses that would produce the following genetic ratios.

    i $3 : 1$ **ii** $1 : 1$ **iii** $1 : 2 : 1$ **iv** $9 : 3 : 3 : 1$ **v** $1 : 1 : 1 : 1$ [10 marks]

  b Explain the inheritance of the human ABO blood groups.   [3 marks]

  c If a group O woman marries a group AB man, state the possible blood groups that their children could have.   [2 marks]

15 Feather colour and shape are two traits in certain breeds of domestic chickens which are controlled by autosomal alleles that are codominant. The following table shows the alleles and phenotypes of both traits.

Trait	Allele	Effect
feather colour	C^B	produces black feathers in homozygotes
	C^W	produces white feathers in homozygotes
feather shape	S^S	produces straight feathers in homozygotes
	S^F	produces frizzled feathers in homozygotes

The heterozygous state for feather colour is grey while the heterozygous state for feather shape is mildly frizzled.

  a Explain what is meant by the following terms:

    i codominant   **ii** homozygote   **iii** heterozygous   **iv** autosomal   [4 marks]

  b A farmer bred a black frizzled rooster with a grey mildly frizzled hen.

    Draw a genetic diagram to show the expected results of this cross. Include in your answer the phenotypic ratios.   [5 marks]

  c Is it necessary to do a test cross to determine the genotypes of the offspring? Explain your answer.   [2 marks]

  d The farmer expressed the wish to have all his chickens to be grey and mildly frizzled. What would you advise him to do? Explain your reason.   [4 marks]

16 a Why is a chi-square test conducted on the results of genetic crosses?   [2 marks]

  b The genes for coat texture and coat colour are found on separate chromosomes. The allele for black coat is dominant to white coat while the allele for rough coat is dominant to smooth coat.

    A test cross was carried out on an individual who showed the dominant traits. The following results were obtained after repeated crosses: 22 rough and black coats, 18 rough and white coats, 25 smooth and black coats, 19 smooth and white coats.

    A chi-square test was conducted on the results.

    i . Differentiate between the terms 'allele' and 'gene'.   [2 marks]

    ii State the null hypothesis of the chi-square test.   [2 marks]

    iii What is meant by 'critical value' of the chi-square test?   [2 marks]

    iv What is the expected phenotypic ratio of the offspring?   [1 mark]

    v Using the results, explain the steps of conducting a chi-square test. The formula is shown: $\chi^2 = \Sigma \dfrac{(O - E)^2}{E}$

       Include in your answer how the critical value is chosen and how a conclusion is drawn.   [8 marks]

Chapter 8
Aspects of genetic engineering

By the end of this chapter you should be able to:

a   outline the principles of restriction enzyme use in removing sections of the genome;

b   explain the steps involved in recombinant DNA technology, including the isolation of genes, cloning of genes and the use of vectors;

c   describe the use of gene technology in the production of human insulin by bacteria;

d   discuss the possible benefits and hazards of gene therapy, including its potential use in the treatment of cystic fibrosis;

e   discuss the medical, agricultural, environmental, ethical and social implications of the use of genetically modified organisms.

We first learned that DNA is the genetic material in the 1950s. Since then, our ability to understand how it works, and then how to manipulate it for our own purposes, has increased at a phenomenal rate. The use of DNA – genes – to produce something that we want is called **gene technology**. It is developing very rapidly. We are becoming more and more able to alter genes within organisms. It is very important to think about what we should do and what we should not do. The fact that we *can* do something does not mean that we *should*.

Genetic engineering

Genetic engineering means using technology to change the genetic material of an organism. It may involve taking genes from an organism of one species and placing them in another, where they are expressed. The DNA that has been altered by this process, and which now contains lengths from two different species, is called **recombinant DNA**.

The complete set of genetic material of an organism (or of a population of organisms) is called its **genome**. Genetic engineering therefore alters the genome of the organism.

The organism to which the new gene has been added is said to be a **recombinant organism**, or a **genetically transformed organism**, or a **transgenic organism**. The term **genetically modified organism** (GMO) is also used. This is not restricted to organisms into which genes from other species have been placed. GMOs can also be organisms that have had their own genes altered using gene technology.

Gene technology is an expensive process, and it still runs a very poor second to conventional selective breeding in terms of numbers of new varieties of crop plants and farm animals that are being produced. There has been considerable opposition to it from many people in many parts of the world, which has greatly slowed down the widespread introduction of genetically modified organisms into food production and some other industries, too. Nevertheless, there are several instances where a product made using genetic engineering is literally a life-saver, such as the human insulin secreted by genetically modified bacteria.

An overview of gene technology

1   The gene that is required is identified. It is either cut out of the chromosomes, made by 'reverse transcription' of mRNA or synthesised directly in the laboratory.

2   Multiple copies of the gene are made using a technique called **PCR**, which stands for **polymerase chain reaction**.

3 The gene is inserted into a **vector** – an organism or structure that is able to deliver the gene into the required cells.

4 The vector inserts the gene into the cells.

5 The cells that have been successfully transformed are identified and **cloned**.

6 The cloned cells are used in industrial or medical situations.

The steps involved in this procedure are described here with particular reference to the production of recombinant bacteria that make insulin. From these bacteria, insulin can be extracted and used by people with diabetes.

Obtaining the DNA coding for insulin

Several different methods are used by the different companies producing human insulin (humulin) using recombinant DNA. It was first achieved by direct synthesis of the DNA.

Insulin is a protein made from two polypeptide chains, A and B, joined by disulphide bridges. It is quite a small protein and its primary structure is precisely known. The genetic code that specifies the sequence of amino acids in the two chains can be deduced from the sequence of amino acids.

The DNA that codes for the A and B polypeptides can be synthesised in a laboratory. Nucleotides can be joined together using the deduced code. However, there needs to be more DNA than just the code for the sequence of amino acids in insulin. Some extra DNA is needed to allow it to be joined to other sections of DNA.

A different approach uses the fact that the gene for insulin is only expressed by beta (β) cells in the islets of Langerhans in the pancreas, and these cells really specialise in doing just that. This means that a very high proportion of the mRNA in these cells has been transcribed from the insulin gene (Figure 8.1).

So mRNA carrying the information for making insulin can be extracted from these cells. The mRNA is then incubated with an enzyme called **reverse transcriptase**. This comes from a group of viruses called **retroviruses** – HIV is an example of one. As the name suggests, this enzyme does something that does not normally happen in

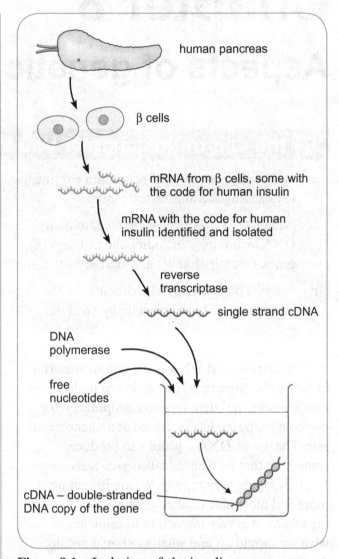

Figure 8.1 Isolation of the insulin gene.

human cells – it makes DNA using RNA as a template. In this instance, complementary single-stranded DNA molecules are made, using the mRNA as a template. These single-stranded molecules are then converted to double-stranded DNA molecules.

Whatever method is used to obtain the DNA coding for insulin, many copies of the DNA are then synthesised by the polymerase chain reaction, PCR. This is described on page 159. Other important techniques often used in separating and identifying lengths of DNA in genetic engineering, such as the use of DNA probes and electrophoresis, are described on pages 157 and 158.

Inserting the DNA into a vector

In order to get the genetic material into a recipient bacterial cell, a go-between called a **vector** has to be used.

One commonly used vector, when the gene is to be inserted into bacteria, is a **plasmid**. A plasmid is a small, circular piece of DNA that occurs naturally in bacteria (Figure 8.2). Plasmids often contain genes that confer resistance to antibiotics. They can be exchanged between bacteria – even between different species of bacteria. (This is a concern for humans, because it means that a person infected with a strain of antibiotic-resistant bacteria may also become a breeding ground for a different species of bacterium that is also resistant to that antibiotic.)

We can make use of this ability of plasmids to get inside bacterial cells. If you can put your piece of human DNA into a plasmid, the plasmid can deliver it into a bacterium.

A length of DNA to be inserted into a plasmid is treated with a **restriction enzyme**. These enzymes are made by bacteria. They are used by the bacterium to attack and destroy DNA that has been inserted into them by viruses. (Viruses called bacteriophages, or phages, can infect bacteria.) There are many different kinds of restriction enzymes, and each kind cuts DNA at a particular base sequence. For example, a restriction enzyme called *Bam*HI always cuts DNA where there is a GGATCC sequence on one DNA strand and, of course, the complementary sequence CCTAGG on the other (Figure 8.3). If you know the base sequences near the ends of the gene you require, then you can use a particular restriction enzyme to cut at these points (Figure 8.4).

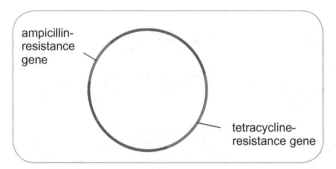

Figure 8.2   Plasmid pBR322.

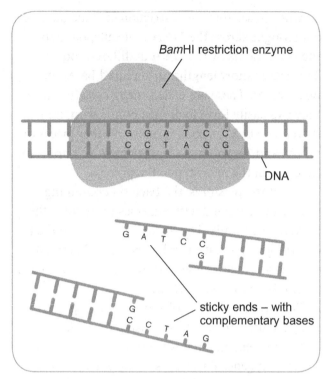

Figure 8.3   Cutting DNA with a restriction enzyme.

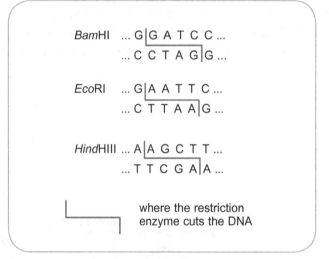

Figure 8.4   Target base sequences of three restriction enzymes commonly used in gene technology.

SAQ

1 Look carefully at the base sequences on the two strands of the DNA that have been cut by each of the restriction enzymes in Figure 8.4. Can you pick out a pattern that is common to all of them?

You can see that the restriction enzyme does not cut straight across the DNA molecule, but cuts the two strands of the DNA at different points. This leaves short lengths of unpaired bases on both pieces. These are called **sticky ends** (because they can easily form hydrogen bonds with, and therefore 'stick' to, similar ends on other pieces of DNA) and, as you will see, they have an important function in the next step in the process.

To get the plasmids, the bacteria containing them can be treated with enzymes to dissolve the bacterial cell walls. The 'naked' bacteria are then centrifuged, so that the relatively large bacterial chromosomes are separated from the much smaller plasmids and the cell debris.

The circular DNA molecule making up the plasmid is then cut open using the same restriction enzyme that was used for cutting out the insulin gene. This leaves sticky ends that are complementary to those on the required gene.

The plasmids and insulin DNA are then mixed up together. Some of the plasmid sticky ends will pair up with some of the insulin DNA sticky ends. The enzyme **DNA ligase** is then used to link together the deoxyribose–phosphate backbones of the DNA molecule, producing a closed circle of double-stranded DNA containing the insulin gene (Figure 8.5).

Not all the plasmids will take up the insulin gene like this. Some of them just join up with themselves again.

Plasmids are not the only kind of vector that can be used. Viruses can also be used as vectors as they, too, have the ability to insert their DNA into other cells. Yet another type of vector are **liposomes**. These are tiny balls of lipids containing the DNA. They have been used in some attempts at gene therapy in humans. Gene therapy is described on pages 163–165.

Getting the plasmids into bacteria

Now the plasmids are mixed with a culture of the bacterium that is to be transformed (Figure 8.6). Calcium ions are added to the mixture, because they affect bacterial cell walls and membranes, making it easier for them to take up the plasmids. A small proportion of the bacteria, perhaps 1%, take up plasmids containing the insulin gene. The rest either do not take up plasmids at all, or take up ones that did not contain the insulin gene.

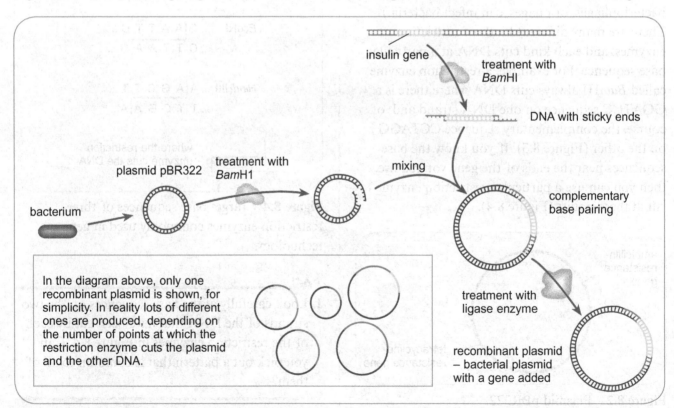

Figure 8.5 Inserting a gene into a plasmid.

To sort out the bacteria that have been transformed (that is, contain the insulin gene) from those which have not, a screening process based on the fact that the plasmid contains antibiotic-resistance genes is used. For example, the plasmid pBR322 contains two resistance genes, one for tetracycline and one for ampicillin. It just so happens that, if the restriction enzyme *Bam*HI is used to cut the DNA of the plasmid, it cuts right through the middle of the tetracycline-resistance gene. So, when the insulin gene is inserted, it inactivates the tetracycline-resistance gene (Figure 8.6).

The bacteria are then grown on agar jelly to which ampicillin has been added. Any that survive must have taken up the plasmid. However, we don't know which have taken up the transformed plasmids (containing the insulin gene) and which have just taken up the unaltered plasmids.

To sort these out, a sample of each colony of bacteria is now placed on another agar plate, this time containing tetracycline (Figure 8.7). This is called **replica plating**. Only the bacteria that have a working resistance gene to tetracycline will survive – and these must be the ones that have *not* taken up the transformed plasmids. So now we can go back to the first plate, and select just the colonies of bacteria that were *not* able to grow on the tetracycline plate. These are the ones that contain the required gene.

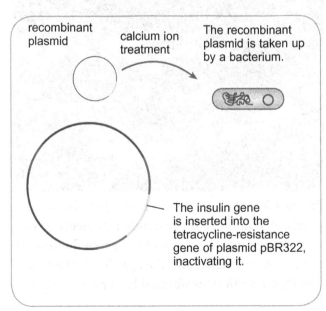

Figure 8.6   Inserting the gene into a bacterium.

recombinant plasmid

calcium ion treatment

The recombinant plasmid is taken up by a bacterium.

The insulin gene is inserted into the tetracycline-resistance gene of plasmid pBR322, inactivating it.

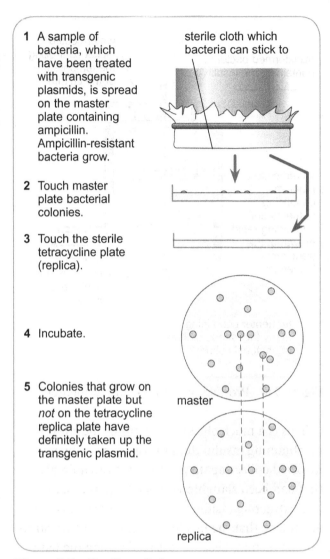

1   A sample of bacteria, which have been treated with transgenic plasmids, is spread on the master plate containing ampicillin. Ampicillin-resistant bacteria grow.

sterile cloth which bacteria can stick to

2   Touch master plate bacterial colonies.

3   Touch the sterile tetracycline plate (replica).

4   Incubate.

5   Colonies that grow on the master plate but *not* on the tetracycline replica plate have definitely taken up the transgenic plasmid.

master

replica

Figure 8.7   Replica plating.

Cloning and production

One selected bacterium, when allowed to reproduce, produces a clone containing many millions of identical cells, all with the insulin gene.

These genetically modified bacteria are now cultured on a large scale, in fermenters (Figure 8.8). They make the two polypeptides of insulin, which are extracted, purified, joined to make insulin and sold for use by people with diabetes.

SAQ

2   Outline the use of each of these enzymes in gene technology:

   a restriction enzymes

   b DNA ligase

   c reverse transcriptase.

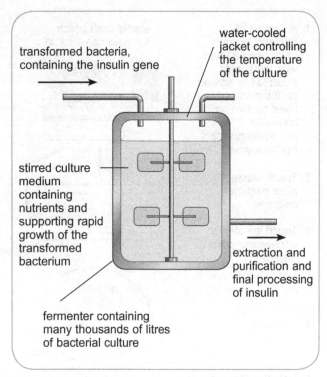

Figure 8.8 Producing insulin using fermenters.

This is a much cheaper and more efficient way of obtaining insulin than the old way – which involved collecting up pancreases from animals that had been slaughtered (usually pigs or cattle) and extracting insulin from them. It also has the advantage that the insulin is absolutely identical to human insulin, because it is made using the base sequence on the human insulin gene.

Other genetic markers

In the technique just described, antibiotic-resistance genes were used to identify the bacteria that had been successfully transformed. However, there has been some concern about this, because it could increase the risk of other, potentially pathogenic, bacteria taking up these genes and becoming resistant to antibiotics. So other markers now tend to be used.

One of the most widely used types of marker is a gene that causes fluorescence. The gene was first found in jellyfish, and it codes for the production of an enzyme that produces a protein that fluoresces bright green in ultraviolet light. The gene for the enzyme is inserted into the plasmids, so all that needs to be done to identify the bacteria that have taken up the plasmids is to shine ultraviolet light on them. The ones that have been transformed will glow green (Figure 8.9).

Promoters

Bacteria contain many different genes, which make many different proteins. But not all these genes are switched on at once. Genes are switched on by **promoters**. If we want the gene that we have inserted into a bacterium to be expressed, then we also have to insert an appropriate promoter for it.

When bacteria were first transformed to produce human insulin, the insulin gene was inserted next to the β galactosidase (lactase) gene, so that they shared a promoter. The promoter switches on the gene when the bacterium needs to metabolise lactose. If the bacteria are grown in a medium containing lactose, they synthesise both β galactosidase and human insulin.

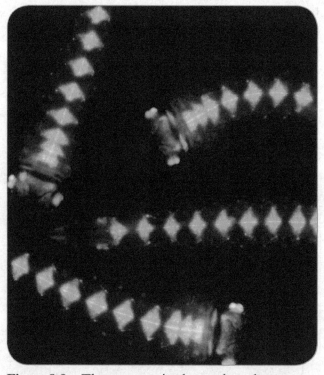

Figure 8.9 These mosquito larvae have been genetically modified for research into their role in transmitting malaria. Genes for a green fluorescent protein were inserted into the eggs from which they developed, so the transformed larvae glow green.

DNA probes

Probes are used in all kinds of gene technology. They help to pick out the required piece of DNA from among a whole collection of different DNA fragments in a mixture.

A probe is a length of single-stranded DNA, often a few hundred base pairs long, that has a base sequence complementary to the one you want to extract. It does not have to be *perfectly* complementary, so it does not matter if you don't know the precise base sequence of the required DNA.

The probe is 'labelled' in some way. It might, for example, be made with nucleotides containing an isotope of phosphorus, ^{32}P, which emits beta radiation that can be detected using X-ray film.

When the probe is mixed with the DNA fragments, it forms hydrogen bonds with any stretches of DNA that are mostly complementary to its own base sequence. If we can see where the probes are, then we also know where these DNA fragments are.

In the photograph below, a grid contains different DNA fragments, which are made up of 20 736 different fragments of human chromosome 17. It has been treated with radioactive probes for genes responsible for breast cancer. Above the grid is an X-ray plate shown after it has been exposed to the grid and it shows some fragments in the grid as being tagged, as they produced dark spots on the X-ray plate above the fragments.

Electrophoresis

One of the techniques used in gene technology – for example, in 'genetic fingerprinting' or in identifying a gene to be transferred to another organism – is called **electrophoresis**. This separates different fragments of DNA according to their sizes.

A tank is set up containing a very pure form of agar called **agarose gel**. A direct current is passed continuously through the gel. DNA fragments are added at one end of the tank. The DNA fragments carry a small negative electric charge. They are pulled through the gel towards the anode (the positively charged electrode). The smaller the fragments, the faster they move.

When the current in the tank is turned off, the gel contains DNA fragments that have ended up in different places. These are not visible straight away. One way of making them visible is to transfer them, very carefully, onto absorbent paper, which is placed on top of the gel.

Now the paper is heated just enough to make the two strands in each DNA molecule separate from one another. Short sequences of single-stranded DNA, called **probes**, are added (page 157). These may be 'labelled' using fluorescence, or using radioactive isotopes. The probes pair up with the DNA fragments on the paper. The positions of the DNA fragments can now be detected, either by shining ultraviolet light onto them or by placing the paper against an X-ray film. The radiation emitted by the probes makes the film go dark. So we end up with a pattern of dark stripes on the film, matching the positions of the DNA fragments in the agarose gel.

The analysis of the patterns made following electrophoresis and labelling of DNA samples taken from different individuals, using the same sections of the genome, is called DNA profiling. DNA profiling can be used to help identify or eliminate possible suspects in criminal investigations, as well as establish family relationships.

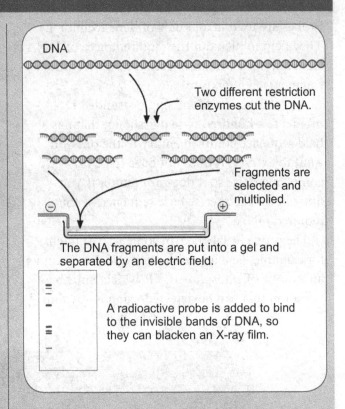

Two different restriction enzymes cut the DNA.

Fragments are selected and multiplied.

The DNA fragments are put into a gel and separated by an electric field.

A radioactive probe is added to bind to the invisible bands of DNA, so they can blacken an X-ray film.

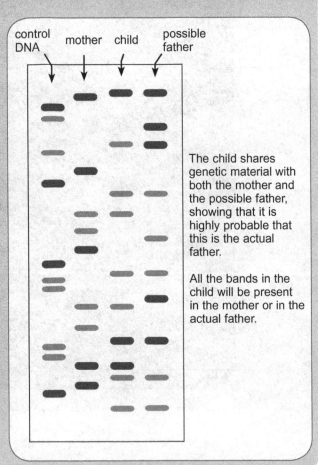

The child shares genetic material with both the mother and the possible father, showing that it is highly probable that this is the actual father.

All the bands in the child will be present in the mother or in the actual father.

The polymerase chain reaction

The polymerase chain reaction, generally known as **PCR**, is used in almost every application of gene technology. It is a method for rapidly producing a very large number of copies of a particular length of DNA.

The illustration shows the various steps in PCR. First, the DNA is denatured, usually by heating it. This separates the DNA molecule into its two strands, leaving bases exposed.

The enzyme **DNA polymerase** is then used to build new strands of DNA against these exposed ones.

However, DNA polymerase will not just begin doing this with no 'guidance' – it needs to know where to start. A **primer** is used to begin the process. This is a short length of DNA, often about 20 base pairs long, that has a base sequence complementary to the start of the part of the DNA strand that you want to copy. The primer attaches to the start of the DNA strand, and then the polymerase will continue to add nucleotides all along the rest of the DNA strand.

Once the DNA has been copied, the mixture is heated again, which once more separates the two strands in each DNA molecule, leaving them available for copying again. Once more, the primers fix themselves to the start of each strand of unpaired nucleotides, and DNA polymerase makes complementary copies of them.

The three stages in each round of copying need different temperatures.

- Denaturing the double-stranded DNA molecules to make single-stranded ones requires a high temperature, around 95 °C.
- Attaching the primers to the ends of the single-stranded DNA molecules (known as **annealing**) requires a temperature of about 65 °C.
- Building up complete new DNA strands using DNA polymerase requires a temperature of around 72 °C. (The polymerases used for this process come from microorganisms that have evolved to live in hot environments.)

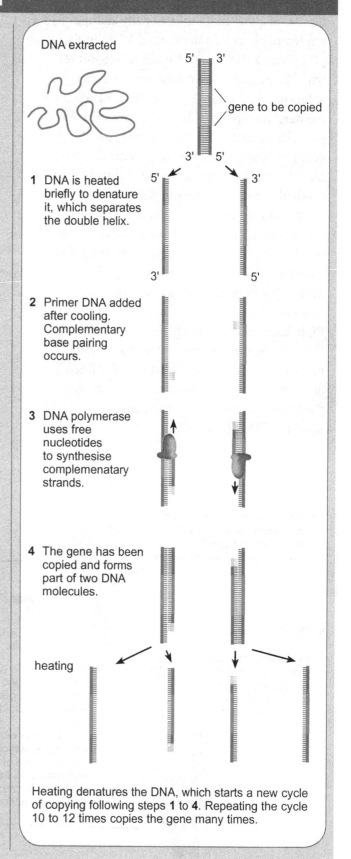

DNA extracted

gene to be copied

1 DNA is heated briefly to denature it, which separates the double helix.

2 Primer DNA added after cooling. Complementary base pairing occurs.

3 DNA polymerase uses free nucleotides to synthesise complemenatary strands.

4 The gene has been copied and forms part of two DNA molecules.

heating

Heating denatures the DNA, which starts a new cycle of copying following steps **1** to **4**. Repeating the cycle 10 to 12 times copies the gene many times.

continued ...

159

Most laboratories that work with DNA will have a machine that automatically changes the temperature of the mixture. You simply place your DNA sample into a tube together with the primers, free nucleotides, a buffer solution and the DNA polymerase, switch on the machine and let it run. The tubes are very small (they each hold about $1\,mm^3$) and have very thin walls, so as the temperature in the machine changes, the temperature inside the tubes also changes very quickly.

You can see how this could theoretically go on forever, making more and more copies of what might originally have been just a tiny number of DNA molecules. A single DNA molecule can be used to produce literally billions of copies of itself in just a few hours. PCR has made it possible to get enough DNA from a tiny sample – for example, a microscopic portion of a drop of blood left at a crime scene. The sequence of bases in the DNA can then be checked against a suspect's DNA (page 158).

Another example of genetic engineering – making Golden Rice™

A lack of vitamin A, **retinol**, in the diet causes health problems to hundreds of thousands of people, in developing countries in South East Asia, for example. Vitamin A is required for the formation of the visual pigment rhodopsin, responsible for the reception of light energy by rod cells in the retina of the eye. Vitamin A deficiency therefore causes night blindness. It also reduces resistance to bacterial and viral infections, and so increases the risk of a person dying from infectious diseases such as measles, or diarrhoea caused by microorganisms.

Most of us get plenty of vitamin A in our diets. It is present in most meat products, especially liver. We can also make retinol from a precursor called **β carotene**, an orange pigment found in plant foods such as carrots. However, people who eat diets based on rice may get insufficient amounts of either β carotene or retinol.

In an attempt to address this problem, by providing an affordable way of improving dietary intake of β carotene to people who eat a rice-based diet, a new variety of rice, called **Golden Rice™**, has been developed through genetic engineering (Figure 8.10).

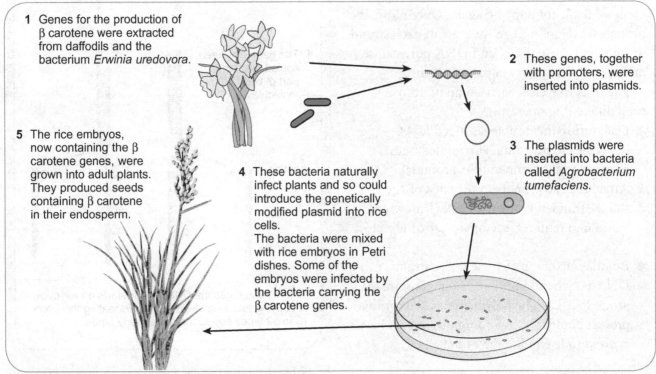

1 Genes for the production of β carotene were extracted from daffodils and the bacterium *Erwinia uredovora*.

2 These genes, together with promoters, were inserted into plasmids.

3 The plasmids were inserted into bacteria called *Agrobacterium tumefaciens*.

4 These bacteria naturally infect plants and so could introduce the genetically modified plasmid into rice cells.
The bacteria were mixed with rice embryos in Petri dishes. Some of the embryos were infected by the bacteria carrying the β carotene genes.

5 The rice embryos, now containing the β carotene genes, were grown into adult plants. They produced seeds containing β carotene in their endosperm.

Figure 8.10 How Golden Rice™ was developed.

The idea was to insert genes into rice plants that would increase the amount of β carotene made by the plants, and that would be expressed in their seeds. The first type of Golden Rice™ that was produced, in 1999–2000, used a gene from daffodils that coded for the production of an enzyme called phytoene synthase, and another from the bacterium *Erwinia uredovora* that coded for the enzyme carotene desaturase. This worked, but the quantities of β carotene that were produced were not high enough to make a significant increase in vitamin A levels in children who ate the rice. The rice only contained about 1.6 µg of β carotene per gram. As a child requires about 300 µg of vitamin A per day, and as you need 12 µg of β carotene to make 1 µg of vitamin A, they would have to eat a lot of rice to get this amount. The low level of β carotene in the Golden Rice™ helped to fuel the criticisms which were already being made of the project – that this was not the way to help people suffering poverty (we should lift them out of poverty, not just give them different food to eat) or that the rice would not be able to help those who need it.

The researchers experimented with other sources of these two genes, and in 2004 produced a better version of Golden Rice™ using genes taken from maize plants and rice itself. This produces up to 31 µg of β carotene per gram of rice (Figure 8.11). In many countries, children eat between 100 g and 200 g of rice per day, so they could get more than half (in some cases all) of their vitamin A requirement by eating Golden Rice™.

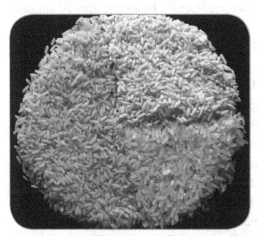

Figure 8.11 Three types of rice – 'ordinary' rice and the first and second versions of Golden Rice™.

Are GMOs safe?

The first genetically modified crops for commercial use were planted in the 1990s, and more varieties have been developed steadily since then. Their introduction immediately sparked controversy. Many people are vehemently opposed to these crops, sometimes because they think there are risks to human health and sometimes because they think there may be damage to the environment. On the other hand, some think such crops are potentially hugely beneficial to humans, and to the environment. In between are the great majority of people, without an extensive scientific background, who do not know which of the many competing claims they see or hear in the media to believe.

So, *are* genetically modified crop plants 'safe'? There are really two issues here. First, could genetically modified crops cause harm to other organisms in the environment? Secondly, is it safe to eat food made from genetically modified plants?

Many genetically modified crops, including maize and cotton, have had genes inserted into them that cause them to make a toxin (poisonous substance) called Bt toxin. This is done by inserting genes from the bacterium *Bacillus thuringiensis*. The toxin is only harmful to insects, not humans. Moreover, it only affects insects once they have taken it into their digestive system, so it kills only insects that are feeding on the crop – that is, insect pests.

However, some harmless – and, indeed, often helpful – insects do feed on pollen, and experiments have found that some of these can be harmed by feeding on pollen of flowers of crops containing Bt toxin. Research has also found that insects feeding on waste material from the crops, such as dead leaves that have fallen onto the ground or into streams, could theoretically be harmed if they ate large enough quantities.

A farmer growing a Bt crop can greatly reduce the amount of insecticide that is sprayed onto the fields, and in some cases may not need to use insecticides at all. This benefits the environment, as insecticides are rarely specific to insect pests, and generally have wide-ranging effects on other insects and perhaps on other groups of organisms as well. Residues of insecticides left on food crops can also cause illness in humans, so a reduction in their use should reduce the risk of harm to people eating foods made from the crops.

One recurrent problem that occurs when insecticides are used to kill insect pests is that the insects may develop resistance to the insecticide, through natural selection (pages 177 to 179). The farmer may therefore need to spray ever-increasing quantities of insecticide in order to keep the pest numbers under control, or he may need to keep changing the type of insecticide that he is using. It was hoped that insects would not develop resistance to Bt toxin so readily, but unfortunately Bt toxin-resistant populations of insect pests of both maize and cotton have already begun to appear. This is requiring new varieties of the GM crops to be produced, containing slightly different versions of the Bt toxin.

Some genetically modified crops have had genes added to them that make them resistant to herbicides (chemicals that kill plants). This means that the farmer can spray the crop with a herbicide which will kill all the weeds growing with the crop, but will not kill the crop plants. The benefits are that it is much easier to control weeds than in non-GM crops, which can increase yields and reduce costs, leading to a potential reduction in the cost of food.

There is no evidence that such GM plants are unsafe to eat, or less nutritious. However, the quantity and quality of research on the environmental impact of these crops is not very great. One worry is that these crops might 'invade' a habitat and reduce the numbers or variety of other, native, plants growing there. Or their genes might somehow spread into other plants, perhaps by wind or pollinating insects. This could change the plants and the ecosystems in which they live in unpredictable ways.

In 2003, the results were published of a large-scale UK study of the effects on biodiversity of growing genetically modified crops. They showed that the effects were fairly small. One GM crop even appeared to increase biodiversity on the farm where it was grown.

In general, the risk of the GM plants, or their genes, spreading into the wider environment appears to be small. After all, the features that have been introduced into the crops would be very unlikely to give them an advantage in a natural situation. There is no reason to expect that they could compete successfully with native plants. Moreover, gene transfer is not likely unless close relatives of the crop plants are growing nearby. For most of our crops – all of the cereals, potatoes, cotton – it is not usual for the crops to be grown in the vicinity of any close, wild relatives.

However, things will be changing in the next few years. Some of the GM crops expected to be introduced in the future *will* have features that could enhance their ability to survive in the wild. For example, varieties of maize are being developed that have the ability to grow in very dry situations, and varieties of rice that are able to grow in salty water. Any genetically modified rice – including Golden Rice™ – could pose a threat, as there are often wild relatives of rice growing right next door to the paddy fields where rice is cultivated, meaning that there is a risk of genes being transferred into the wild plants. It will be important to carry out large-scale field trials on these new GM crops, if safety is to be ensured and public fears allayed.

On the whole, people have more positive views towards GM crops when they can see a benefit to the consumer. The first GM crops had no obvious benefit to the consumer – they benefited only the grower or retailer. This undoubtedly contributed to the public opposition to their introduction. Crops like Golden Rice™, however, may be more acceptable, if benefits to health can be clearly demonstrated.

It is important to remember that a genetically modified plant is really not that different from one that has been produced by traditional selective breeding. Both of them contain genes that are different from those found in natural populations of the wild plants from which the crop plant has been developed. We could, for example, develop drought-resistant varieties of maize by selective breeding over many generations, and end up with plants that are very similar to those produced by genetic engineering. We need to be just as careful in our use of plants bred traditionally as we are with those produced by gene technology.

3 Summarise the arguments for and against the use of GM crops as two sets of bullet points, under the headings 'Potentially beneficial' and 'Potentially harmful'. You may like to look on the internet to find more points to add to your list. Do take care to use only sites that are likely to be providing reliable information. Do not put too much faith in information provided by sites where a group of people are attempting to promote a biased point of view for which there is little or no evidence.

Gene therapy

Gene therapy is the treatment of a disease by manipulating the genes in a person's cells.

Attempts have been made to treat **cystic fibrosis** by gene therapy. It is worth looking at the story in some detail, as it shows how a thorough knowledge of the biology underlying a disease, and of genetic engineering techniques, can open up new possibilities for treating previously untreatable diseases. It also shows how unexpected problems can greatly lengthen the time taken to put a new idea into practice.

Cystic fibrosis is a genetic disease in which abnormally thick mucus is produced in the lungs and other parts of the body. A person with cystic fibrosis is very prone to bacterial infections in the lungs because it is difficult for the mucus to be removed, and bacteria can breed in it. The thick mucus adversely affects many other parts of the body. The pancreatic duct may become blocked, and people with this disease often take pancreatic enzymes by mouth to help with digestion. Around 90% of men with cystic fibrosis are sterile because thick secretions block ducts in the reproductive system.

In Chapter 7, we saw that cystic fibrosis is caused by a recessive allele of the gene that codes for a transporter protein called **CFTR**. This protein sits in the plasma membranes of cells –

for example, in the alveoli – and allows chloride ions to pass out of the cells. The recessive alleles (there are several different faulty ones) code for an incomplete or faulty version of this protein, which does not act properly as a chloride ion transporter.

In a healthy person, the cells lining the airways and in the lungs pump out chloride ions through the channel in the membrane formed by CFTR (Figure 8.12). This results in a relatively high concentration of chloride ions outside the cells. This reduces the water potential below that of the cytoplasm of the cells. So water moves out of the cells by osmosis, down the water potential gradient (page 63). It mixes with the mucus there, making it thin enough for easy removal by the sweeping movements of cilia.

However, in someone with cystic fibrosis, this does not happen. Much less water moves out of the cells, so the mucus on their surfaces stays thick and sticky. The cilia, or even coughing, can't remove it all.

The *CFTR* gene

The gene that encodes the CFTR protein is found on chromosome 9. The most common defective allele is the result of the deletion of three bases. The CFTR protein made using this allele is therefore missing one amino acid. The machinery in the cell recognises that this is not the right protein, and does not place it in the plasma membrane.

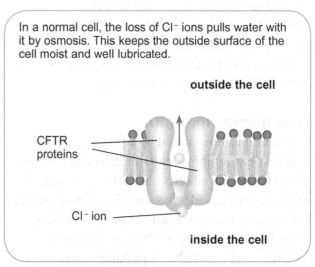

In a normal cell, the loss of Cl⁻ ions pulls water with it by osmosis. This keeps the outside surface of the cell moist and well lubricated.

outside the cell

CFTR proteins

Cl⁻ ion

inside the cell

Figure 8.12 The CFTR protein forms channels for chloride ions in the plasma membrane.

Because the faulty *CFTR* alleles are recessive, someone with one faulty allele and one normal allele is able to make enough CFTR protein to remain healthy. This makes it a good potential candidate for gene therapy. We don't need to remove the genes that are already there – we just need to get a correct, dominant allele into the cell, and it should – in theory – be able to make enough of the CFTR protein to allow the cell to work properly.

In practice, all attempts to do this have so far run into major difficulties. Trials in the UK began in 1993. The normal allele was inserted into liposomes, which were then sprayed as an aerosol into the noses of nine volunteers. The hope was that the liposomes would be able to move through the lipid layers in the plasma membranes of the cells lining the respiratory passages, carrying the gene with them. The trial succeeded in introducing the gene into a few cells lining the noses of the volunteers, but the effect only lasted for a week because these cells have only a very short lifespan and are continually replaced.

Researchers in the USA tried a different vector. In a trial involving several people with cystic fibrosis, they introduced the gene into normally harmless adenoviruses and then used these to carry the gene into the passages of the gas exchange system. The gene did enter some of the cells, but some of the volunteers experienced unpleasant side-effects as a result of infection by the virus. Because of this, the trials were stopped.

Work has not been completely abandoned, however, and research continues into other possible ways of introducing working copies of the correct *CFTR* allele into human cells.

Gene therapy for SCID

Another condition caused by a person's genes is SCID, severe combined immunodeficiency disease. This is caused by a faulty allele of a gene coding for an enzyme called adenosine deaminase (ADA). This enzyme is essential for the proper working of the immune system. Without it, the immune system is unable to fight off pathogens, and without treatment a child born with this condition will almost certainly die in infancy or childhood.

Several methods of gene therapy have been trialled to treat this disease. They involved the removal of some of the patient's T cells (white blood cells) and the insertion of the correct allele into them using a vector such as a **retrovirus**. (Retroviruses are viruses that contain RNA. When they have inserted their RNA into a host cell, the host cell uses the RNA to make a complementary length of DNA.) The cells that have successfully taken up the allele are then cloned to produce large numbers of them, which can then be replaced into the patient's body.

This technique is still in need of refinement. For example, several patients who appeared to have been successfully treated have gone on to develop leukaemia (a cancer involving white blood cells). This has happened because the new allele has, by chance, been inserted in a position where it affects genes that normally control the cell cycle. However, as these patients would have died anyway, the risk of cancer in this instance is generally seen as an acceptable, albeit very unfortunate, one in the treatment of this rare and inevitably fatal disease.

SAQ

4 Summarise the potential benefits and hazards of gene therapy.

5 Use the internet to find examples of diseases, other than cystic fibrosis and SCID, for which attempts have been made at treatment using gene therapy. What features do all of these diseases have in common?

Somatic and germ line gene therapy

The two examples of gene therapy that we have looked at – for SCID and for cystic fibrosis – involve **somatic gene therapy**. This means that the cells that are being genetically modified are body cells. They are not involved in reproduction. If the genes in them are modified, the effect stops there, in that person. The modified genes will not be passed on to any offspring that person has (Figure 8.13).

Germ line gene therapy would involve changing the genes in cells that would go on to form gametes, and therefore possibly zygotes. If this were done, then *all* of the cells in the new organism would carry the genetic modification. At the moment, this kind of gene therapy for humans is banned in most countries, including the UK and the USA. However, some people think this type of therapy could bring huge benefits. You may like to think about what these could be, and consider your opinion on whether or not germ line therapy should be allowed to take place.

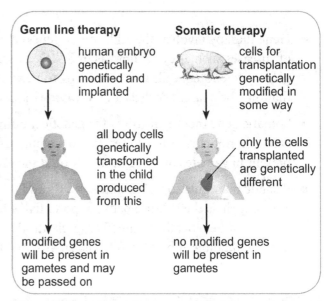

Figure 8.13 Germ line and somatic gene therapy compared.

Summary

- Genetic engineering involves the production of recombinant DNA, which contains genes inserted from a source other than the organism itself. The genes need to be inserted in such a way that they will be expressed in the genetically transformed organism, which may mean that a promoter is also added to the DNA.

- Restriction enzymes make staggered cuts in DNA at points where particular base sequences are present. The short lengths of unpaired bases that they leave are known as sticky ends. If another piece of DNA is cut with the same kind of restriction enzyme, then the two lots of sticky ends will be complementary to one another and able to join by hydrogen bonding between the bases.

- Plasmids may be used as vectors to carry DNA into bacterial cells. The plasmids are cut open with restriction enzymes, and then the required DNA is added to them. DNA ligase links the new DNA with the rest of the DNA in the plasmid. The plasmids can be taken up by bacterial cells, which are therefore transformed by the addition of the new DNA. If the DNA has come from another organism, the bacteria are now said to be transgenic.

- Viruses and liposomes can also be used as vectors.

- Many plasmids contain genes that confer resistance to antibiotics, and can be passed from one bacterium to another. These resistance genes can be used as markers to check which bacteria have successfully taken up plasmids containing the required gene. Genes that code for green fluorescent protein are being increasingly used as markers.

- Human insulin is now produced by transgenic bacteria, which have been genetically engineered to contain and express the human insulin gene.

- Golden Rice™ has been genetically engineered to contain genes that produce large amounts of β carotene in the grains. It is hoped that growing and eating this rice will help some of the many people in developing countries who suffer from vitamin A deficiency.

continued …

- Gene therapy involves the addition of genes to human cells that could cure or reduce the symptoms of diseases such as cystic fibrosis or SCID. There are difficulties to be surmounted in getting enough genes into enough cells for the therapy to be worthwhile, and also in avoiding some of the side-effects that may be caused, such as an increased risk in developing cancer.

- Somatic gene therapy involves the genetic modification of body cells, whereas germ line therapy (which is not currently allowed in most countries) involves genetic modifications that would be present in gametes or a zygote, and would therefore be passed on to subsequent generations.

- Genetic manipulation of animals, plants and microorganisms raises important ethical and social issues, such as the balance between potential suffering to other organisms and benefits to humans. The use of genetically modified crop plants also raises issues relating to human health, and to damage that might be caused to the environment or to other species.

Questions

Multiple choice questions

1   For what purpose are restriction enzymes used in the production of recombinant DNA?
   A forming DNA from mRNA
   B cutting open the circular DNA of plasmids
   C joining pieces of DNA to form recombinant DNA
   D breaking down bacterial cell walls to release plasmids

2   Vectors in gene technology are molecules that:
   A are covalently bound to and carry foreign DNA into cells.
   B cut foreign DNA.
   C protect bacterial cells from invasion by foreign DNA.
   D hydrolyse DNA.

3   Bacteria protect themselves from invading viruses by:
   A using ligases to cut viral DNA upon entry.
   B using restriction enzymes to cut viral DNA upon entry.
   C joining with vectors to carry viral DNA.
   D joining viral DNA and their own DNA using ligases.

4   The following are four steps in genetic engineering:
   I   cloning
   II   obtaining DNA/DNA cleavage
   III screening
   IV   production of recombinant DNA.

   Which of the following identifies the correct sequence of the steps involved in genetic engineering?
   A I, II, II, IV
   B II, IV, III, I
   C II, IV, I, III
   D IV, II, III, I

continued ...

5 Restriction enzymes cut DNA molecules only at specific target sites with particular base sequences. The target sites for three restriction enzymes are shown below.

Restriction enzyme	Target sites
*Bam*HI	G GATCC
	CCTAG G
*Eco*RI	G AATTC
	CTTAA G
*Hpa*II	CC GG
	GG CC

Which enzymes would be most suitable for cutting the following segment of DNA?

  C C G A A T T G G A T C C G G T A G
  G G C T T A A C C T A G G C C A T C

A *Bam*HI and *Eco*RI

B *Bam*HI and *Hpa*II

C *Eco*RI and *Hpa*II

D *Bam*HI only

6 One of the steps in genetic engineering is the screening of the recombinant DNA to determine if the foreign DNA has been taken up. In this process, which of the following is used?

A antibiotics

B viruses

C a vector

D dyes

7 The R plasmid shown below is used as a vector in the process of genetic engineering. If *Bam*HI was used as the restriction enzyme, which of the following is the **best** method to identify the transformed plasmid?

A It would not survive if exposed to ampicillin.

B It would not survive if exposed to tetracycline.

C It would survive in ampicillin but not tetracycline.

D It would survive in tetracycline but not ampicillin.

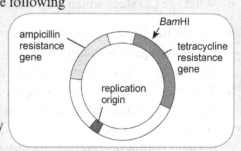

8 Experimental treatments of cystic fibrosis by gene therapy have involved:

A inhaling a spray that delivers normal DNA to the lungs.

B modifying cells outside the body and then transplanting them back in again.

C inserting a gene into the DNA of the egg of the mother.

D both somatic and germ line therapy.

9 Which of the following correctly describes gene therapy?

A It is the treatment of a genetic disorder by adding a normal gene to replace a faulty one.

B It can only be done inside the body.

C Recessive genes are removed and replaced with normal dominant genes.

D Normal dominant genes are replaced.

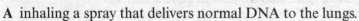

continued ...

Structured questions

10 Insulin used to be obtained from the pancreases of animals. Genetic engineering is now used to produce insulin.

    a The first step in the genetic engineering procedure is to obtain the gene responsible for producing insulin. Identify **two** ways by which this may be done.   [2 marks]

    b The diagram below shows some steps in the genetic engineering procedure.

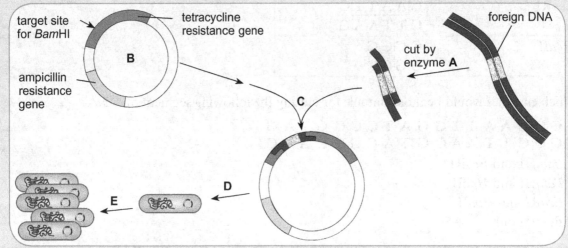

      i Name the type of enzyme used to cut the plasmid and foreign DNA.   [1 mark]

      ii What is the specific enzyme **A** used to cut the foreign DNA?   [1 mark]

      iii Draw the plasmid **B** after it has been cut by the enzyme *Bam*HI.   [2 marks]

      iv Explain the appearance of the foreign DNA after it has been cut by enzyme **A**.   [2 marks]

      v What is the name and function of the enzyme used at **C**?   [2 marks]

      vi In the step at **D**, the recombinant DNA is inserted into the bacteria. How is this achieved?   [2 marks]

      vii What is the name of step **E**?   [1 mark]

    c What are the steps which follow **E** in order to produce insulin?   [2 marks]

11 **a** One of the steps in genetic engineering is the screening of the cloned bacteria for the desired gene. In addition to the use of genetic markers for antibiotic resistance, suggest one other way in which the modified bacteria can be identified.   [1 mark]

    b The steps below show how the gene is added to a plasmid. The manufactured plasmid, pBR322, has genes for resistance to the antibiotics tetracycline and ampicillin as shown by the black areas. A restriction endonuclease is introduced that cuts the plasmid within the tetracycline resistance gene; and an insulin gene is added.

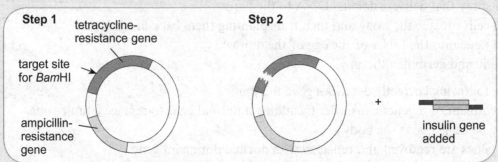

      i Draw the **three** possible outcomes of **Step 2**.   [6 marks]

continued ...

ii   The products of **Step 2** are inserted into the bacterium, *E. coli*, by heat shock, cloned and then screened to determine the colonies with the desirable insulin gene. What genetic marker is used to identify the desirable colonies?   [1 mark]

iii The colonies obtained after cloning the recombinant bacterium are exposed to the antibiotics ampicillin and then tetracycline.

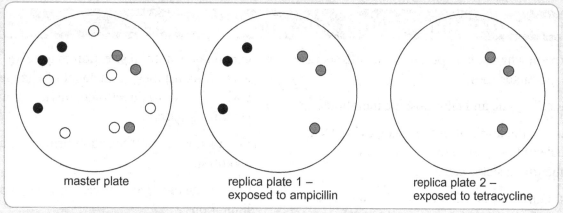

master plate

replica plate 1 –
exposed to ampicillin

replica plate 2 –
exposed to tetracycline

Using your answer for **b i**, explain the results shown above.   [7 marks]

Essay questions

12 a What are restriction enzymes?   [3 marks]

b Insulin is produced by genetic engineering.
Describe the steps involved in producing insulin by this method.   [7 marks]

c Antibiotic markers are used to identify the recombinant bacterium with the desired insulin gene.
Give **one** advantage and **one** possible problem of using this type of genetic marker.   [2 marks]

d Give **three** advantages of genetically produced insulin.   [3 marks]

13 a Explain what is meant by the terms 'recombinant DNA' and 'transgenic organism'.   [4 marks]

A BBC news item showed a grieving family in India. The father of the family had committed suicide because his genetically modified cotton crop had failed and he was unable to purchase more of the seeds to replant. This led to him being unable to meet his financial obligations. Genetically modified crops have an impact on both the environment and humans.

b Discuss the social impact of genetically modified organisms. (Include in your answer both the benefits and hazards.)   [6 marks]

c Discuss the ethical impact of genetically modified organisms on today's society.   [5 marks]

14 Attempts have been made to treat cystic fibrosis, which is caused by a recessive allele, by using gene therapy.

a Explain what is meant by the term 'gene therapy'.   [2 marks]

b Why is it theoretically easier to treat diseases caused by recessive alleles by gene therapy, rather than those caused by dominant alleles?   [2 marks]

c Explain how gene therapy works, using the example of cystic fibrosis.   [6 marks]

d Discuss the possible benefits and hazards of gene therapy.   [5 marks]

Chapter 9
Variation and natural selection

By the end of this chapter you should be able to:

a explain why sexually produced organisms vary in characteristics;

b describe gene and chromosome mutations;

c discuss the implications of changes in DNA nucleotide sequence for cell structure and function in sickle cell anaemia;

d explain how mutation brings about genetic variation, with reference to sickle cell anaemia and Down's syndrome;

e explain why heritable variation is important to selection;

f explain how environmental factors act as forces of natural selection, with reference to antibiotics and *Biston betularia*;

g explain how natural selection may be an agent of constancy or an agent of change, with reference to directional, disruptive and stabilising selection;

h discuss how natural selection brings about evolution;

i discuss the biological species concept and its limitations;

j explain the process of speciation, including the role of isolating mechanisms (reproductive, geographic, behavioural and temporal) and with reference to allopatric and sympatric speciation.

Variation

Each living organism is classified into a particular species – a group of organisms with similar characteristics that can breed together to produce fertile offspring. For example, you belong to the species *Homo sapiens*. A Jamaican ackee tree belongs to the species *Blighia sapida*.

Although all the individuals in a species have similar characteristics, they are not identical with one another. They show **intraspecific variation**. This variation has two basic causes – the organisms' genes, and their environment.

Genetic variation

In Chapter 6, we saw how genes with different alleles can cause variation, through the processes involved in sexual reproduction. Meiosis causes variation through **independent assortment** and **crossing over**. **Random fertilisation** introduces genetic variation between offspring.

Genetic variation can also be brought about by

changes in the DNA in a cell. This is called **mutation**. Genetic variation is inherited.

Environmental variation

Some of the variation that we see between individuals is not caused by their genes. For example, one person might cut their hair short, while another has long hair. Two people with similar combinations of alleles for height, that would allow both of them to grow tall, might end up being very different heights, because they ate different diets when they were young. Two plants with identical genes may have very different sizes and colours of leaves, because one is growing in the shade, or in soil that is low in nitrate or magnesium ions, while the other is growing in the sunshine, or in soil that is rich in ions.

Differences like this arise during an organism's lifetime. They do not normally affect the DNA, and so they are not passed on to the offspring. Environmental variation cannot be inherited.

Mutations

The processes of DNA replication, transcription of the code onto mRNA and the translation of this to an amino acid sequence are all very carefully quality-controlled by the cell. Nevertheless, things do sometimes go wrong.

Occasionally, the structure of a DNA molecule is damaged. There are many possible causes – it most often happens when the DNA is being copied. Despite the fact that DNA polymerase will not normally allow a 'wrong' base to be used, just occasionally a different one does creep in.

A random, unpredictable change in a DNA molecule is called a **mutation**.

Types of mutation

Mutations can be categorised into two main groups. **Gene mutations** are changes in the sequence of nucleotides in the DNA that makes up a gene. **Chromosome mutations** are changes in the structure or number of chromosomes in a cell.

Figure 9.1 shows three different kinds of gene mutation that might take place.

Substitution of one base for another quite often has no effect – a **silent mutation**. This is because the DNA code is **degenerate**, meaning that each amino acid is coded for by more than one triplet (Table 5.1, page 102). For example, GAA and GAG both code for the amino acid leucine.

Deletion, however, is almost certain to make a big difference. Deletion involves the loss of one base pair from the DNA molecule. Because the bases are read as triplets, if one pair goes missing then the whole sequence is read differently. This is called a **frame shift** (Figure 9.2).

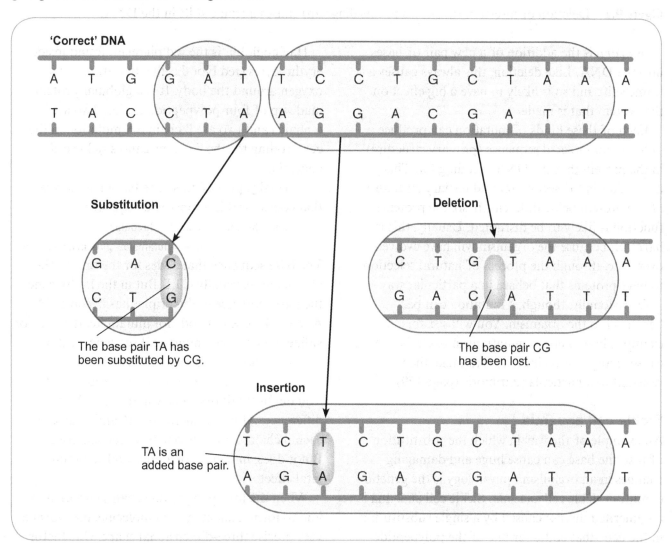

Figure 9.1 Mutations.

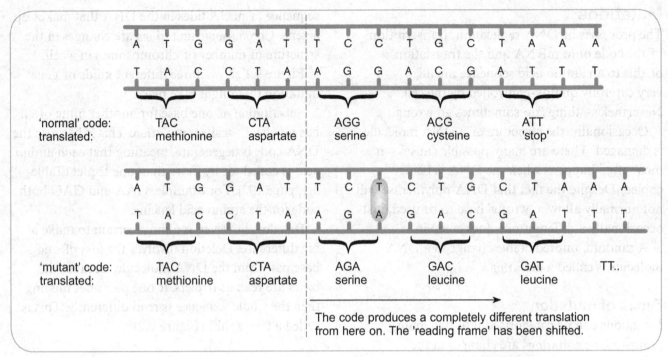

'normal' code:	TAC	CTA	AGG	ACG	ATT
translated:	methionine	aspartate	serine	cysteine	'stop'

'mutant' code:	TAC	CTA	AGA	GAC	GAT TT..
translated:	methionine	aspartate	serine	leucine	leucine

The code produces a completely different translation from here on. The 'reading frame' has been shifted.

Figure 9.2 Deletion or insertion (as shown in this diagram) causes a frame shift in the DNA.

Insertion is the addition of a new pair of bases into the DNA. Like deletion, this always causes a frame shift and so is likely to have a big effect on the protein that is made.

Each of these kinds of mutation can produce a different amino acid sequence (primary structure) in the protein that the DNA is coding for. This may result in the secondary and tertiary structure of the protein being different. If so, the protein's function is likely to be disrupted. Usually, this is harmful, because the organism will have evolved over time, through the process of natural selection, to have proteins that behave in a particular way.

Occasionally, though, a mutation can be beneficial to the organism. You will see, for example, how random mutations in bacteria can cause changes in their DNA that make them resistant to a particular antibiotic (page 179).

Sickle cell anaemia

An example of the way in which the substitution of just one base can cause huge and damaging changes in an organism's physiology is the genetic condition **sickle cell anaemia**. Sickle cell anaemia is an inherited disease caused by a single substitution in the gene that codes for one of the polypeptide chains in haemoglobin.

Haemoglobin is the red pigment, found inside erythrocytes (red blood cells), that transports oxygen around the body. It is a globular protein made up of four polypeptide chains. Two are α chains and two are β chains. A mutation in the gene coding for the β chains causes sickle cell anaemia.

Normally, part of this gene has a base sequence that codes for this amino acid sequence:

– valine – histidine – leucine – threonine
– proline – glutamate – glutamate – lysine –

The base sequence that codes for the first of the glutamates is usually CTT. But in the faulty gene the base sequence in this triplet has become CAT. And CAT does not code for glutamate. It codes for valine. So now the amino acid sequence will be:

– valine – histidine – leucine – threonine
– proline – valine – glutamate – lysine –

You might think that this would not make much difference. After all, there are 146 amino acids in each β chain, and only one has been changed. But it does, in fact, have a huge and sometimes fatal effect.

When the four polypeptide chains curl up and join to form a haemoglobin molecule, they form a very precise three-dimensional shape. One factor influencing this shape is that some amino acids have side chains that are hydrophilic, while others are hydrophobic. The polypeptides tend to curl up so that most of the hydrophobic amino acids are in the middle of the molecule, well away from the watery cytoplasm inside the erythrocyte. The hydrophilic side chains tend to be on the outside, where they interact with water molecules.

Glutamate has a side chain that is hydrophilic. In the 'correct' version of the haemoglobin molecule, it lies on the outside and helps to make the haemoglobin soluble. Valine, however, has a side chain that is hydrophobic. So in the 'incorrect' version, the haemoglobin molecule has a hydrophobic side chain on its outer surface, where there should be a hydrophilic one.

The valine side chains cannot interact with water, but they *can* interact with each other. Most of the time, this does not happen. But if the oxygen level in the blood falls, then the valines form bonds between themselves that stick haemoglobin molecules together. Long fibres of stuck-together haemoglobin molecules are produced. As the fibres form inside the erythrocytes, they pull the cell out of its usual biconcave shape. Some cells become sickle shaped (Figure 9.3).

In this state, the erythrocytes are not only useless but also dangerous. The fibres of haemoglobin cannot carry oxygen – hence the name 'anaemia' for this disease. Moreover, these misshapen cells cannot pass through capillaries. They cause blockages, which are very painful and can do serious damage to tissues. When this happens, a person is said to be having a sickle cell crisis.

Chromosome mutations

Chromosome mutation can be defined as a random change in the number (Figure 9.4) or structure of chromosomes (Figure 9.5). It is most likely to occur during meiosis I, when it is easy for things to go wrong as the paired chromosomes line up on the crowded equator at metaphase and are pulled apart in anaphase. Errors can result in the chromosomes not being shared equally between the daughter cells.

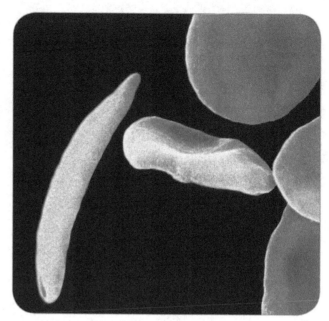

Figure 9.3 The cell at the left of this SEM is a sickled erythrocyte (× 5000).

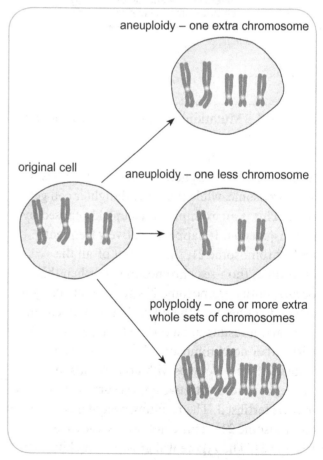

Figure 9.4 Mutations of chromosome number.

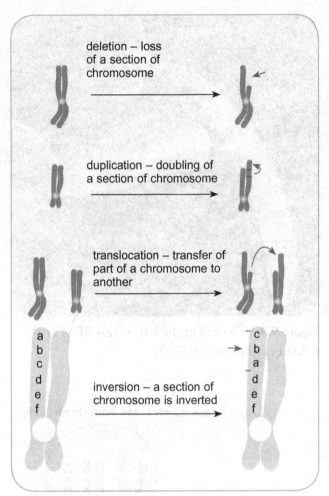

Figure 9.5 Mutations of chromosome structure.

Down's syndrome

Sometimes, one daughter cell gets both copies of a chromosome, while the other daughter cell gets none. This is an example of **aneuploidy** caused by **non-disjunction**. It happens relatively frequently with chromosome 21, the shortest of all the autosomes (non-sex chromosomes). When this occurs, both chromosome 21s go into one daughter cell and neither into the other. If this happens in an ovary, it results in an oocyte (female gamete) with either no chromosome 21 or with two copies instead of one. Oocytes with no chromosome 21 die, but those with two copies survive and may be fertilised. The resulting zygote has three chromosome 21s. This condition is known as **trisomy 21**. The zygote will grow into a child with **Down's syndrome**.

About 4% of cases of Down's syndrome are caused in a different way. During meiosis, part of the long arm of chromosome 21 breaks off and attaches to another chromosome. This is called **translocation**.

It is not known why having three copies of this chromosome produces Down's syndrome. This condition occurs in around 1 in 700 births. Its frequency greatly increases with the age of the mother, a result of the higher chance of mutation occurring during the formation of oocytes in older ovaries. Paternal age can also be a factor.

Children with Down's syndrome can have physical characteristics including eyes that slant upwards and a flattened nose bridge (Figure 9.6). They often have cheerful and friendly personalities, though there are mental health issues of varying severity. There are also health problems such as heart defects or muscle weakness, but these can often be kept well under control with appropriate medical treatment. Nevertheless, most people with Down's syndrome have a relatively short life expectancy.

Figure 9.6 A person with Down's syndrome has an extra chromosome 21 in their cells.

Klinefelter's and Turner syndromes

These two conditions are both the result of non-disjunction. Unlike Down's syndrome, however, these involve the sex chromosomes.

In the case of Klinefelter's syndrome, the mutation happens during oogenesis in the female parent. The two X chromosomes fail to separate as they should. One daughter cell gets both of them while the other gets none. This results in an oocyte with the genotype XX (it should be X). It has 24 chromosomes instead of 23.

When this oocyte is fertilised by a sperm carrying a Y chromosome, a zygote is produced with the genotype XXY. The number of chromosomes in the zygote is 47 – that is, 2n + 1.

Turner syndrome is caused when a faulty oocyte, with no X chromosomes, is fertilised by a sperm carrying an X chromosome. The resulting zygote has the genotype X0. It has 45 chromosomes instead of 46.

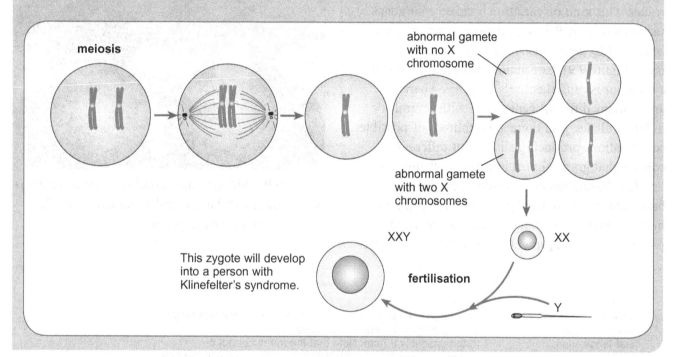

Discontinuous and continuous variation

Variation in some characteristics is very 'cut and dried'. For example, a person has one of four blood groups in the ABO system. There are no in-betweens – everyone is either A, B, AB or O. This kind of variation, where there are relatively few clearly defined groups to which an individual can belong, is called **discontinuous variation** (Figure 9.7).

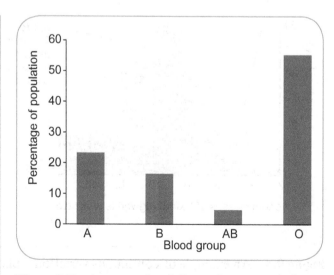

Figure 9.7 An example of discontinuous variation – ABO blood groups.

Discontinuous variation is almost always caused by genes, with little or no environmental influence. Usually, just one or two genes are involved, each of them having only a few alleles. Human ABO blood groups, for example, are controlled by a single gene with three alleles.

However, for most characteristics, variation is not so clear-cut. For example, human skin or eye colour is impossible to categorise into clearly defined colours. Leaf length in the Oxford ragwort can range between around 2 mm to 180 mm, with any length possible between these two extreme values. This kind of variation is called **continuous variation** (Figure 9.8). Continuous variation may be caused by genes, or by the environment, or both (Figure 9.9). Continuous variation in human eye colour is caused entirely by genes. There are so many different genes, each with several different alleles, that there are hundreds of possible combinations producing all sorts of different colours that grade almost imperceptibly into one another. Variations in skin colour, on the other hand, are caused partly by genes (again there are many of these with many different alleles) and partly by the environment, in particular the degree

of exposure to sunlight. Variations in leaf length on a single plant, however, must be caused entirely by the environment (for example, the degree of shading on the leaf) because all the cells in the plant were produced by mitosis from a single zygote and so contain exactly the same genes.

SAQ

1 Suggest an example of discontinuous variation in humans other than ABO blood groups.

Figure 9.9 Most of the variation between humans is continuous variation, and is influenced by the environment as well as genes.

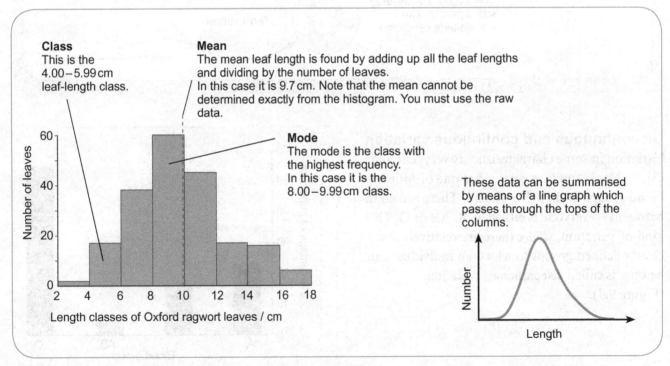

Class
This is the 4.00–5.99 cm leaf-length class.

Mean
The mean leaf length is found by adding up all the leaf lengths and dividing by the number of leaves.
In this case it is 9.7 cm. Note that the mean cannot be determined exactly from the histogram. You must use the raw data.

Mode
The mode is the class with the highest frequency.
In this case it is the 8.00–9.99 cm class.

These data can be summarised by means of a line graph which passes through the tops of the columns.

y-axis: Number of leaves
x-axis: Length classes of Oxford ragwort leaves / cm

line graph y-axis: Number
line graph x-axis: Length

Figure 9.8 An example of continuous variation – plant leaf length.

When a characteristic is influenced by the combined effect of many genes, this is known as **polygenic** inheritance. Polygenic characteristics tend to show continuous variation and a normal distribution. There are many human characteristics that are influenced by a large number of genes, each having a small effect. The tendency towards obesity is one example. The environment often contributes towards these characteristics. Obesity, for example, is also affected by exercise and what we eat.

SAQ _____

2   For each of these examples of variation between sunflower plants, suggest whether they are caused by genes alone, environment alone, or interaction between genes and environment:
   a the height of the plant
   b the colour of the flower petals
   c the diameter of a mature flower
   d the percentage of seeds that develop after fertilisation.

Charles Darwin and the theory of natural selection

In 1856, a startling new theory was put forward by Charles Darwin and – quite independently – Alfred Russel Wallace. Darwin is by far the more famous of these two brilliant scientists, perhaps because his publications developed his theory more fully, and were widely read and discussed during the latter half of the 19th century. His book *On the Origin of Species* is still in the bestseller lists today.

Darwin was a thinker and experimenter. He made observations of the world around him, and then developed logical theories about how and why things happened. He worked on many different areas of biology, but he is best known for his theories about how living organisms may have evolved over time.

Darwin proposed a mechanism called **natural selection** to explain how organisms might change over time. His theory grew out of four observations and three logical deductions from them.

Observations:
● All organisms over-reproduce – far more offspring are produced than are required to keep the population at a steady size.
● Population numbers tend to remain fairly constant over long periods of time.
● Organisms within a species vary.
● Some of these variations are inherited.

Deductions:
● There is competition for survival – the 'struggle for existence'.
● Individuals with characteristics that best adapt them for their environment are most likely to survive and reproduce.
● If these characteristics can be inherited, then the organisms will pass the characteristics on to their offspring.

Darwin argued that, if this happened over a long period of time, then the characteristics of a species could gradually change, as better-adapted individuals were more likely to survive and pass on their adaptations to their offspring. Gradually, the species would become better and better adapted to its environment.

Overproduction

Almost all organisms have the reproductive potential to increase their populations. Rabbits, for example, produce several young in a litter, and each female may produce several litters each year. If all the young rabbits survived to adulthood and reproduced, then the rabbit population would increase rapidly. Figure 9.10 shows what could happen.

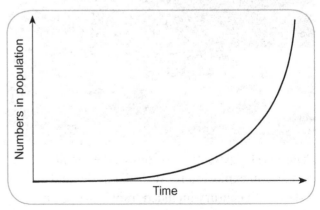

Figure 9.10   If left unchecked, numbers in a population can increase exponentially.

This sort of population growth actually did happen in Australia in the 19th century. In 1859, twelve pairs of rabbits from Britain were released on a ranch in Victoria, as a source of food. The rabbits found conditions to their liking. Rabbits feed on low-growing vegetation, especially grasses, of which there was an abundance. There were very few predators to feed on them, so the number of rabbits soared. Their numbers became so great that they seriously affected the availability of grazing for sheep (Figure 9.11).

Such population explosions are rare in normal circumstances. Although rabbit populations have the potential to increase at such a tremendous rate, they do not usually do so.

As a population of rabbits increases, various **environmental factors** come into play to keep down the numbers. These factors may be **biotic factors** – that is, caused by other living organisms – such as predation, competition for food or infection by pathogens; or they may **abiotic factors** – that is, they are caused by non-living components of the environment – such as water supply or nutrient levels.

For example, the increasing number of rabbits eats an increasing amount of vegetation, until food is in short supply. The larger population may allow the populations of predators, such as foxes, stoats and weasels, to increase (Figure 9.12). Overcrowding may occur, increasing the ease with which diseases such as myxomatosis (Figure 9.13) can spread. Myxomatosis is caused by a virus that is transmitted by fleas. The closer together the rabbits live, the more easily fleas, and therefore viruses, will pass from one rabbit to another.

These environmental factors act to reduce the rate of growth of the rabbit population. Of all the rabbits born, many will die from lack of food, be killed by predators or die from myxomatosis. Only a small proportion of young will grow to adulthood and reproduce.

Figure 9.12   Stoats are predators of rabbits.

Figure 9.11   Attempts to control the rabbit population explosion in Australia in the mid-to-late 19th century included 'rabbit drives', in which huge numbers were rounded up and killed. Eventually, myxomatosis brought numbers down.

Figure 9.13   Myxomatosis is a deadly disease of rabbits, but some have developed resistance to it.

Natural selection

What determines which will be the few rabbits to survive, and which will die? It may be just luck. However, some rabbits will be born with a better chance of survival than others. Variation within a population of rabbits means that some will have features that give them an advantage in the 'struggle for existence'. The ones that are best adapted to their environment are most likely to survive and reproduce.

One feature that varies is coat colour. Most rabbits have alleles that give the normal agouti (brown) colour. A few, however, have darker coats. Such darker rabbits will stand out from the others and are more likely to be picked out by a fox. They are less likely to survive – at least, in their normal environment – than brown rabbits. The chances of a dark rabbit surviving long enough to reproduce and pass on its genes for coat colour to its offspring are less than the chances for a normal brown rabbit. Brown rabbits are better adapted to their environment.

Predation by foxes is an example of a **selection pressure**. Selection pressures increase the chances of some genetic variations being passed on to the next generation and decrease the chances for others. The effect of this is **natural selection**. Natural selection increases the frequency of certain characteristics within a population, at the expense of others. The characteristics that best adapt an organism to its environment are most likely to be passed on to the next generation.

In Trinidad streams, guppy populations are separated by waterfalls. Research has shown that there is strong natural selection on any population in a pool with high populations of predators. Here, the guppies become mature at an earlier age, are relatively small at maturity and produce litters more frequently. If guppies from such populations are moved to a pool with fewer predators they evolve into the larger, longer-lived forms of guppies over seven to eleven generations.

Antibiotic resistance

The development of resistance by bacteria to antibiotics and other medicinal drugs is a good example of natural selection – and one that has great significance for us.

Antibiotics are chemicals produced by living organisms, which inhibit or kill bacteria but do not normally harm human tissue. Most antibiotics in general use are produced by fungi. The first antibiotic to be discovered was penicillin, which was first used during World War II to treat a wide range of infectious diseases caused by bacteria. Penicillin prevents cell wall formation in bacteria.

If someone takes penicillin to treat a bacterial infection, bacteria that are susceptible to penicillin will not be able to grow or reproduce (Figure 9.14). In most cases, this will be the entire population of bacteria. However, by chance, there may be among them one or more individuals that are resistant to penicillin. One example is found in the populations of the bacterium *Staphylococcus*, where some bacteria produce an enzyme, penicillinase, which inactivates penicillin.

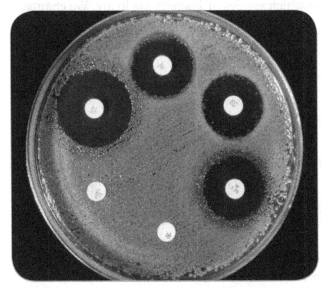

Figure 9.14   The green areas on the agar jelly in this Petri dish are colonies of the bacterium *Escherichia coli*. The white discs are pieces of card impregnated with different antibiotics. Where there is a clear area around a disc, the antibiotic has prevented the bacteria from growing. You can see that this strain of *E. coli* is resistant to the antibiotics on the discs at the bottom left and has been able to grow right up to the discs.

These individuals have a tremendous selective advantage. Bacteria that are not resistant are killed, while those with resistance can survive and reproduce. Bacteria reproduce very rapidly in ideal conditions, and even if there was initially only one resistant bacterium, it might produce ten billion descendants in 24 hours. A large population of penicillin-resistant *Staphylococcus* would result.

Such antibiotic-resistant strains of bacteria are constantly appearing. One of the most worrying is MRSA, which stands for methicillin-resistant *Staphylococcus aureus*. This bacterium, normally harmless, is capable of infecting people whose immune systems are not strong – perhaps because they have another illness. Many people who were already ill have picked up MRSA infections while in hospital and have died as a result. This bacterium has become resistant to almost all antibiotics, so infections are very difficult to treat. Figure 9.15 illustrates how the number of deaths from *Staphylococcus aureus* changed between 1993 and 2005, in England and Wales.

By using antibiotics, we change the environment in which species of bacteria are living. We change the selection pressures. Individual bacteria that are lucky enough to have genes that make them better adapted to the new environment win the struggle for existence, and pass on their advantageous genes to their offspring. The more we use antibiotics, the greater the selection pressure we exert on bacteria to evolve resistance to them.

Alleles for antibiotic resistance often occur on **plasmids** (page 153), small circles of DNA other than the main bacterial 'chromosome'. Plasmids are quite frequently transferred from one bacterium to another, even between different species. So it is possible for resistance to a particular antibiotic to arise in one species of bacterium and be passed to another.

Insecticide resistance

Just as natural selection has led to the development of populations of bacteria that are resistant to antibiotics, so it has led to the development of resistance to insecticides in some populations of insects.

There are many kinds of insect that we would like to see fewer of. Mosquitoes transmit malaria, a major cause of death, especially among children, in many tropical and subtropical parts of the world. Insects eat our stores of food, and damage crops. A wide range of insecticides has been developed to try to keep populations of insect pests to a reasonably low level.

Almost 20% of the insecticides that are used in the world are aimed at getting rid of insects that damage cotton plants. Cotton is a major crop in countries including USA, Australia, India, Pakistan and China (Figure 9.16 and Figure 9.17).

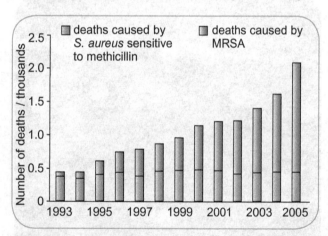

Figure 9.15 Deaths from *Staphylococcus aureus* between 1993 and 2005, in England and Wales.

3 a Using Figure 9.15, describe the changes in the numbers of deaths from non-antibiotic-resistant *Staphylococcus aureus* and methicillin-resistant *S. aureus* between 1993 and 2005, in England and Wales.
   b Suggest explanations for these changes.

Figure 9.16 Cotton plants produce heads of seeds surrounded by silky fibres of cellulose.

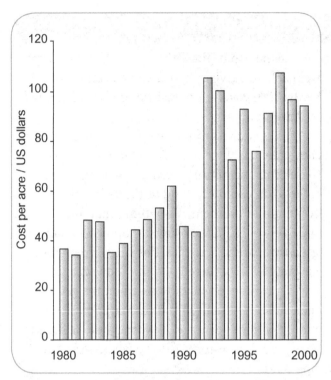

Figure 9.17 Costs of treating cotton against insect pests in the state of Mississippi, USA, between 1980 and 2000.

SAQ

4 a Using Figure 9.17, describe how the costs of controlling insect pests of cotton in Mississippi changed between 1980 and 2000.

 b Suggest at least two other possible explanations for these changes.

The major pest of cotton is the cotton boll worm, *Helicoverpa armigera* (Figure 9.18 and Figure 9.19). This is a species of moth, and it is the caterpillars that cause all the damage. They feed not only on cotton, but also on crops of maize, groundnuts (peanuts) and sorghum. Resistance to several different insecticides has developed, and growers have sometimes fallen back on using highly toxic chemicals that not only kill boll worms but also other beneficial or harmless organisms.

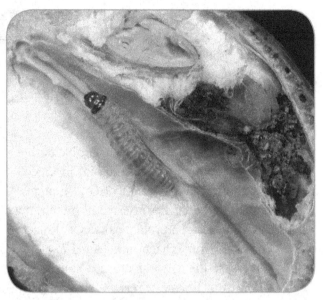

Figure 9.18 The cotton boll worm, *Helicoverpa armigera*.

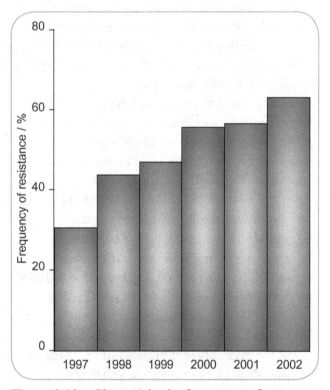

Figure 9.19 Changes in the frequency of resistance of cotton boll worms to a commonly used insecticide, in Australia.

SAQ

5 Explain how the increase in the percentage of boll worms showing resistance, illustrated in Figure 9.19, could have developed.

GM cotton to beat the boll worm

Today, there is a lot of pressure on growers to use fewer chemicals. But they still have to control cotton boll worms – if they do not, then they could easily lose their entire crop. This did actually happen in Mississippi in 1999.

In an attempt to get around this problem, the multinational company Monsanto has produced a genetically engineered variety of cotton plant. The plants have had genes inserted into them that enable them to make a protein called Cry1Ac. The gene came from a bacterium called *Bacillus thuringiensis*, so the cotton plants are sometimes called Bt cotton. The protein is a toxin which kills insects that eat the cotton, by attaching to receptors on the plasma membranes of the cells lining the insects' guts and destroying the gut wall. This particular variety of the toxin only affects butterflies and moths (and their caterpillars), because only they have the receptors to which the toxin can bind.

The cotton is marketed as INGARD®. Seeds are expensive – much more so than unmodified cotton – but growers should still gain good profits because they should not need to use insecticides, and should get high yields because less of the crop will be lost. The GM cotton could be good for the environment, because it should only harm caterpillars that actually eat the cotton plants.

However, natural selection is still at work, and cotton boll worms are already becoming resistant to the toxin in the Bt cotton. Researchers in Australia investigated how resistance could arise. They took a large number of boll worms collected from different places, and fed them on crystals of the Cry1Ac protein and also spores from *Bacillus thuringiensis*. This gave the boll worms a much higher dose of the toxin than they would normally get if they ate the Bt cotton plants. After seven days, the researchers took the survivors and transferred them to a diet that did not contain the toxin, so that they could be sure to live long enough to produce the next generation. They repeated this procedure for 28 generations. The graph shows their results. You can see that, to begin with, few of the boll worms were able to survive exposure to the toxin. But by the 12th generation, resistance was beginning to develop.

The researchers found that the resistance was caused by a lack of one of the binding sites on the plasma membranes of the boll worms' cells – the toxin could not bind to the cells, and so could not harm them. They think that one of the boll worms in the original population had a gene that caused it to lack this site (or perhaps a new mutation occurred), so this individual and its offspring survived and passed on the gene to their offspring. In each successive generation, more and more of the boll worms were resistant ones whose ancestor was the original resistant boll worm.

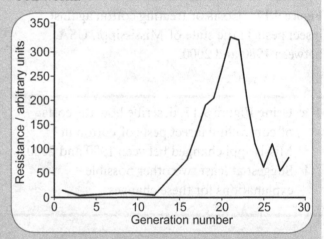

Interestingly, however, after generation 23 the level of resistance fell. The researchers think that there may be a price that the boll worms have to pay for being resistant – perhaps lacking the receptor site disadvantages them in some way. However, these boll worms were still more resistant than the original colony, quite enough to cause problems in fields of Bt cotton.

So what can be done, if insect pests can even develop resistance to toxins in genetically modified cotton? At the moment, growers should use at least two different weapons against boll worms – two different insecticides, for example. It is much less likely that a boll worm will, by chance, have genes that make it resistant to both.

Industrial melanism

One of the earliest well-documented examples of natural selection in action involved a rather dull-looking moth, called *Biston betularia*, in the UK. This moth, whose common name is the peppered moth, flies by night and rests on tree branches during the day. Various birds eat it, and the speckling on the moth's wings acts as camouflage against the tree bark, reducing the chances of a moth being seen by a bird and eaten.

Until 1849, all the specimens of *B. betularia* that had been seen in the UK had pale wings with dark markings. However, in 1849 a melanic (black) specimen was caught near the industrial city of Manchester (Figures 9.20 and 9.21). In the ensuing years, more and more black moths were caught. It seemed that, at least in some areas of the UK, the proportion of melanic *B. betularia* was increasing, at the expense of the normal pale speckled variety.

We now know that the difference between these two forms of the moth is caused by a single gene. The normal, pale speckled colour is produced by a recessive allele, and the melanic form by a dominant allele. Until the 1960s, the frequency of the dominant allele (and therefore of the melanic moths) continued to increase in areas close to industrialised areas of the UK. In other parts of the country, the recessive allele (and therefore the pale moths) continued to be the more common.

It was found that the selection pressure causing the change in allele frequency was predation. In areas where the air was polluted, lichen was unable to grow on the tree bark and deposits of soot made the bark darker. This meant that the melanic moths were better camouflaged than the pale ones. The melanic moths were more likely to survive, reproduce and pass on their alleles to their offspring.

As air pollution from industry has decreased, the advantage has swung back in favour of the pale form of the moth again. Today, there are very few melanic *B. betularia* moths in the UK. Although mutations of the allele do continue to occur fairly frequently, any moth that has the rare dominant form of the allele is at a selective disadvantage, and not likely to survive long enough to reproduce.

Stabilising and directional selection

Most of the time, in most populations, the individuals are already well adapted to their environment. The alleles present in the population are the ones that confer the most advantageous characteristics. If the environment remains fairly stable, then the same alleles will be selected for in every successive generation. Nothing changes. This is called **stabilising selection** (Figure 9.22).

Figure 9.20 Pale (normal) and melanic forms of the peppered moth, *Biston betularia*, on light and dark tree bark.

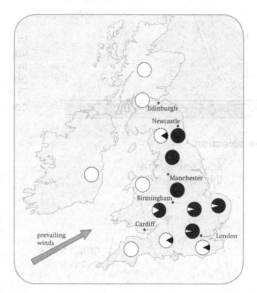

Figure 9.21 The distribution of the pale and dark forms of the peppered moth, *Biston betularia*, in the early 1960s. The ratio of dark to light areas in each circle shows the ratio of melanic to pale moths in that part of the country.

However, if there is a change in the environment, this might result in a change in the selection pressures on the population. A variation that previously was not advantageous may begin to confer better survival value than another, resulting in **directional (evolutionary) selection**. Or perhaps a completely new, advantageous variation arises, by mutation. The change in the environment, or the appearance of a new allele, can bring about a change in the genetically determined characteristics of subsequent generations of the species, which we call **evolution**.

There is a third possible effect of selection. Sometimes, the variations in the middle of the range may be at a selective disadvantage compared to both of the extremes. In this case, the result of natural selection may be a population that contains two different forms, at either end of a spectrum for a particular characteristic. This is called **disruptive selection**.

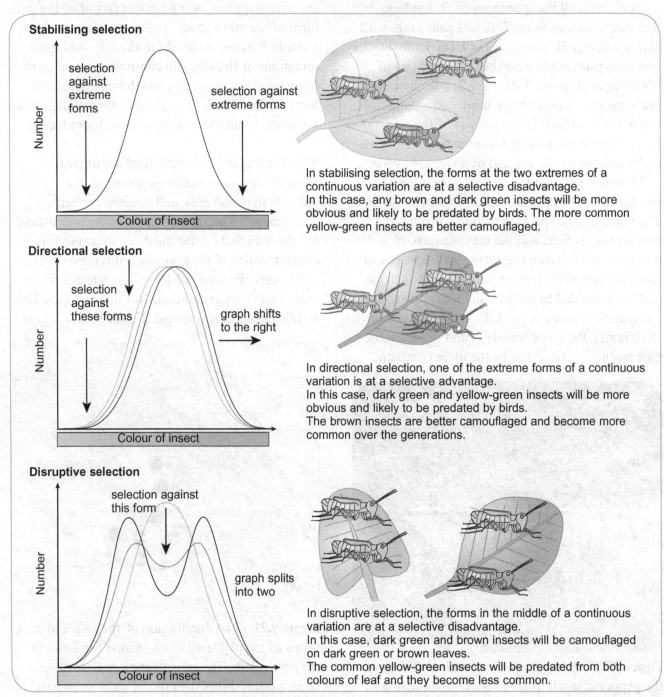

Stabilising selection

In stabilising selection, the forms at the two extremes of a continuous variation are at a selective disadvantage.
In this case, any brown and dark green insects will be more obvious and likely to be predated by birds. The more common yellow-green insects are better camouflaged.

Directional selection

In directional selection, one of the extreme forms of a continuous variation is at a selective advantage.
In this case, dark green and yellow-green insects will be more obvious and likely to be predated by birds.
The brown insects are better camouflaged and become more common over the generations.

Disruptive selection

In disruptive selection, the forms in the middle of a continuous variation are at a selective disadvantage.
In this case, dark green and brown insects will be camouflaged on dark green or brown leaves.
The common yellow-green insects will be predated from both colours of leaf and they become less common.

Figure 9.22   Stabilising and directional selection.

Speciation

Speciation is the formation of a new species. It is not easy to define exactly what we mean by a species, but one generally accepted definition is that it is a group of organisms, with similar **morphology** and **physiology**, which can interbreed with one another to produce fertile offspring. So, to produce a new species from an existing one, some of the individuals must:

- become morphologically or physiologically different from the members of the original species;
- no longer be able to breed with the members of the original species to produce fertile offspring.

The splitting apart of this 'splinter group' from the rest of the species is known as **isolation**. Sometimes, the organisms are separated by a physical barrier, such as a mountain range or river, and this is called **geographical isolation**. However, it is not until the two groups are so different from each that they can no longer interbreed that they are said to show **reproductive isolation**, and have become different species. There is no gene flow between two different species – each is effectively genetically isolated from the other.

Allopatric and sympatric speciation

Speciation is thought to often begin when a geographical barrier separates two populations of a species. The two groups then evolve along different lines, either because of different selection pressures in the two geographically separated areas, or because of genetic drift. If the barrier then breaks down and the two populations come together again, they may have changed so much that they can no longer interbreed, and can be said to be two different species (Figure 9.23).

The production of a new species as a result of

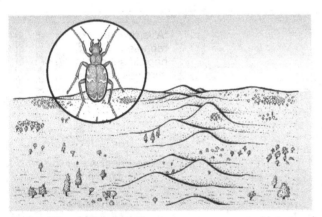

A population of tiger beetles becomes separated into two by the formation of a mountain chain.

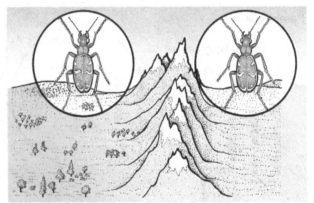

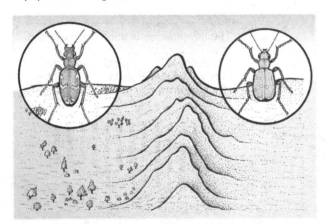

Natural selection pressures differ on the two sides of the mountain chain due to factors such as level of rainfall.

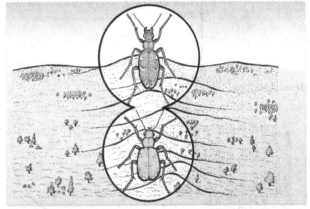

Over a long period the two populations diverge and may become different species.

Figure 9.23 Geographical isolation leading to speciation.

geographical separation is known as **allopatric speciation**.

However, new species can sometimes evolve without being geographically separated. This is called **sympatric speciation**. For example, in North America and Canada there is a fly called *Rhagoletis pomonella*, whose maggots feed on various fruits such as apples or hawthorn berries. It looks as though the maggots used to feed on hawthorn berries, but then some of them began to feed on apples instead, soon after apples were introduced into North America in the mid-19th century. Adult flies that grew up feeding on apples tend to lay their eggs on apples, and to mate with other flies that grew up on apples. A similar pattern occurs in the flies that feed on hawthorn berries. Possibly these two populations will eventually become so distinct that they can no longer interbreed, and we can say that the apple fly and the hawthorn fly are two distinct species.

Sympatric speciation by polyploidy

New species can arise more quickly than this by polyploidy, sometimes in a single generation. Speciation by polyploidy is especially important in plants where it has been involved in the production of about 15% of all plant species. This kind of sympatric speciation includes some of the best documented examples of speciation we have observed.

Imagine a plant with 8 chromosomes in each of its diploid cells. If, however, during the first mitotic division of a newly fertilised egg the chromosomes fail to separate, the daughter cell will contain 16 chromosomes (Figure 9.24). Mitosis will occur as normal in such a cell and the whole body of the plant is made up of the products of this cell. This new plant is a tetraploid – each cell contains 4 sets of chromosomes. As all the chromosomes come from the same plant, it is an **autotetraploid**.

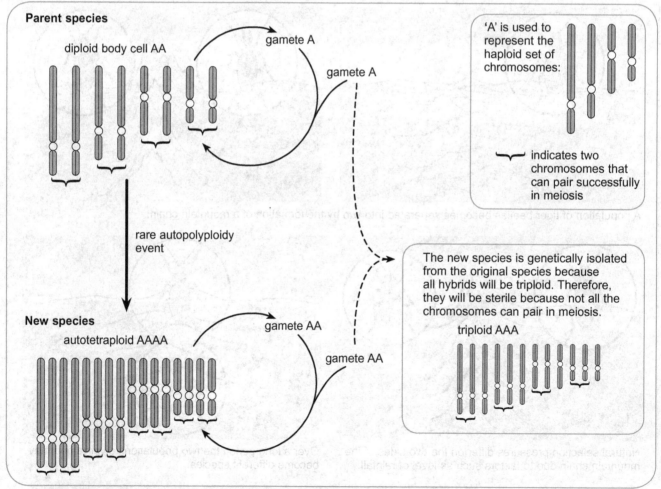

Figure 9.24   Formation of a new species from one parent – autopolyploidy.

Tetraploid plants are often more vigorous than diploid plants and will successfully produce gametes by meiosis. Although its cells have 16 chromosomes, every chromosome has a partner to pair with. The gametes will each contain 8 chromosomes.

An autotetraploid is likely to be fertilised by a gamete from a diploid plant. A gamete from the autotetraploid contains 8 chromosomes and a gamete from the original diploid plant contains 4 chromosomes. The product is an individual with 12 chromosomes. This is a triploid, containing three sets of chromosomes.

The plant will be healthy and vigorous. However, it has problems with meiosis. As there are three chromosomes of each homologue, they can't pair successfully and no fertile gametes will be produced. The plant is sterile and so is genetically isolated from the parent.

Many plants can multiply asexually, so this triploid plant can probably spread and produce a population. It is a unique clone and it may be distinct enough to be called a new species. This is **autopolyploid speciation**.

Polyploidy can also result in a new species arising from two *different* species in **allopolyploid speciation** (Figure 9.25). In the first instance, this involves the production of a hybrid from the infrequent event of a gamete from one species combining with a gamete of a different species. Hybrids are generally vigorous plants but they suffer a similar problem to autotriploids during meiosis. As the hybrid has chromosomes from two different species, no chromosome has a homologue to pair with. Meiosis is unsuccessful and the hybrid will be sterile but may survive if the plant reproduces asexually.

However, a very rare event might occur in

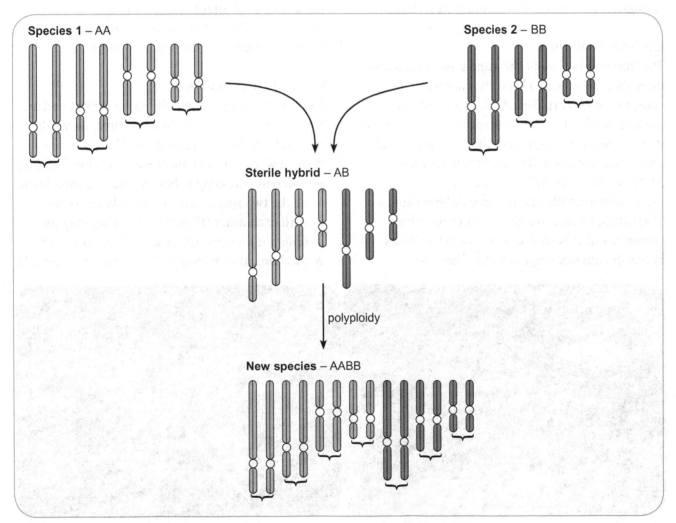

Figure 9.25   Formation of a new species from two parents – allopolyploidy.

one of these hybrid plants. It may produce a few viable gametes by polyploidy. This could be due, for example, to a failure of the spindle forming in meiosis I, producing gametes with the diploid number of chromosomes instead of the haploid number. If both a male and female gamete can be produced in the same way in the male and female parts of a flower of that plant and they fertilise, a viable zygote can be produced. It is a tetraploid and will be a fertile plant. However, it is incapable of interbreeding with either of the original parents. This will definitely be a new species.

Reproductive isolation

What is it that prevents two groups breeding, even when they are living in the same place? There are many different factors that can prevent interbreeding between two closely related species. One way of classifying them is to group them into **ecological**, **temporal** and **reproductive barriers**.

Ecological barriers

Two species may live in the same area at the same time, yet rarely meet. This is the case with the apple and hawthorn flies. At the moment, these are still classified as the same species, but there is a strong case for the argument that they are already reproductively isolated from one another and so could be said to be different species.

Another example involves two different species of crayfish, *Orconectes virilis* and *Orconectes immunis*, which both live in freshwater habitats in North America (Figure 9.26). They look very similar to one another, but *O. virilis* lives in streams and lake margins, while *O. immunis* lives in ponds and swamps. They rarely meet and do not interbreed. It seems that *O. virilis* is less good at digging than *O. immunis*, and cannot easily burrow into the mud when a pond dries up, so it is less able to survive summer drying than *O. immunis*. *O. immunis* is able to live perfectly well in the lakes and streams where *O. virilis* lives, but *O. virilis* is more aggressive and drives *O. immunis* out of crevices where it tries to shelter.

Temporal barriers

Two species may live in the same place and even share the same habitat, but not interbreed because they are not active at the same time of day, or do not reproduce at the same time of year. For example, the spectacular flowering shrubs *Banksia attenuata* and *Banksia menziesii* both live in the same area of Western Australia (Figure 9.27). *B. attenuata* flowers in summer, but *B. menziesii* flowers in winter, so they cannot interbreed.

Reproductive barriers

Even if two species share the same habitat and are reproductively active at the same time, they still may not interbreed successfully. There are many ways in which this can be prevented. They include:

- different courtship behaviour, so that individuals of the two species are not stimulated to mate with each other (Figure 9.28) – they may make different movements, or sing different songs;
- mechanical problems with mating – for example,

Figure 9.26   a *Orconectes virilis*; **b** *Orconectes immunis*, sometimes called the nail polish crayfish.

Figure 9.27 **a** *Banksia attenuata*; **b** *Banksia menziesii*.

one may be much smaller than the other or have different shapes or sizes of reproductive organs;

- gamete incompatability, so that even if mating takes place successfully the sperm cannot fertilise an egg;
- zygote inviability, so that even if a zygote is produced it dies;
- hybrid sterility, so that even if a zygote develops successfully, the resulting hybrid cannot produce gametes and so cannot reproduce.

The species concept

The definition of a species that we gave on page 185 involves reproductive isolation between two groups. This is sometimes known as the **biological species concept**. The biological species concept is a useful one in trying to work out how new species can evolve. However, it has one major limitation – it can only be used for organisms that reproduce sexually. We cannot use it to determine whether groups of asexually reproducing organisms belong to the same or different species. Nor can we use it to classify extinct organisms that are only known as fossils, old bones or skins.

But we do still classify all organisms into a particular species. Each species is given a unique binomial, a name made up of its genus and its species – for example, *Homo sapiens*. Even organisms that do not reproduce sexually, or where we don't know enough about them to determine whether or not they can interbreed with other species, are classified in this way. Biologists use a variety of different methods to decide whether two groups of organisms belong to the same species or to different ones, often largely based on their morphology. The species **taxon** can be used even when we cannot apply the biological species concept to a group of organisms.

For example, birds that are clearly very closely related in an evolutionary sense, but live in different parts of the world and have different colouring, may be classified as different species even if nothing is known about whether or not they are able to breed together. This is sometimes known as the **phylogenetic (evolutionary) species concept**. It is based on the fact that we can clearly see a difference between the two groups – we can tell them apart – and we are certain that they must have evolved from a common ancestor. We do not need to know how *far* they have evolved from one another. It is enough to know that they are clearly two distinct groups each with their own distinctive characteristics.

Figure 9.28 A mallard drake will only mate with a female who displays appropriate courtship behaviour. **a** A pair of mallards displaying to one another; **b** although a pintail female looks very like a mallard female, her courtship behaviour will only interest a pintail male.

Care has to be taken where there are very clearly distinguished forms of the same species. In some species of wild bee there are forms that are so different in appearance that you might think they are different species. But study reveals that they are fully interbreeding. The occurrence of distinctly different forms of one species is known as **polymorphism**.

Using the phylogenetic species concept often means that many more groups are classified as separate species than using the strict biological species concept. Conservationists sometimes make use of this to make their case stronger. For example, using the biological species concept there are 101 bird species that are endemic to Mexico (are found nowhere else). Using the phylogenetic species concept, there are 249.

So which is the better concept to use – the biological species concept, or the phylogenetic species concept? It depends on what you are using it for. The biological species concept gives a clear-cut definition, which can be applied rigorously and in the same way for different groups of organisms. But it is limited because it can only be used for sexually reproducing organisms. And even in those that do, we often don't have enough information to determine whether or not there is complete reproductive isolation. It is also impossible to use on fossils.

However, because the phylogenetic species concept is not so rigorous, different people might make different decisions about whether particular groups of organisms are species or not, which can be confusing. But it does at least allow us to make a decision, which we might not be able to do at all if we stuck rigidly to the biological species concept.

Summary

- There is variation between individuals of a species. This may be caused by genes, by the environment or by both.

- Mutation is a random change in the genes or chromosomes of an organism. It is most likely to happen during DNA replication or during meiosis. Sickle cell anaemia is an example of a condition caused by gene mutation, and Down's syndrome is an example of a condition caused by chromosome mutation.

- If variation is caused by genes, then individuals with particular alleles, or combinations of alleles, may be better adapted to their environment than other individuals. These better-adapted organisms are said to have a selective advantage. They are more likely to survive and reproduce, passing on their alleles (including the advantageous ones) to their offspring. Over time, this can cause a shift in the allele frequency in the population.

- If a population of organisms is already well adapted to its environment, then natural selection may not bring about any change in allele frequency. This is known as stabilising selection. However, if the environment changes or a new allele appears by mutation, then natural selection may result in a shift in allele frequencies, known as directional selection. Natural selection can also favour two extremes within a population, resulting in disruptive selection.

- Speciation is the formation of a new species. The biological species concept defines a species as a group of organisms that are morphologically and physiologically similar, and that cannot breed with other groups to produce fertile offspring. This can only be determined for groups that reproduce sexually.

- Speciation involves the reproductive isolation of two or more populations, which therefore become new species. It can be allopatric or sympatric. Isolating mechanisms may include geographical separation, differences in ecology, timing of reproductive activity, behaviour or reproductive incompatibility.

Questions

Multiple choice questions

1 Which of the following statements could **not** be used to describe a species? It is a group of organisms:

  A sharing unique structural and functional characteristics.

  B capable of reproducing with one another to produce viable offspring.

  C sharing the same ecological niche.

  D all with identical DNA.

2 Two species of tree crickets live in a forest and arose from a common ancestor. Their habitats overlap. The two species never interbreed because the calls of the males are different and they do not attract the females of the other species. This type of speciation arose because of:

  A mechanical isolation.

  B behavioural isolation.

  C temporal isolation.

  D ecological isolation.

3 Sickle cell anaemia is a genetic disease in humans. Individuals homozygous for the sickle cell allele generally die in infancy from severe anaemia. In areas where malaria is present, individuals who are homozygous for the normal allele succumb to malaria and the heterozygotes have an advantage.

  What type of natural selection is occurring in the malarial areas?

  A stabilising

  B disruptive

  C directional

  D directional then disruptive

4 The diagram shows a summary of Darwin's theory of natural selection.

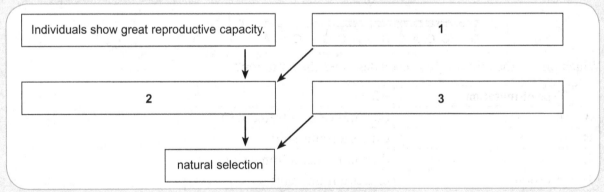

Which statements should be placed in boxes **1**, **2** and **3**?

	Populations remain constant.	Variation is shown in all populations.	There is a struggle for existence.
A	3	1	2
B	1	3	2
C	2	3	1
D	3	2	1

continued ...

5 Which of the following graphs shows a population which has been subjected to disruptive selection?

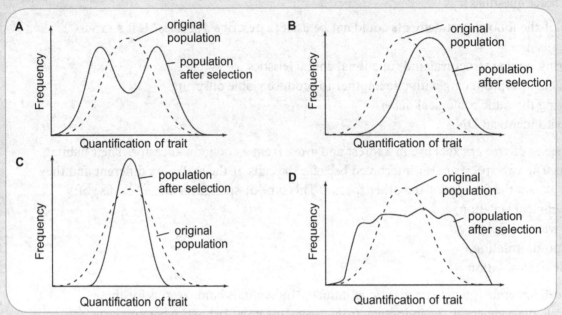

6 Down's syndrome is caused by a chromosome mutation. It arises from:

    A non-disjunction of chromosome 21 in mitosis during the formation of the egg.

    B non-disjunction of chromosome 21 in meiosis during the formation of the egg.

    C non-disjunction of chromosome 21 in mitosis during the formation of the sperm.

    D non-disjunction of chromosome 21 in meiosis during the formation of the sperm.

7 The diagram below shows a form of gene mutation.

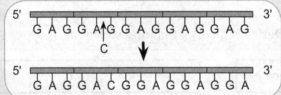

Which one of the following correctly describes the mutation?

	Type of mutation	Effect
A	inversion	does not cause a frame shift
B	deletion	causes a frame shift
C	substitution	does not cause a frame shift
D	insertion	causes a frame shift

continued ...

8 The diagrams below show four types of chromosomal mutation.
Which of the diagrams represents a translocation?

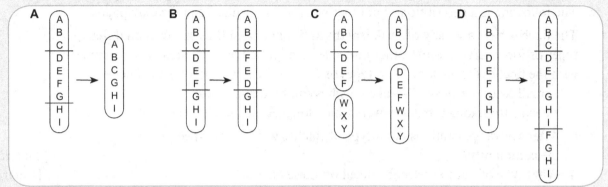

9 Which of the following introduces new variation into a gene pool?
A crossing over   B random fusion of gametes   C gene mutation   D chromosomal inversion

10 Two male and two female finches were blown to an island 200 miles from their home.
This island was uninhabited by finches. They reproduced and a new population of finches
was established. The birds were unable to make the journey back to their original homeland.
When a finch from the original homeland was added to the new population, it could not
interbreed with them. What type of evolutionary event had occurred?
A sympatric speciation   B allopatric speciation   C migration   D artificial selection

Structured questions

11 a Name **three** types of natural selection. [3 marks]
  b Read the passage below on different types of natural selection and answer the
  questions that follow.
  In the Galapagos Islands there are three basic groups of finches. They have a
  variety of beak sizes. One of the groups is the ground finches which range in size
  from small to large. These birds eat a variety of seeds. In 1977, there was a severe
  drought and the vegetation died. Their food source was scarce. The birds quickly
  depleted the small, soft seeds. Only large, tough seeds were left. The large birds
  with deep, strong beaks were equipped to open the hard seeds. Smaller finches died.
  i What selection pressure was operating during the drought year? [1 mark]
  ii If the drought continued for several generations, what type of natural selection
    would act on these finches? [1 mark]
  iii The graph below shows the population of ground finches before the drought.

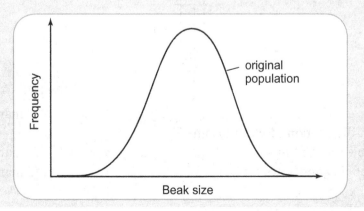

continued ...

Draw a graph to show the population several generations after the drought. [2 marks]

    iv What is the effect of the drought on the median of the curve? [1 mark]

c The following is a description of another population of finches in the Galapagos.

The finches have a variety of beak lengths and shapes and thus can feed on different types of food. After a while, other bird species migrated into the area. They competed with the finches for the food. Two types of food sources were left for the finches:
- small seeds which small-beaked finches can hold and crack
- nectar from long tubular flowers which long-beaked finches can reach.

    i After several generations, what type of finches would you expect in the community? [2 marks]

    ii What type of natural selection acted on these finches? [1 mark]

    iii After several generations, what shape of graph of frequency against beak size would you expect? [1 mark]

d Copy the graph in **b iii** then draw another graph to show what you would expect if:

    i stabilising selection had occurred over several generations [2 marks]

    ii no selection pressure was present over several generations. [1 mark]

12 a What do you understand by the term 'mutation'? [1 mark]

  b Identify **two** types of mutation. [2 marks]

  c One of the ways chromosome mutation occurs is by changes to its structure.
There are different ways this can be achieved.
The diagrams below show one chromosome. Copy and complete each one to show how each type of chromosome mutation may arise.

    i deletion   **ii** duplication

    iii inversion   **iv** translocation

[7 marks]

  d Down's syndrome results from non-disjunction of chromosome 21.
Make an annotated diagram to show how this arises. [3 marks]

  e What is the difference between allopoidy and autoploidy? [2 marks]

continued ...

13 When evolutionary biologist John Endler began studying Trinidad's wild guppies
in the 1970s, he observed a wide variation among guppies from different streams,
even among guppies living in different parts of the same stream. He observed:

- Males from one pool sported vivid blue and orange splotches along their sides,
  while those further downstream carried only modest dots of colour near their tails.
- There were differences in the distribution of guppy predators.
- There were differences in the colour and size of gravel in different stream locations.

He formulated different hypotheses to explain the differences in colour and size of the
males and then simulated situations to test these hypotheses.

a What is a possible hypothesis that could be tested, based on the above observations? **[2 marks]**

b In one simulation, he proceeded as follows:

One hundred guppies, all of which were very drab in colouration at the beginning of the
test, were placed in a pool. In the same pool about 30 rivulus were placed, one of the
least voracious of all guppy predators. These predator fish, which typically occupy
the highest reaches of guppy streams, generally eat only young guppies and do not
distinguish between males and females.

The following results were obtained:

After five generations

number of guppies:	227
number of generations:	5
number of weeks:	112
male colour types:	
brightest:	42%
bright:	7%
drab:	17%
drabbest:	34%

After twelve generations

number of guppies:	232
number of generations:	12
number of weeks:	407
male colour types:	
brightest:	95%
bright:	1%
drab:	2%
drabbest:	2%

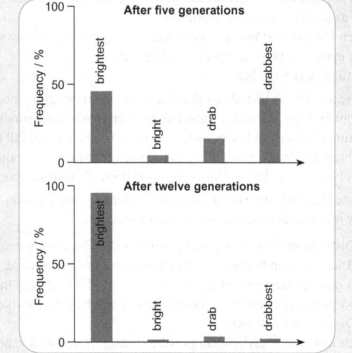

i Explain why the guppy population increased after five generations. **[2 marks]**
ii Describe the composition of the guppy population after 5 and 12 generations. **[5 marks]**
iii Suggest a reason for the change in composition of the guppy population. **[2 marks]**

c Another simulation was done as follows:

One hundred guppies, an even mix of all four guppy types (brightest, bright,
drab, and drabbest) were placed in the pool. Different types of guppy predators
were also placed in the pool: 30 rivulus, 30 acara, and 30 cichlids. Cichlids,
generally only found in the largest, deepest stretches of guppy streams, are the
most voracious of all guppy predators; each may eat as many as four or
five guppies per day.

continued ...

i What do you expect to happen to the numbers of guppies after 5 generations? [1 mark]

The bar graph below shows the results of the simulation after 12 generations.

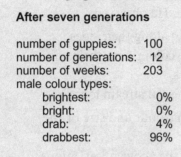

After seven generations

number of guppies: 100
number of generations: 12
number of weeks: 203
male colour types:
    brightest: 0%
    bright: 0%
    drab: 4%
    drabbest: 96%

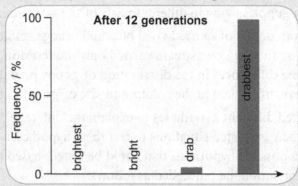

ii Which type of guppy was most prevalent in the pool after 12 generations?
Give a reason for your answer. [2 marks]

d What type of natural selection was operating in both simulations? [1 mark]

Essay questions

14 a Biological evolution can be defined as gradual changes in the genetic composition
of a population with the passage of each generation. Discuss how natural selection
acts as the mechanism of evolution. [7 marks]

b When Darwin and Wallace proposed the origin of species by natural selection,
they made several observations and deductions.

A situation is given below:

There are 3 types of rodents that live on a pine tree: small, medium and large.
To obtain food, the rodents must jump from tree to tree. Small rodents have difficulty
making the jump and large ones tend to be too heavy and fall through the trees as
the branches break. Each year the rodents reproduce and about 400 offspring of
each type are produced. Only about one-third of them survive to reproductive age.

Relate Darwin's observations and deductions to this situation. [5 marks]

c Why is heritable variation important in selection? [3 marks]

15 a i Distinguish between a gene mutation and a chromosome mutation. [2 marks]

ii Briefly explain how gene and chromosome mutations arise. [4 marks]

iii A gene mutation which involves a deletion of a base may have a greater effect
on the polypeptide than substitution of one base for another.
Discuss this statement. [4 marks]

b Sickle cell anaemia is an example of a disease caused by a gene mutation.
Discuss the effect of this mutation on the structure and functions of haemoglobin. [5 marks]

16 a Explain what is meant by the term 'species' according to the biological species concept. [3 marks]

b What are the limitations of the definition in **a**? [3 marks]

c i 'Isolating mechanisms can lead to speciation.' Discuss this statement. [6 marks]

ii A species of wren arrived in an isolated island in the Pacific ocean thousands of
years ago. The island had no wren population previously. Recently, some migrant
wrens arrived on the island and were able to interbreed with the resident wrens.
The offspring however were poorly adapted and produced no offspring.
Explain, with reasons, whether the local wrens and the recently migrated wrens
were the same species. [3 marks]

Chapter 10

Asexual reproduction and vegetative propagation

By the end of this chapter you should be able to:

a explain the term 'asexual reproduction', including a discussion of binary fission, budding, asexual spore formation, fragmentation and one example of asexual reproduction in plants, including the role of meristems;

b describe and explain the principles and the importance of vegetative propagation as exemplified by the production of cuttings and tissue culture, including hormone stimulation;

c discuss the advantages and disadvantages of asexual reproduction;

d discuss the genetic consequences of asexual reproduction.

Reproduction

Reproduction is the production of new organisms from parent organisms. It is necessary because organisms do not live for ever, so that young must be produced to maintain the existence of the species.

There are two fundamental types of reproduction. In **asexual reproduction**, one parent produces one or more new individuals through mitosis. The new individuals are therefore genetically identical to their parent, and to each other. In **sexual reproduction**, one or two parents produce haploid gametes, which then fuse together (fertilisation) to produce a diploid zygote. At some stage in a life cycle involving sexual reproduction, meiosis is involved, and this – as we have seen – results in genetic variation.

In this chapter, we will look at various examples of asexual reproduction. Sexual reproduction is described in Chapters 11 and 12.

Asexual reproduction in prokaryotes – binary fission

Prokaryotes, such as bacteria, reproduce by splitting one cell into two, a process known as **binary fission**. This is not the same as mitosis, because bacterial DNA is circular (unlike the linear chromosomes of eukaryotes) and bacteria do not produce spindle fibres.

Figure 10.1 shows binary fission in a bacterium. First, the circular DNA molecule is duplicated, by semiconservative replication. Often, the DNA is attached to the plasma membrane during this process. The two, genetically identical, circles of DNA are then separated. In some bacteria, this is done by the two DNA circles becoming attached to polymer 'brushes' that form at opposite ends of the cell and pull the two DNA molecules apart. In others, a protein polymer forms between the two DNA molecules, pushing them apart. The parent cell grows in length, and produces a new plasma membrane and cell wall about half way along. This allows it to divide into two complete cells, each with a genetically identical copy of the original cell's DNA.

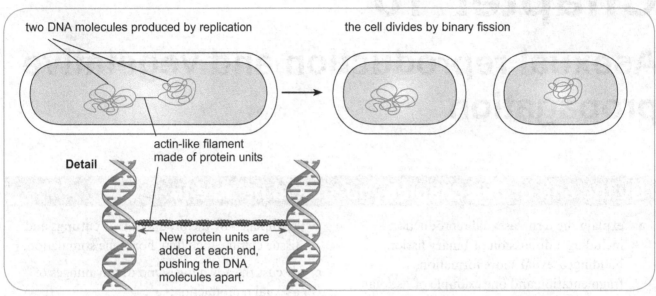

Figure 10.1 Binary fission in a bacterium.

Telomeres and immortality

If organisms were immortal, then reproduction would not be necessary. Why can't organisms live for ever? One answer may be a feature of chromosomes called **telomeres**.

Each chromosome in our cells – and also in the cells of all other eukaryotes so far studied, including protoctists, fungi, plants and other animals – has repetitive sequences of nucleotides at each end called telomeres. In vertebrates, these chromosome ends have many repeats of the nucleotide sequence TTAGGG. In a young human cell, there may be many thousands of nucleotides in each telomere. The DNA in the telomeres does not code for proteins – it does not contain genes.

When DNA is replicated during interphase, the enzymes involved are not able to work right to the very end of the chromosome. Consequently, between 50 to 200 bases at the 3' end remain unpaired. If these were part of functional genes, then each time the cell divided it would lose some genes. The telomeres prevent this happening. Each time the DNA is duplicated, part of each telomere is lost. This doesn't affect the rest of the chromosome, so all the genes can continue to function normally.

However, each time the DNA is duplicated – that is, each time the cell divides – another piece of each telomere is lost. Eventually, the telomeres become so short that the cell can no longer continue to divide without losing important parts of its DNA. In humans, cells containing chromosomes with telomeres shorter than this critical length will not continue to divide. This means that it becomes difficult – perhaps impossible – for tissues to repair themselves as they could in a younger person. There is some evidence that the progressive shortening of telomeres contributes to certain features of ageing, and to the limited lifespan of most organisms.

However, some of our cells do contain an enzyme called telomerase, and this enzyme is able to build new telomeres at the end of chromosomes. Although every cell contains the gene for telomerase, it is only switched on in a small number of cells, including those that are responsible for producing gametes, and in cells in a very young embryo. This means that reproduction 'resets' telomere length. A human baby starts its life with chromosomes with long telomeres, which gradually get shorter and shorter as the person ages.

Asexual reproduction in fungi

Examples of asexual reproduction in fungi include budding in yeast (Figure 10.2) and spore production in *Penicillium* (Figure 10.3).

You may be able to grow some *Penicillium* fungus if you leave a citrus fruit in a warm, damp place for a while. The fungus looks like a blue-green powder on the surface of the fruit (Figure 10.4).

Most fungi are made up of long, thin threads called **hyphae**. All of the hyphae, which branch and intertwine with one another, make up the body of the fungus, called the **mycelium**. The mycelium lies within or on the surface of whatever the fungus is growing and feeding on. In *Penicillium*, each hypha is made up of a chain of haploid cells divided by cross walls. The fungus grows by producing new cells at the ends of its hyphae.

Some hyphae grow upwards, into the air, and these are known as aerial hyphae. At the tips

Figure 10.4   *Penicillium* growing on a lemon. The blue-green colour is due to the millions of tiny spores that it has produced.

of these hyphae, branches called **conidiophores** ('bearers of conidia') are formed. These produce strings of blue-green spores, by mitosis. The spores are called **conidia**, or conidiospores. As the parent cells are haploid, the conidia are also haploid.

The spores are protected by a waterproof covering and are easily spread by air currents. If a spore lands on a suitable substrate, it will germinate and then divide by mitosis to produce a new mycelium.

Penicillium is also able to produce another type of spore, called ascospores, by a sexual process.

Fragmentation in starfish

Many different kinds of organism are able to grow new individuals simply by splitting parts off from their bodies. We don't find it surprising that plants can do this, but it is also a process used by numerous animal groups, including cnidarians, annelid worms and starfish.

Starfish belong to the phylum Echinodermata. They have bodies with a five-fold (pentameral) symmetry. Many starfish have five or ten arms that radiate out from a central disc, in which there is a single opening to the digestive system, on the animal's lower surface. Starfish move using hundreds of tube feet that are connected to a series of tubes containing liquid – a kind of hydraulic

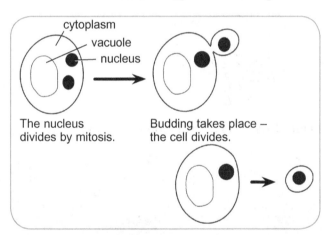

Figure 10.2   Budding in yeast.

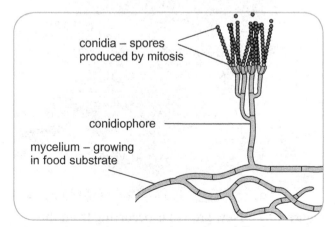

Figure 10.3   The production of conidia (asexual spores) in the fungus *Penicillium*.

system in which muscle contraction causes pressure changes that cause the tube feet to move.

Although starfish generally reproduce sexually, some can also reproduce asexually by allowing one arm, or part of an arm, to break off. The parent grows a new arm to replace the lost one, and the part that has broken off grows into a complete new individual (Figure 10.5).

Figure 10.5 A new starfish, which will end up with six arms, is growing from an arm that has broken off from another individual. All of the new cells are produced by mitosis, so the new animal will be genetically identical to its parent.

Vegetative reproduction in ginger

Plants are almost all able to reproduce sexually, through gametes that are produced in flowers, and this is described in Chapter 11. However, many plants can also reproduce asexually. There is a multitude of different ways in which they do this, many of which are used by humans to propagate plants that we grow as crops or in gardens.

Ginger, *Zingiber officinale*, can reproduce using its underground stems, called **rhizomes**. These swollen structures contain food reserves, which the plant uses to tide it over during periods when it cannot grow and photosynthesise.

A plant stem has a series of places from which leaf stalks (petioles) grow, and these are called nodes. The length of stem between nodes is an internode. Buds, called axillary buds, occur in the axils (angles) of the leaf stalks. Each bud contains a group of cells that are able to undergo mitosis for growth. Such a group of cells is known as a **meristem**.

If a piece of rhizome is broken off, a new plant is able to grow from the buds that occur along the rhizome (Figure 10.6 and Figure 10.7). The roots that grow are **adventitious roots**, because they grow straight out of a stem.

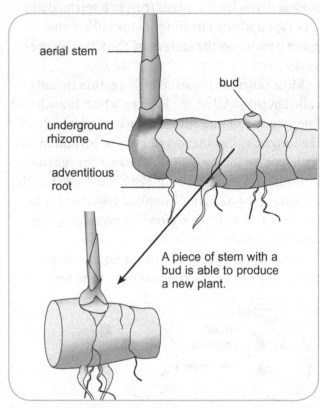

Figure 10.6 Vegetative reproduction in ginger.

Figure 10.7 New growth is occurring from this ginger rhizome. If the branches are broken off and planted, they will grow into two new plants.

Meristems

In animals, including humans, cell division and growth can occur almost all over the body. However, this is not the case in plants, where most cell division takes place in meristems.

Figure 10.8 shows the position of the main meristems in a plant, and Figure 10.9 shows some of these in more detail. If you were able to produce and stain a root tip squash when you were studying mitosis, then you have already seen some cells in the root tip meristem. The cells that divide by mitosis at meristems are undifferentiated. Differentiated (specialised) plant cells cannot divide. Meristematic cells are small, with relatively small amounts of cytoplasm compared to the size of their nucleus, and they have very small vacuoles. They tend to be packed closely together.

Some of the new cells that are produced at a meristem remain as meristematic cells, ready to divide to produce more new cells. Some of them,

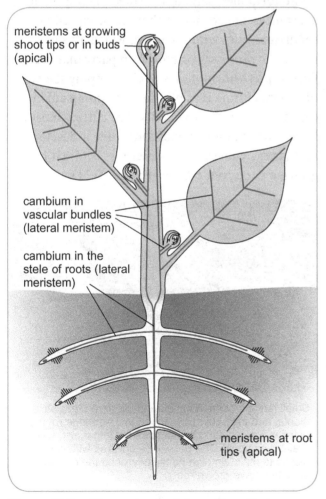

Figure 10.8 The main meristems in a plant.

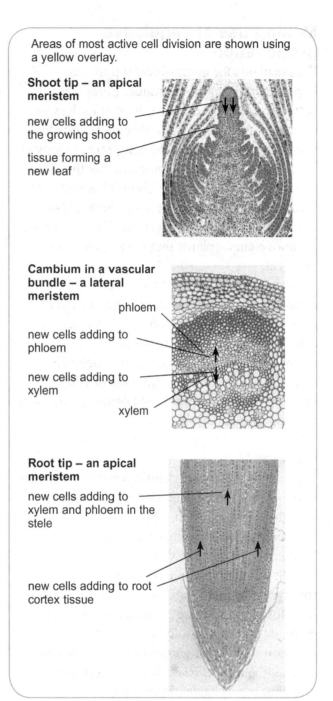

Areas of most active cell division are shown using a yellow overlay.

Shoot tip – an apical meristem

new cells adding to the growing shoot

tissue forming a new leaf

Cambium in a vascular bundle – a lateral meristem

phloem

new cells adding to phloem

new cells adding to xylem

xylem

Root tip – an apical meristem

new cells adding to xylem and phloem in the stele

new cells adding to root cortex tissue

Figure 10.9 The structure of the apical and lateral meristems in a plant.

however, change in size, shape and structure to become specialised for a particular function. This is called **differentiation**. It involves the switching on of a particular set of genes that causes the cell to produce a set of proteins which, in turn, affect its structure and metabolism. For example, some may become xylem vessels or phloem sieve tubes, while others may develop chloroplasts and become photosynthetic.

Making use of vegetative propagation

We use plants for many different purposes – in particular as food, to produce fibres (e.g. cotton) or to make our environment look more attractive. In general, when we breed plants for these purposes, we want the new plants to look as similar as possible to the originals. We therefore tend to propagate many of them using methods involving asexual reproduction, which is also known as vegetative propagation. These methods produce **clones** – plants that are all genetically identical.

Some of these methods have been used for hundreds of years. For example, humans must have propagated plants from **cuttings** a very long time ago. More recently, **tissue culture** has provided a way of producing thousands of genetically identical plants from tiny amounts of starting material. Tissue culture is used commercially on a very large scale.

Cuttings

For many plants, taking cuttings is extremely easy to do – you simply break a piece from the plant and push it into the soil. The piece of the plant – the cutting – eventually grows new roots and becomes a complete new plant.

Today, the production of cuttings is done on a small scale by many gardeners, and on a massive scale for plants that are grown as commercial crops, or that can be sold in large numbers as decorative plants for gardens and in homes. Sugar cane, *Saccharum officinarum*, is an example of a crop that is propagated by cuttings.

To propagate sugar cane, stems or pieces of stem are laid on the ground (Figure 10.10). The stem is cut into several pieces, the length of which varies according to different customs. Each piece of stem must contain at least one node and bud – usually, they contain three or more. The pieces of stem are laid horizontally in rows in a field and then covered with a thin layer of soil. Within days, new shoots grow upwards from the buds, and new underground stems (rhizomes) grow through the soil and produce roots. Other plants may not grow from cuttings as readily as sugar

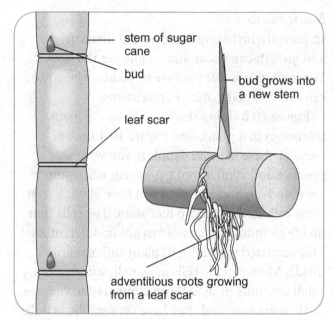

Figure 10.10 Propagating sugar cane by cuttings of sections of the stem.

cane, and for these **plant hormones** may be used. Plant hormones, sometimes called plant growth substances, are chemicals that are produced in small quantities within a plant, which affect metabolic processes or growth in particular areas. One of these hormones is **auxin**. Dipping the base of a stem cutting into powder containing auxin can stimulate it to grow roots. The cutting must be kept in warm, moist conditions while it develops, as it is very vulnerable to water loss before it has grown a good root system (Figure 10.11).

Small lengths of stem have been cut from *Gazania* plants, and their cut ends dipped into hormone rooting powder. Each cutting is then inserted into moist, well-drained compost. The pot will be covered with clear plastic to maintain high humidity around the cuttings until they have grown roots.

Figure 10.11 Planting stem cuttings.

Stems are not the only parts of plants from which cuttings can be taken. For example, *Passiflora* (passion flower) can be propagated from root cuttings, and *Saintpaulia* (African violet) from leaf cuttings (Figure 10.12).

Figure 10.12   African violets (*Saintpaulia* sp.) can be propagated from leaf cuttings. The cutting on the left has been dipped into rooting hormone, which has stimulated it to produce roots much more quickly than the untreated leaf on the right.

Tissue culture

Tissue culture is a technique that allows very large stocks of a particular variety of plant to be built up quickly and relatively cheaply. It can be carried out at any time of year, in any country, as it takes place inside a laboratory. Plants are only moved outside or into the soil in glasshouses when they are well developed.

When a plant is wounded, cells close to the wound divide by mitosis to produce a **callus** – a fairly shapeless bundle of cells that seals the wound and prevents the entry of fungi or other potential pathogens. Tissue culture involves stimulating cells to behave in a similar way to callus cells. Plant hormones are then used to stimulate the bundle of cells to differentiate to form the variety of different cell types which form themselves into root, stem and leaf tissues.

Plant tissues differ from animal tissues in the ability of cells to produce other cell types. Mammalian tissues have limited ability to do this. In an adult, this ability is restricted to a few cell types like stem cells. However, plant tissues are **totipotent** – all the genetic information in some cells in any tissue can be used to produce different tissues.

Tissue culture begins with the removal of a small group of cells from a plant (Figure 10.13). We have seen that not all plant cells are able to divide by mitosis, and those that do are called **meristematic cells**. However, a few meristematic cells are found in all plant tissues. Experience has enabled knowledge to be built up of the best source of cells to use in different plant species and varieties. The group of cells that is removed is called an **explant**. Care must be taken to ensure that the explant does not contain any cells infected by viruses.

There are various ways in which the explant can now be treated. A common procedure is to immerse it in a well-aerated solution containing a balance of hormones, in particular **auxin** and **cytokinin**, which stimulate cell growth and division. The solution also contains nutrients needed by the cells (which will not be able to photosynthesise) – such as sucrose as an energy source, and inorganic ions such as potassium, magnesium and nitrate – and is sterile. It is very important to maintain sterile conditions throughout the procedure, as the

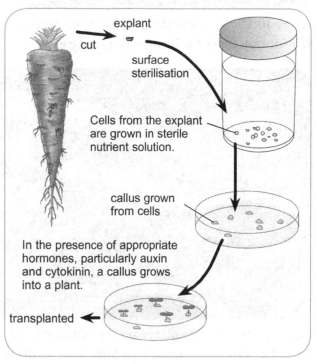

Figure 10.13   Plant tissue culture.

tiny bundles of plant cells are highly vulnerable to attack by fungi or bacteria. The explants are usually dipped into a disinfectant solution immediately after removal from the parent plant, and sterile techniques are used during their handling and transfer into the sterile nutrient solution.

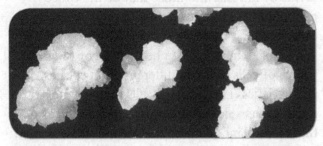

Figure 10.14   Calluses derived from onion cells.

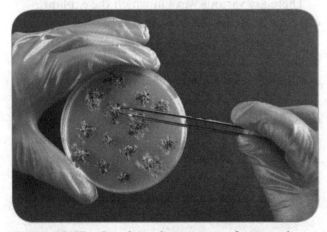

Figure 10.15   Sundew plants grown from explants. Why do you think the handler is wearing gloves?

The explants can be kept in this solution for long periods of time, and these time periods can be extended almost indefinitely if a few cells are taken out every now and then and grown in a fresh culture solution. The undifferentiated cells in the explants divide repeatedly by mitosis, producing a callus (Figure 10.14). Every now and then, some of the cells can be harvested for growing on into new plants. A complete callus could be used, or it could be subdivided to produce several smaller ones.

The calluses are placed onto sterile agar jelly (support medium) containing nutrients and a particular mix of plant growth substances that stimulates them to grow roots and shoots (Figure 10.15). The particular growth substances that are used, and their relative concentrations, have been found by trial and error but they generally include auxins and cytokinins, which affect cell division and cell differentiation. The mixture varies from species to species and may be changed at different stages during growth – one might be used to stimulate production of roots, and then a different mix to stimulate production of shoots.

When the plant is large enough, it is transplanted into sterile soil. From now on, it should grow into a complete new plant (Figure 10.16). The biggest danger is infection by fungi, and it is important to maintain sterile conditions until the plant can fend for itself.

Figure 10.16   Orchid plants are very difficult to propagate by any means other than tissue culture. Now they can be produced on a huge scale, as at this orchid nursery in Bangkok, Thailand. It is also easy to airfreight the small plants, still in their sterile environment, to other countries.

Advantages and disadvantages of asexual and sexual reproduction

Many organisms are able to reproduce either asexually or sexually, and do both at different stages in their life cycles. This must mean that both methods have advantages to the organism, and that perhaps the different methods each come into their own under particular circumstances.

When thinking about the advantages and disadvantages of sexual and asexual methods of reproduction, it is important to be clear in your mind about whether you are discussing their value to *the species of organism* or to a human who is propagating that organism for their own purposes. We will deal with the first of these situations – the biological advantages and disadvantages to a particular organism of each of these methods of reproduction. We will concentrate on plants, as it is in this group of organisms that many species are able to reproduce both asexually and sexually. However, you may also like to think about other groups of organisms, and whether or not each of the features discussed here also applies to them.

Asexual reproduction

The fundamental feature of asexual reproduction is that it produces clones – groups of genetically identical organisms. The great advantage of this is that if an individual is already well adapted to its environment, then all of its offspring will have identical genotypes and will be equally well adapted. Plants therefore tend to use asexual reproduction when they are spreading across an area in which they are already growing successfully. However, as we shall see, there are circumstances in which genetic stability is not what is required.

Asexual reproduction can also be especially useful if a plant has just colonised a new area. It is possible that only one individual has arrived there, so if it is going to reproduce it has to do it on its own. This could mean that asexual reproduction is a good option, although it is important to remember that sexual reproduction can also take place with only one parent, if that parent is able to produce both male and female gametes (as many plants do).

Another advantage of asexual reproduction is that the young plants can use their parent as a source of nutrients until they have developed their own strong root and shoot systems. Only then do they need to break away from the parent and begin to live an independent life.

Sexual reproduction

The fundamental feature of sexual reproduction is that it produces genetic variation among the offspring. As we saw in Chapter 9, it is this variation that provides the foundation for change through natural selection.

Imagine, for example, that a new fungal disease arrives in an area. The plants of a particular species are susceptible to it, and many are infected and die. However, one or two of them just happen to have alleles that confer resistance to the infection, and these survive and reproduce, allowing the species to continue. If this species had only reproduced asexually, then every individual may have been susceptible, so the entire species could have been wiped out by the fungus.

This is not to say that absolutely no variation occurs with asexual reproduction. You saw in Chapter 9 that mutations can bring about random changes in genes, and these can happen at any time. However, sexual reproduction also provides for variation through independent assortment, crossing over and random fertilisation, none of which occur during asexual reproduction.

In plants, asexual reproduction generally produces new plants that grow in close proximity to the parent. Sexual reproduction, however, produces seeds inside fruits (page 221) and these are generally adapted to be dispersed away from the parent. This reduces competition with the parent plant and with each other, and also increases the chance of the species colonising new areas. The genetic variability between the seeds means that some of them may even be able to grow successfully in an environment that is not the same as the one in which the parent plant was growing.

Another big advantage of seeds is that they are often able to survive through long periods of adverse conditions, such as extreme cold or drought. In climates where there are long, dry summers or cold winters, seeds can lie dormant through these times of year, only germinating when suitable environmental conditions return.

One potential disadvantage of sexual reproduction is that two parents may be needed, and that male gametes must somehow travel to a female gamete before fertilisation can take place. Many plants, however, have flowers that produce both male and female gametes, so they are able to fertilise themselves if no male gametes from elsewhere manage to reach their female gametes.

This situation is rarer in animals, and there have been several cases of the numbers of animals of an endangered species dropping so low that it became almost impossible for a last remaining male or female to find a mate.

Saving the black robin

The black robin, *Petroica traversi*, lives on the remote Chatham Islands in the South Pacific. In 1970, there were only 20 black robins left, all on a tiny island called Little Mangere. That year, only one chick was produced and survived. The chick was female, and she was caught by conservationists and taken to another, larger, island, Mangere, in the hope that a new population could begin to develop there. She was named Old Blue.

By 1980, there were only five black robins on Little Mangere, and it was thought they were probably too old to breed. However, just in case, they were all moved to Mangere – and one of them did indeed become a mate for Old Blue. Old Blue laid two eggs, both of which were taken away by the conservationists and placed in the nest of a much commoner bird, a tom tit, on a third island, called South East. The tom tit looked after the eggs as if they were her own, and successfully raised two black robin chicks. Old Blue continued to lay more eggs, some of which were also removed and reared in the same way.

Once the black robin chicks reared by the tom tits were large and strong enough, some were returned to Mangere Island and some released on South East, to live their own lives. Today, there are more than 250 black robins in existence, all descended from the single pair of black robins on Mangere. Without the intervention of humans, this species would almost certainly be extinct today.

SAQ

1 Use the information on pages 205 to 206 to construct a list of bullet points summarising:
  a the advantages of asexual reproduction
  b the disadvantages of asexual reproduction
  c the advantages of sexual reproduction
  d the disadvantages of sexual reproduction.

2 Construct a similar list, but this time thinking out for yourself the advantages and disadvantages, *to a human*, of propagating plants (or animals) either asexually or sexually.

Summary

- Asexual reproduction involves the production of new individuals by mitosis from a single parent, producing groups of organisms that are genetically identical, called clones.

- Prokaryotes, including bacteria, reproduce by binary fission. The circular DNA is duplicated and the two circles pushed or pulled to either end of the cell. A new plasma membrane and cell wall grows across the centre of the cell, which splits into two.

- A spore is a cell or group of cells surrounded by a protective coat. Fungi, for example, can produce spores by mitosis. Spores can survive adverse conditions, and can be transported over quite long distances, helping to disperse the fungus.

- Some species of multicellular organisms, including some plants and animals, can reproduce asexually by splitting off part of their bodies. Starfish, for example, can grow an entire new animal from an arm shed from a parent individual.

- Many plants commonly reproduce asexually. For example, ginger grows new plants from buds along its underground stems (rhizomes).

- In plants, cell division only occurs in areas called meristems. These are found, for example, in the apex (tip) of the shoot and root, and in the buds in the leaf axils.

- Humans use vegetative propagation to generate large numbers of new, genetically identical plants. Cuttings can be taken from roots, shoots or leaves. In some species, these grow roots very readily, but in others it may be helpful to dip the lower end of the cutting into auxin to stimulate root production.

- Tissue culture takes very small pieces of tissue from a parent plant and induces new growth by exposing it to different plant hormones at different stages. Very large numbers of plants can be produced from very small amounts of tissue.

- Asexual reproduction is advantageous in that it produces genetically identical individuals that will be as well adapted to their environment as their parent. However, in some circumstances this may not be an advantage – for example, if a new environmental factor (e.g. a new disease) arises, or if the plant is spreading into a different environment.

Questions

Multiple choice questions

1 Many unicellular organisms reproduce by the processes of:
  A fission and budding.
  B regeneration and ovulation.
  C cuttings and seed formation.
  D sperm and egg formation.

2 A mass of dividing, undifferentiated cells in tissue culture is called an:
  A an explant.   B a callus.   C an embryo.   D a meristem.

continued ...

3 Some plants can produce new individuals by cuttings and without seed formation. This type of asexual reproduction is called:

    A budding.   **B** vegetative reproduction.   **C** spore formation.   **D** fragmentation.

4 Which of the following **best** describes asexual reproduction?

Asexual reproduction is the formation of new individuals from:

    A a single individual with the involvement of gametes.

    B a single individual without the involvement of gametes.

    C two individuals with the involvement of gametes.

    D two individuals without the involvement of gametes.

5 The diagram to the right shows a potato which is developing into a plant. What method of reproduction does this represent?

    A sporulation.

    B binary fission.

    C vegetative propagation.

    D fragmentation.

6 Which of the following correctly describes the diagram to the right?

	Type of reproduction	Example	Example of organism
A	sexual	seed formation	cactus
B	sexual	spore formation	mushroom
C	asexual	spore formation	bread mould
D	asexual	seed formation	*Hibiscus*

7 Which of the following pairs of statements about the advantages and disadvantages of asexual reproduction is **true**?

	Advantage	Disadvantage
A	Allows for rapid production of new individuals.	Lacks genetic variation to adapt to a new environment.
B	Large amount of spores are produced.	Less competition with parent plants for resources.
C	Only one parent is required.	Formation of new gametes.
D	Lacks genetic variation to adapt to a new environment.	More competition with parent plants for resources.

8 Plant cells can be grown in tissue culture and regenerate new plants because the individual cells:

    A have all the DNA needed to make a new plant.

    B can differentiate into seeds.

    C are meristematic.

    D can only express certain genes.

9 Micropropagation or tissue culture is described as:

    A growing plants in greenhouses.

    B culturing whole plants from explants.

    C growing crops on a small scale.

    D culturing whole plants from spores.

continued ...

10 The diagram gives an overview of tissue culture.

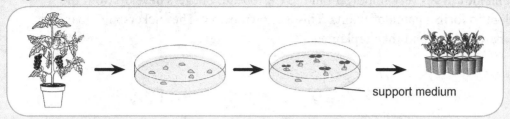

The most commonly used semi-solid support for the plantlets is:

A auxin.

B cytokinin.

C agar.

D callus.

11 Which of the following is **not** an example of asexual reproduction?

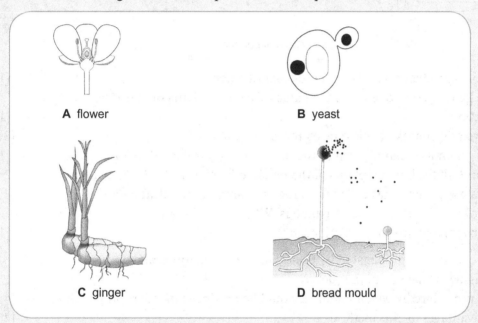

A flower

B yeast

C ginger

D bread mould

Structured questions

12 Tissue culture propagation refers to test-tube methods of culturing whole plants asexually from very small pieces of tissue called explants cut from parent plants. These are placed in a sterile culture in a petri dish containing a special culture medium.

   a What property must the small piece of plant tissue have in order to develop?   [1 mark]

   b Name **three** parts of a plant from which small pieces of tissues can be obtained.   [3 marks]

   c Name **four** substances that the special culture medium contains.   [4 marks]

   d Why is it necessary for the plant tissue cultures to be sterile?   [2 marks]

   e This process is used to produce pineapple plants. How can the process of tissue culture increase the supply of these plants?   [3 marks]

   f As with any process there are advantages and disadvantages. Identify one advantage and one disadvantage of tissue culture or micropropagation.   [2 marks]

continued ...

13 The banana industry is very important in the Caribbean. Suckers develop from the parent plant to form a patch of plants. This is shown below. The suckers are separated from the parent plant and then replanted.

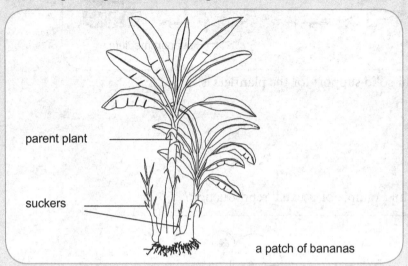

parent plant

suckers

a patch of bananas

a   What type of reproduction is shown in the diagram above?   [1 mark]
b   What is **one** genetic consequence to the patch of banana plants of this type of reproduction?   [1 mark]
c   What are **two** advantages of this type of reproduction?   [2 marks]
d   What is the generalised name given to the patch of plants which develop?   [1 mark]
e   Do you expect all the banana plants in the patch to be identical? Give a reason.   [2 marks]
f   The Guyana government has called for a regional approach to deal with the Black Sigatoka, a deadly disease of bananas. Why would disease progress quickly among the banana plants in the Caribbean?   [1 mark]

Banana plants can also be propagated by tissue culture or micropropagation. Meristematic tissue is taken from the sucker.
g   Outline the procedure by which plantlets would be produced by micropropagation.   [4 marks]
h   Give **two** advantages of this method.   [2 marks]

Essay questions

14 a   In the process of tissue culture, meristematic tissue is grown on a culture medium to form a callus. What do you understand by the terms 'meristematic' and 'callus'?   [6 marks]
b   Discuss the process of cloning plants from tissue culture including the effects of auxins and cytokinins.   [9 marks]

15 a   Differentiate between asexual reproduction and vegetative reproduction.   [2 marks]
b   Living organisms show a range of methods of asexual reproduction. Discuss, with examples:
     i   binary fission
     ii spore formation
     iii fragmentation
     iv cuttings from plants.   [9 marks]
c   Discuss the advantages and disadvantages of asexual reproduction.   [3 marks]
d   Suggest a reason why mammals do not reproduce asexually.   [1 mark]

Chapter 11

Sexual reproduction in the flowering plant

Sexual reproduction

Sexual reproduction is a type of reproduction that involves specialised sex cells, called **gametes**. Gametes are usually haploid. During sexual reproduction, the nuclei of two gametes fuse together in a process called **fertilisation**. The product of fertilisation is a diploid cell called a **zygote**. The zygote divides repeatedly by mitosis to produce an **embryo**, which eventually develops into a mature individual.

In most species of organism, the gametes are of two types. One is relatively small and moves. This is the male gamete. The other type is relatively large and does not move. This is the female gamete.

The production of gametes usually involves meiosis, which introduces genetic variation between the gametes. This variation is increased when gametes containing different combinations of alleles fuse together.

In this chapter, we will look at sexual reproduction in a major group of plants, the flowering plants or **angiosperms**. Sexual reproduction in humans is covered in Chapter 12.

Flowering plant reproduction – an overview

It may be helpful to step back and take a broad overview of sexual reproduction in a flowering plant, before we go into the various stages in more detail.

The reproductive organ of a flowering plant is the flower (Figure 11.1). Here, the male gametes and female gametes are made, and fertilisation takes place.

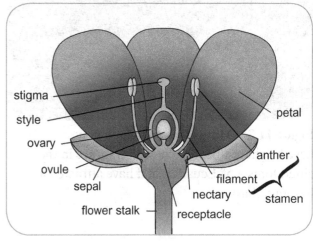

Figure 11.1   The basic structure of a flower.

The male gametes consist of haploid nuclei, inside pollen grains that are produced in the anthers of the flower. (It is important to realise that the pollen grains are not themselves the male gametes, but *contain* the male gametes.) The female gametes are haploid nuclei inside a structure called an embryo sac, found inside an ovule.

Unlike the male gametes of animals, the male gametes of flowering plants are not motile – that is, they cannot move by themselves. Plants therefore depend on outside agencies to transport their male gametes, inside pollen grains, from one flower to another. This is called **pollination**, and it results in the deposition of pollen grains onto a stigma.

The pollen grain then germinates, producing a tube that grows towards an ovule and penetrates an embryo sac. The male gametes inside the pollen grain pass down the tube and fuse with the female gametes inside the embryo sac. This produces a zygote, which divides by mitosis to produce an embryo plant. The ovule becomes a seed, containing this embryo.

Figure 11.2 Many bird-pollinated flowers have bright red petals. These flowers are in the cloud forests of Ecuador, and have attracted a hummingbird.

Flower structure

If you look carefully at a flower, you should be able to see that the different parts are arranged in concentric rings, called **whorls**, attached to the top of the flower stalk. The part to which they are attached is the receptacle. These parts are always arranged in the same order, shown in Figure 11.1.

Sepals and petals

The outer whorl is the **calyx** and is made up of the **sepals**. In many flowers, these are dull in colour, and their main function is to protect all the other parts of the flower when it is still a bud. The next whorl is the **corolla**, made up of **petals**. In insect-pollinated or bird-pollinated flowers these are often brightly coloured and scented, as their function is to attract insects or birds to the flower. The petals advertise the presence of the flower from a distance, and the colour and scent of the petals may determine the kind of insect or other animal which is attracted (Figures 11.2 and 11.3). For example, bird-pollinated flowers are often red. The dark reddish-brown, strangely scented flowers of *Stapelia* smell like rotten meat, and attract flies. In many flowers, nectaries secrete a fructose-rich fluid called nectar, which also helps to attract insects and birds for pollination.

Figure 11.3 Stapelias are often called carrion flowers, because of their smell. They attract flies looking for dead animals in which to lay their eggs, which instead become pollinators for the flowers.

Male and female parts of the flower

The next whorl in from the petals is made up of the male parts of the flower and is called the **androecium**. It contains several – often very many – **stamens**. Each stamen has a stalk called a **filament**, which supports an **anther**. Inside the anthers, the male gametes are formed inside pollen grains. This is described in the next section.

Finally, in the centre of the flower, the female parts of the flower are found. These make up the **gynoecium**. There are one or more **ovaries**, each containing one or more **ovules**, inside which the female gametes develop inside an **embryo sac**. At the top of each ovary is a **style** which supports a **stigma**. The stigma has the function of capturing pollen grains, one of the first stages in the series of events which will bring the male gametes to the female gametes.

Not all flowers have both male and female parts – some are just male, and some are just female. We will look at some examples of these, and their significance, on pages 219–220.

Formation of gametes

Male gametes

Male gametes are formed inside pollen grains. Pollen grains are formed inside the anthers (Figure 11.4 and Figure 11.5).

Each anther contains four compartments called **pollen sacs**. The wall of each pollen sac contains several layers of cells. One of the outer layers is made up of cells with thickened walls and is called the **fibrous layer**; this helps to liberate the pollen grains when they are ripe. The innermost layer is called the **tapetum**. The cells in this layer help to provide nutrients to the developing pollen grains.

In the centre of each pollen sac, diploid **pollen mother cells** divide by meiosis, each producing four haploid cells. In some species, these stay together in a group of four called a **tetrad**, but in others they separate. Each of the haploid cells develops a tough, protective wall around itself, becoming a **pollen grain**. The wall is made up of two layers, an outer, very tough, waterproof **exine**, containing sporopollenin, and an inner **intine**. In places, the

exine is absent, leaving a thin area in the wall called a **pit**. The form and structure of the exine varies from species to species, and it is possible to identify a plant just by looking at its pollen grains. They often have spikes or knobs to help them to stick to the bodies of insects or birds.

The haploid nucleus inside each pollen grain divides by mitosis, forming two haploid cells separated by a very thin cell wall. One of the haploid nuclei is called the **generative nucleus**, and the other is the **tube nucleus**. (At a later stage, after the pollen has landed on a stigma, the generative nucleus will divide by mitosis to produce two **male gamete nuclei**.)

When the pollen grains are fully formed, the anthers split open in a process called **dehiscence**. They split along a line between the two pollen sacs on either side (Figures 11.5 and 11.6), exposing the pollen grains on the surface. Before following the pollen grains further, we will look at how the female gamete-forming structures develop to a similar stage.

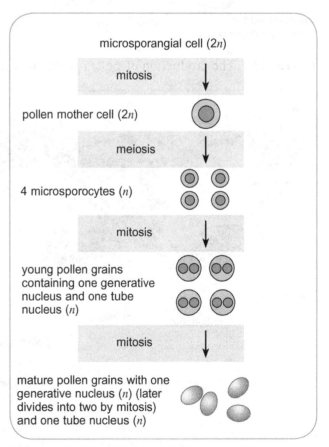

Figure 11.4 Stages in the production of pollen grains.

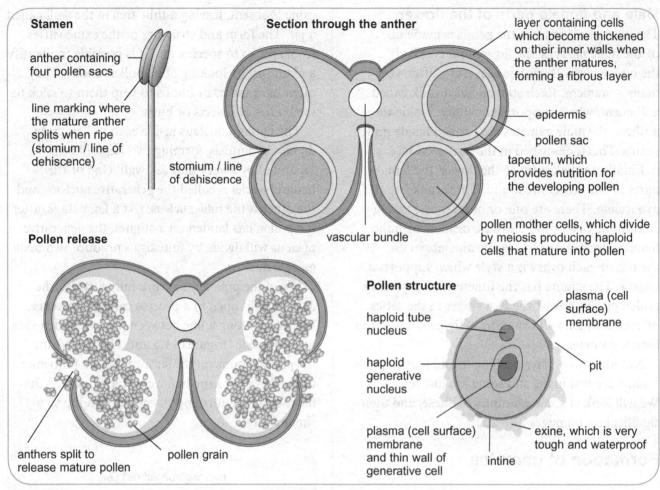

Stamen

anther containing four pollen sacs

line marking where the mature anther splits when ripe (stomium / line of dehiscence)

stomium / line of dehiscence

Section through the anther

layer containing cells which become thickened on their inner walls when the anther matures, forming a fibrous layer

epidermis

pollen sac

tapetum, which provides nutrition for the developing pollen

vascular bundle

pollen mother cells, which divide by meiosis producing haploid cells that mature into pollen

Pollen release

anthers split to release mature pollen

pollen grain

Pollen structure

haploid tube nucleus

haploid generative nucleus

plasma (cell surface) membrane and thin wall of generative cell

intine

plasma (cell surface) membrane

pit

exine, which is very tough and waterproof

Figure 11.5 The production of pollen grains in an anther.

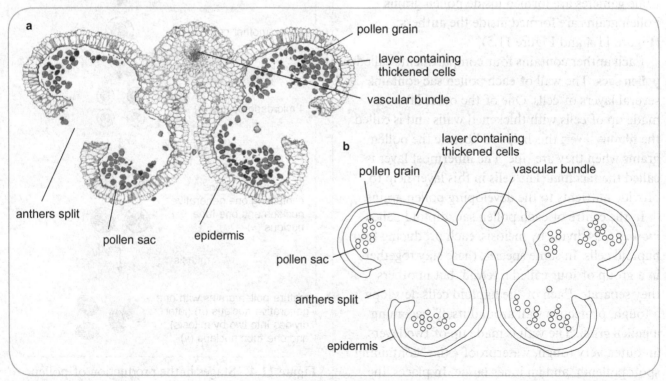

a

pollen grain

layer containing thickened cells

vascular bundle

anthers split

pollen sac

epidermis

b

layer containing thickened cells

pollen grain

vascular bundle

pollen sac

anthers split

epidermis

Figure 11.6 **a** Electron micrograph of a mature lily anther; **b** drawing of the electron micrograph in **a**.

1 Construct a flow diagram to show how the male gametes are produced in a flower.

Female gametes

The female gametes are produced inside structures called **embryo sacs** which develop inside the ovules from megasporangial cells (Figure 11.7).

Ovules are found inside ovaries. Each ovule is connected to the ovary by a stalk called the **funicle**. The ovule has an outer covering, or **integuments**, surrounding a tissue made up of relatively undifferentiated cells called the **nucellus**. At one end of the ovule, the integuments do not quite meet, leaving an opening called the **micropyle**. The other end of the ovule, furthest from the micropyle and nearest to the funicle, is called the **chalaza**.

Inside each ovule, a large, diploid, **megaspore mother cell** develops. This cell divides by meiosis to produce four haploid cells (megaspores). All

but one of these degenerate, and the one surviving haploid cell then develops into an **embryo sac**.

The embryo sac absorbs nutrients from the nucellus and grows larger. Its nucleus divides by mitosis three times, forming eight haploid nuclei. Two of these nuclei are found near the centre of the embryo sac. The other six arrange themselves at the ends of the embryo sac, three at one end and three at the other. They usually develop cell membranes. The three haploid cells at the end nearest the chalaza are called **antipodal cells**. One of the haploid cells at the end nearest to the micropyle is a little larger than the other two. This is the **female gamete**, the **egg cell**. The other two cells at this end are called **synergids**. The two nuclei in the middle may fuse together to form a single diploid nucleus called the **primary endosperm nucleus**. Thus, a mature embryo sac usually contains six haploid nuclei and one diploid nucleus (Figure 11.8).

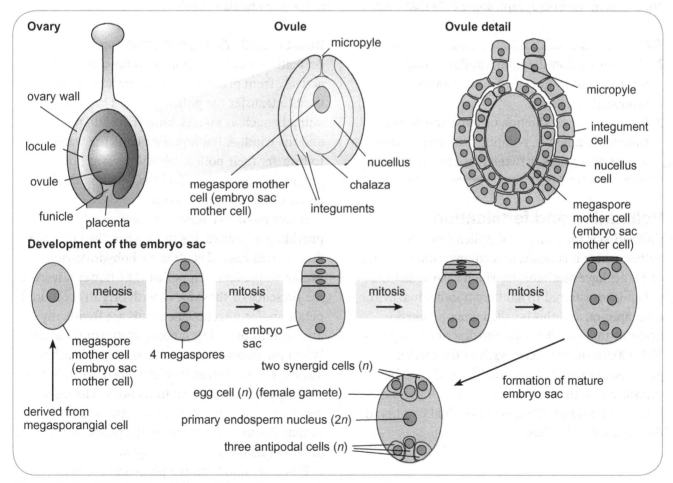

Figure 11.7 The development of the embryo sac (megaspore).

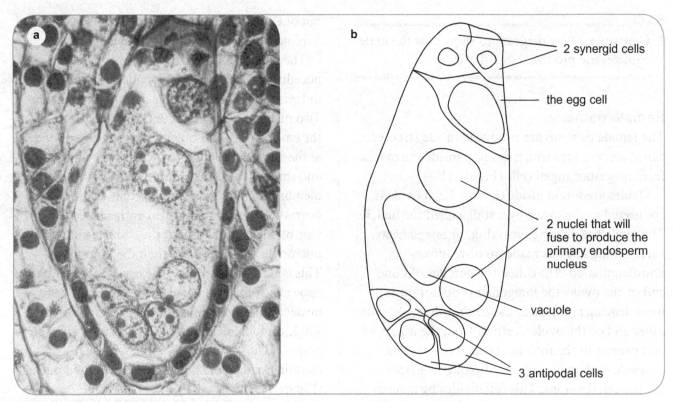

Figure 11.8 **a** Light micrograph of a mature embryo sac; **b** drawing of the electron micrograph in **a**. Note that in the micrograph some cells that overlie the embryo sac are also visible.

2 Construct a flow diagram, similar to that for SAQ **1**, to show how the female gamete is produced in a flower.

3 Compare the production of male and female gametes in a flower, pointing out similarities as well as differences between these two processes.

Pollination and fertilisation

Pollination is the transfer of pollen from the anther, where it is made, to a stigma. Many species of plants have mechanisms which ensure that the pollen is transferred to a different individual of the same species, and this is called **cross-pollination**. Some of these mechanisms are described on pages 219–220. In other species, such as the garden pea (*Pisum sativum*), it is usual for pollen to be transferred to the stigma of the same flower, or at least to a flower on the same individual plant, and this is called **self-pollination**.

Insect- and wind-pollination

As neither plants nor pollen grains can move actively from place to place, other agents are used to transfer the pollen grains. These include animals, such as insects, birds, bats and mice, and the wind. A few aquatic plants use water to transfer their pollen. We will look at how pollination happens in an insect-pollinated flower and a wind-pollinated flower.

Insect-pollinated flowers attract insects by providing a 'reward' for them when they visit the flower. This reward is often carbohydrate-rich nectar, or protein-rich pollen. The flower advertises the presence of these foods with brightly coloured petals and/or a scent. The petals are frequently arranged to provide a landing platform for insects. When the insect arrives at the flower, it brushes against the anthers as it collects its reward. Some of the pollen grains stick to its body. The insect flies away, and will often go straight to another similar flower. Here, some of the pollen grains may brush off its body onto the stigma.

It is clear that both the plant and the insect benefit from this arrangement. Insects and flowers have evolved together over millions of years, and this has resulted in some very elaborate pollination mechanisms. Some flowers, such as orchids, can only be pollinated by one species of insect (Figure 11.9).

Figure 11.9 A wasp is pollinating this orchid. The orchid produces a smell which is a sex hormone (pheromone) that makes the wasp want to mate with the flower as though it were a female wasp. In the process, the wasp passes pollen from one orchid to another.

Wind-pollinated flowers have no need to attract insects, so they do not waste resources in producing large, brightly coloured petals or nectar. They are usually relatively small flowers, and are often held on long stalks so that they can easily catch the wind. Figure 11.10 shows the structure of a typical grass flower. The filaments are long and dangle freely out of the flower; they are very flexible and move easily in the wind, shaking the pollen free from the anthers. Huge quantities of tiny, lightweight pollen grains are produced, and you can often see clouds of this pollen floating up from flowering grasses if you brush against them. The stigmas are long and feathery, and protrude from the flower. Their large exposed surface increases their chance of catching pollen grains floating on the wind.

Fertilisation

Pollination results in the arrival of pollen grains on to the stigma. The pollen grains stick to the surface of the stigma, absorb water and begin to germinate. This normally only happens if the pollen grain is on the stigma of the same species of flower and – in some species – if it is on the stigma of a different flower of the same species. You can read more about this on page 219.

The contents of the pollen grain push out through one of the pits in the wall, forming a **pollen tube**. The tube grows down through the style towards an ovule. The tube nucleus remains close to the tip of the tube as it makes its way through the style. Digestive enzymes are secreted from the tip of the tube, which is probably directed towards the ovule by chemicals that the ovule secretes (chemotaxis). As the tube grows, the generative nucleus divides by mitosis, forming two haploid **male gametes**.

In most plants, the pollen tube enters the ovule through the micropyle and then the locule close to the ovary wall. In a few species, the pollen tube may digest its way in through the chalaza. Once the tube has penetrated the ovule, the tube nucleus degenerates as its role is completed. Behind it, the two male gametes make their way into the embryo sac.

As you would expect, one of the male gametes fuses with the egg cell, forming a **diploid zygote**.

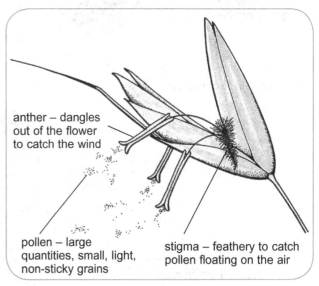

anther – dangles out of the flower to catch the wind

pollen – large quantities, small, light, non-sticky grains

stigma – feathery to catch pollen floating on the air

Figure 11.10 Structure of a wind-pollinated flower.

But, in plants, a **double fertilisation** takes place inside the embryo sac. The other male gamete fuses with the diploid nucleus in the centre of the embryo sac, forming a **triploid nucleus** (that is, one possessing three complete sets of chromosomes). This triploid nucleus is called the **endosperm nucleus** (Figure 11.11).

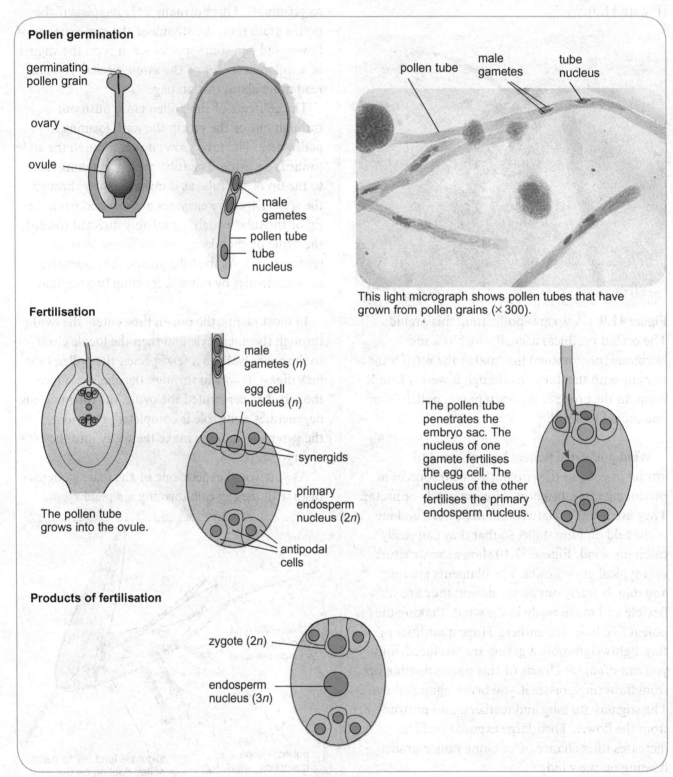

Pollen germination

germinating pollen grain

ovary

ovule

male gametes

pollen tube

tube nucleus

pollen tube

male gametes

tube nucleus

This light micrograph shows pollen tubes that have grown from pollen grains (× 300).

Fertilisation

male gametes (n)

egg cell nucleus (n)

synergids

primary endosperm nucleus (2n)

antipodal cells

The pollen tube grows into the ovule.

The pollen tube penetrates the embryo sac. The nucleus of one gamete fertilises the egg cell. The nucleus of the other fertilises the primary endosperm nucleus.

Products of fertilisation

zygote (2n)

endosperm nucleus (3n)

Figure 11.11 Fertilisation.

Outbreeding and inbreeding

Many species of plant have mechanisms which make it almost impossible for self-fertilisation to take place. In these species, fertilisation can only occur between different plants of the same species, and sometimes only between unrelated plants of the same species. This is called **outbreeding**. In some plant species, however, self-fertilisation or breeding between closely related individuals is possible. This is called **inbreeding**.

Outbreeding increases the amount of genetic variation in the population. Unrelated individuals are more likely to possess different alleles of genes than closely related individuals, so outbreeding maintains a relatively large number of different alleles in the population. Inbreeding over several generations, on the other hand, can result in almost every individual in the population possessing the same alleles.

Many reasons have been put forward as to why genetic variation is 'desirable' within a population. We have already seen that this increases the likelihood that at least some individuals will have resistance against any particular parasite or pathogen, and that variation is necessary in order for evolution to take place.

In many species, inbreeding can lead to a condition called 'inbreeding depression', in which individuals are weak and less likely to survive. This is the result of increased homozygosity – they are more likely to have the same two alleles of any gene. Some of these alleles will be recessive and harmful, and inbreeding increases the chances of individuals possessing two recessive alleles of a gene.

Outbreeding mechanisms

1 Some species of plant are **dioecious**. This means that male flowers and female flowers are found on separate plants. As each plant produces either male gametes or female gametes, but not both, self-pollination (and hence self-fertilisation) is impossible.

   Examples of dioecious plants include nutmeg (*Myristica fragrans*), marijuana (*Cannabis sativa*), pawpaw (*Carica papaya*) (Figure 11.12)

Figure 11.12 This pawpaw flower has only female parts, and you can see several stigmas. All the flowers on this plant are female. Male flowers will be found on other pawpaw plants.

and mamoncillo (*Melicoccus bijugatus*), which also has a number of other common names including chenet and guaya.

2 Some species of plant in which the flowers have both male and female parts are able to prevent self-pollination by ensuring that these two parts do not mature at the same time. In **protandry**, the stamens mature before the stigmas. This means that, when the pollen is ripe and ready to be taken from the stamen, the stigma of that flower cannot receive it. This situation is found in rosebay willowherb or fireweed (*Epilobium angustifolium*) (Figure 11.13). In **protogyny**, the stigmas become receptive before the pollen is released. Examples of protogynous species include avocado (*Persea americana*) and soursop (*Annona muricata*).

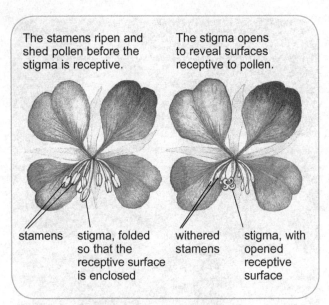

The stamens ripen and shed pollen before the stigma is receptive.

The stigma opens to reveal surfaces receptive to pollen.

stamens | stigma, folded so that the receptive surface is enclosed | withered stamens | stigma, with opened receptive surface

Figure 11.13 In rosebay willowherb, self-pollination is prevented by protandry.

3 Some species have separate male and female flowers on the same plant, and are said to be **monoecious**. Examples include *Zea mays* (Figure 11.14) and pumpkin (*Cucurbita*). If these two flower types do not mature at the same time, then self-pollination is prevented. Castor oil (*Ricinus communis*) has its female flowers carried above the male ones, so that pollen cannot fall onto the stigmas.

Figure 11.14 Maize (*Zea mays*): **a** male flowers at the top of the stem; **b** female flowers lower down.

4 Some species have flowers that have two different forms, a situation known as **heterostyly**. Examples include red cordia (*Cordia sebestena*) and primrose (*Primula vulgaris*) (Figure 11.15). The two types of flowers are called pin and thrum. Pin flowers have a long style, holding the stigma well above the anthers. Thrum flowers have a short style, so that the stigma is below the anthers. This makes it awkward for pollen to be transferred from anther to stigma on a pin flower.

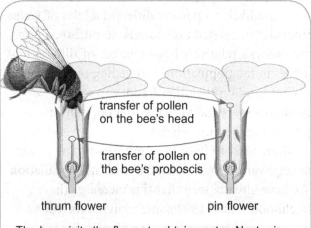

transfer of pollen on the bee's head

transfer of pollen on the bee's proboscis

thrum flower   pin flower

The bee visits the flower to obtain nectar. Nectar is produced at the base of the petals. The bee has to thrust its head into the flower so that the proboscis can reach the nectar. Any pollen attached to the proboscis can pollinate a thrum flower's stigma. Pollen is deposited on the bee's head by the thrum flower which can pollinate the stigma of a pin flower.

Figure 11.15 Prevention of self-pollination in primroses.

5 Even if self-pollination does occur, there are often mechanisms that prevent self-fertilisation taking place. For example, if pollen from the same flower, or from a different flower on the same plant, lands on a stigma, the pollen grain may not be stimulated to germinate and grow a pollen tube. In many species, including grasses, tobacco (*Nicotiana tabacum*) and many of the cabbages (*Brassica* spp.), the compatability of pollen and stigma is determined by a gene called S, which has many different alleles. This gene codes for a protein that is involved in cell recognition, and pollen grains and stigmas with different alleles of this gene produce different varieties of the protein. If a pollen grain lands

on a stigma with the same alleles as its own, it either will not germinate at all, or the pollen tube will not grow all the way through to the style. This situation is called **self-incompatibility**. In other cases, the pollen may germinate and the male gametes may travel down the tube, but either fertilisation does not occur, or the zygote that is produced does not develop. This is known as **self-sterility**.

Development of a seed

Once fertilisation has taken place, the ovule is called a seed. It remains attached to the parent plant and continues to receive nutrients from it. Figure 11.16 shows how a seed develops.

The diploid zygote divides by mitosis to form an **embryo plant**, attached to the wall of the ovule by a large **basal cell** and a little column of cells called the **suspensor**. The embryo develops a **radicle**

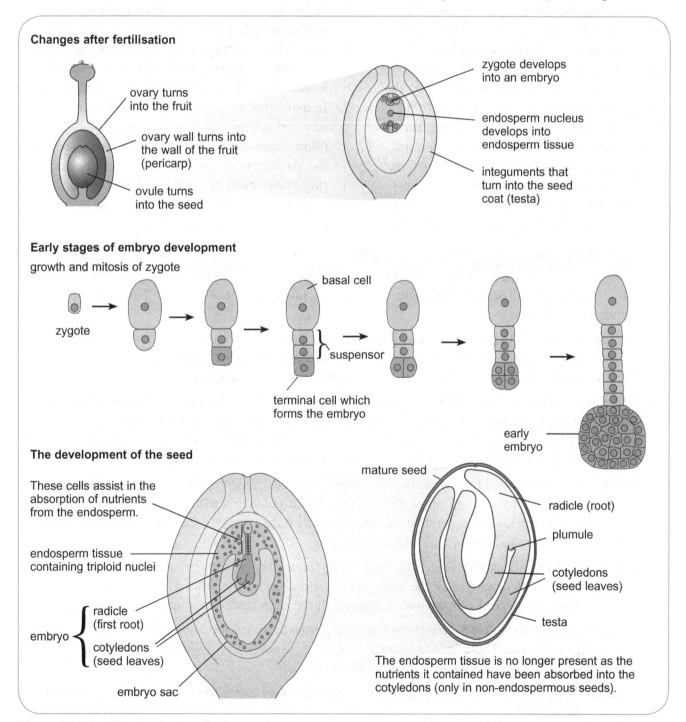

Changes after fertilisation

ovary turns into the fruit

ovary wall turns into the wall of the fruit (pericarp)

ovule turns into the seed

zygote develops into an embryo

endosperm nucleus develops into endosperm tissue

integuments that turn into the seed coat (testa)

Early stages of embryo development

growth and mitosis of zygote

zygote

basal cell

suspensor

terminal cell which forms the embryo

early embryo

The development of the seed

These cells assist in the absorption of nutrients from the endosperm.

endosperm tissue containing triploid nuclei

embryo
{ radicle (first root)
cotyledons (seed leaves)

embryo sac

mature seed

radicle (root)

plumule

cotyledons (seed leaves)

testa

The endosperm tissue is no longer present as the nutrients it contained have been absorbed into the cotyledons (only in non-endospermous seeds).

Figure 11.16 The development of a seed.

or embryo root, a **plumule** or embryo shoot, and two **cotyledons**. In some species, the cotyledons will become the first leaves, the 'seed leaves', of the young plant when the seed germinates, while in others the plumule is the first part to reach the light and photosynthesise. In many species, the cells of the cotyledons build up large stores of food within the seed, such as starch, which will be used by the embryo in the early stages of germination.

The triploid endosperm nucleus also divides by mitosis, forming a tissue called the **endosperm**, which surrounds the developing embryo. The function of the endosperm is to provide nourishment for the embryo. In some seeds, the endosperm has completed its function within a few days of fertilisation and disappears. In others, such as cereal grains, it remains as the main storage tissue to provide nutrients during germination.

While all this is happening, the integuments of the ovule are developing into the **testa** of the seed. This involves thickening and toughening, as waterproof substances such as lignin are laid down in the cell walls. The small gap in the integuments, the micropyle, remains as a tiny hole in the testa.

The wall of the ovary also undergoes changes after fertilisation. The ovary becomes a **fruit** and its wall becomes the **pericarp** of the fruit. The seeds are, of course, contained within the fruit, and the fruit is often adapted to disperse the seeds away from the parent plant. The pericarp may become fleshy and sweet-tasting to attract animals, or it may develop hooks and spines to stick to their hair. Wind-catching projections may form, or the fruit may become dry and hard, later splitting forcefully and throwing the seeds in all directions. In many species, other parts of the flower are involved in these developments. In dandelions (*Taraxacum* spp.), for example, the calyx becomes the 'parachute' of the fruit. In general, however, the various parts of the flower have completed their roles by now and they wither and fall off as the seeds develop inside the fruits.

Summary

- Flowers are organs of sexual reproduction. The male gametes are made in the anthers, and the female gametes in the ovules.

- In the pollen sacs of the anther, pollen mother cells divide by meiosis, producing four haploid cells, each of which develops a tough outer wall and becomes a pollen grain. Each haploid nucleus divides by mitosis to produce two haploid nuclei, the tube nucleus and the generative nucleus.

- In the ovule, an embryo sac mother cell (megaspore mother cell) divides by meiosis to produce four haploid cells, three of which disintegrate. The fourth develops into an embryo sac. Its nucleus divides by mitosis to produce eight haploid nuclei. One of these is the female gamete. Two nuclei in the centre of the embryo sac fuse to produce a diploid primary endosperm nucleus.

- Pollen may be transferred from an anther to a stigma by birds, insects or the wind. If a pollen grain lands on a suitable stigma, it produces a pollen tube which grows down through the style to the ovule. The tube nucleus and generative nucleus move down the tube, and the tube nucleus divides by mitosis to produce two haploid male gamete nuclei.

- The tube nucleus disintegrates. One of the male nuclei fuses with the female nucleus to produce a diploid zygote, which develops into the embryo. The other male nucleus fuses with the diploid primary endosperm nucleus to produce a triploid endosperm nucleus, which develops into the endosperm.

- Many plants have mechanisms to prevent inbreeding, which can lead to homozygosity and inbreeding depression. These include being dioecious, protandrous, protogynous, or showing heterostyly, self-incompatibility or self-sterility.

- After fertilisation, the ovule develops into a seed and the ovary into a fruit. The embryo develops a radicle, plumule and cotyledons.

Questions

Multiple choice questions

1 The diagram to the right illustrates a pollen grain.

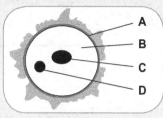

Which of the labelled structures produces the male gametes?

2 The diagram to the right shows a
cross-section of an anther.

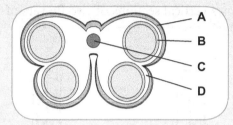

Which of the labelled structures nourishes the developing pollen grains?

3 Double fertilisation of an ovule of a flowering plant results in:
A one cotyledon and one endosperm.
B one embryo and one endosperm.
C one seed and one cotyledon.
D a fruit and a seed.

4 Which of the following is the **correct** sequence of events in a pollen sac?
A pollen mother cells → meiosis → two haploid cells → meiosis → two pollen grains per cell
B pollen grain → meiosis → pollen tube nucleus → meiosis → generative nucleus
C pollen mother cell → meiosis → tetrad of pollen cells → mitosis → two haploid nuclei per cell
D pollen grain → mitosis → pollen tube nucleus → mitosis → generative nucleus

5 The function of the endosperm is:
A seed coat formation.
B pollen formation.
C nourishment of the embryo.
D direction of the growth of pollen tube.

6 The structure that encloses and transports the male gametes to another plant is called a:
A seed. B pollen grain. C fruit. D embryo.

7 The diagram below represents the formation of the embryo sac in a flowering plant.

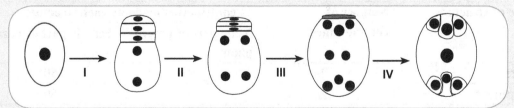

Which steps represent the process of mitosis during the formation
of the embryo sac?
A I only B I and II only C II and III only D I, II, III and IV

continued…

8 Papaya is a plant species which has its male and female flowers in separate plants. Which of the following correctly identifies the type of plant and pollination which occur in the papaya plant?

	Type of plant	Type of pollination
A	monoecious	self
B	dioecious	self
C	monoecious	cross
D	dioecious	cross

9 The diagram below shows vertical sections of two flowers of primrose.

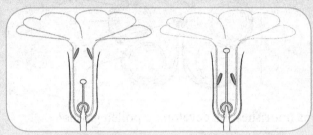

What is the **most** likely type of pollination and an advantage and disadvantage of this type of pollination?

	Type of pollination	Advantage	Disadvantage
A	cross	maintenance of useful characteristics	more genetic variation
B	self	maintenance of useful characteristics	more genetic variation
C	cross	increased heterozygosity and genetic variation	needs a pollinating agent
D	self	increased heterozygosity and genetic variation	needs a pollinating agent

10 The diagram below shows an ovule from a dicotyledonous plant, immediately after self-fertilisation.

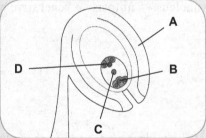

Which of the following correctly identifies the structure, the number of chromosomes and contribution made by each parent?

	Name of structure	Number of chromosomes	% contribution made by each structure	
			Male part of parent plant	Female part of parent plant
A	zygote	$2n$	50	50
B	integument	$2n$	0	50
C	endosperm nucleus	$3n$	33.3	66.7
D	antipodal cell / nucleus	$2n$	0	100

continued ...

Structured questions

11 The micrograph represents a cross-section through an anther of lily (*Lilium*).
Study the figure and answer the questions below.

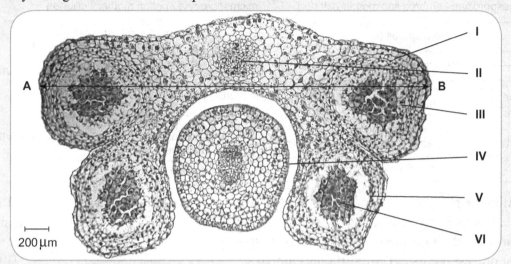

200 µm

a   Identify the structures labelled **I** to **VI** in the micrograph above.   [3 marks]
b   Calculate the actual distance between **A** and **B** shown on the micrograph of the
    anther. Show your working and give your answer to the nearest micrometre (µm).   [2 marks]
c   In the micrograph there are microspore mother cells (pollen mother cells).
    Briefly describe how these cells develop into pollen grains.   [4 marks]
d   What are the genetic implications of the processes that occur during the
    formation of the pollen grains?   [2 marks]
e   Make a labelled drawing of a mature pollen grain.   [4 marks]

12 a Identify the structures **I** to **IX** in the mature carpel shown below.   [4 marks]

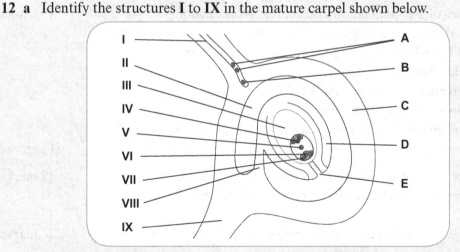

b   Copy the diagram and add the labels 'ovule' and 'embryo sac'.   [2 marks]
c   How are the functions of structures **VIII** and **IX** similar?   [1 mark]
d   Explain briefly how structures **A** and **B** arise and their genetic composition.   [3 marks]
e   What is the role of **E** in the process of fertilisation?   [1 mark]
f   State what happens to structures **C** and **D** after fertilisation.   [2 marks]
g   Draw on your diagram the path taken by the pollen tube to the stage
    immediately before fertilisation.   [2 marks]

continued ...

13 The diagrams below show the development of a pollen grain and an embryo sac.

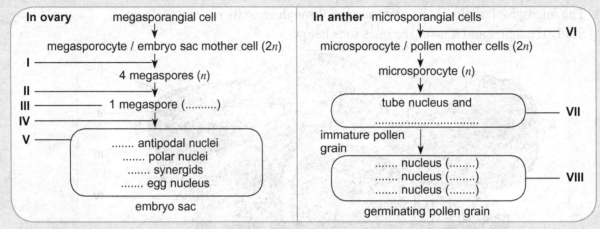

a i Identify the process which occurs at **I** and **VI** in the diagrams above. [1 mark]

ii Explain what happens at **II**. [2 marks]

iii What is the chromosome content of the megaspore at **III**? [1 mark]

iv What process occurs at **IV** to form the embryo sac? [1 mark]

v Give the number of each type of nucleus at **V**. [3 marks]

vi What is the name of the other nuclei formed at **VII**? [1 mark]

vii What are the names of nuclei formed during germination of the pollen grain at **VIII**? [2 marks]

b Using named examples of flowering plants, describe two mechanisms which ensure cross pollination. [4 marks]

Essay questions

14 Plants are involved in asexual reproduction, self-fertilisation and cross-fertilisation. All these mechanisms have advantages and disadvantages for the plants.

a Compare the processes of self-pollination and asexual reproduction. [3 marks]

b Listed below are descriptions of four plants. Suggest the type of reproduction or fertilisation found in each plant, giving reasons for your answer.

i This plant is widely scattered and rare.

ii This plant grows in an environment that is stable.

iii This plant grows in a changing environment.

iv This plant has over time been subjected to a variety of fungal diseases and yet it flourishes. [12 marks]

15 a Explain the difference between pollination and fertilisation. [2 marks]

b Using annotated diagrams, explain the sequence of events from pollination to fertilisation. [7 marks]

c Discuss the development of embryo sac after fertilisation. [6 marks]

16 a Using an annotated diagram, describe the structure of the anther. [4 marks]

b Describe with the aid of diagrams how the two male gametes are produced from a pollen mother cell. [5 marks]

c State **two** advantages and **two** disadvantages of cross pollination. [4 marks]

17 a Using annotated diagrams, describe the structure of a young ovule. [4 marks]

b Describe the formation of an embryo sac in a flowering plant. [5 marks]

c State **two** advantages and disadvantages of self-pollination. [4 marks]

Chapter 12
Sexual reproduction in humans

By the end of this chapter you should be able to:

a   describe the structure of the male and female reproductive systems;

b   make drawings from prepared slides of the mammalian ovary and testis;

c   explain gametogenesis, including the difference between the secondary oocyte and ovum;

d   compare the structure of the secondary oocyte and the sperm, and discuss how their structures suit their functions;

e   explain how hormones regulate gametogenesis;

f   discuss the importance of hormones in the control of the menstrual cycle, with reference to negative feedback;

g   describe how and where fertilisation and implantation normally occur;

h   discuss how knowledge of human reproductive anatomy and physiology has been applied to the development of contraceptive methods;

i   discuss the functions of the amnion;

j   discuss the possible effects of maternal behaviour on foetal development, including the role of nutrition, alcohol abuse, use of legal and illicit drugs, and cigarette smoking.

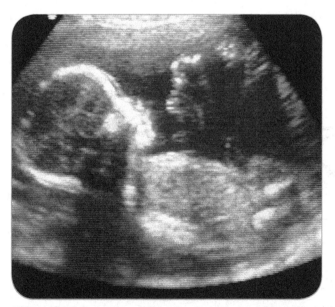

Figure 12.1   A foetus within the mother's uterus as seen by ultrasound. The placenta, through which the foetus obtains its oxygen and nutrients, is the tree-like structure above it.

Human reproduction

Humans, like all mammals, can only reproduce sexually (Figure 12.1). In this chapter, we will look at how the male and female reproductive systems work, and then consider some of the ways in which couples can prevent unwanted pregnancies.

Reproduction in humans involves:

● gametogenesis, which is the production of male and female gametes;

● fertilisation, which is the fusion of the nuclei of a male and a female gamete;

● pregnancy and birth.

The structure of the urogenital system

Figure 12.2 and Figure 12.3 show the organs of the reproductive system in females and males. As the organs of the urinary system (kidneys and bladder) are closely associated with the reproductive organs, this is sometimes known as the **urogenital system**.

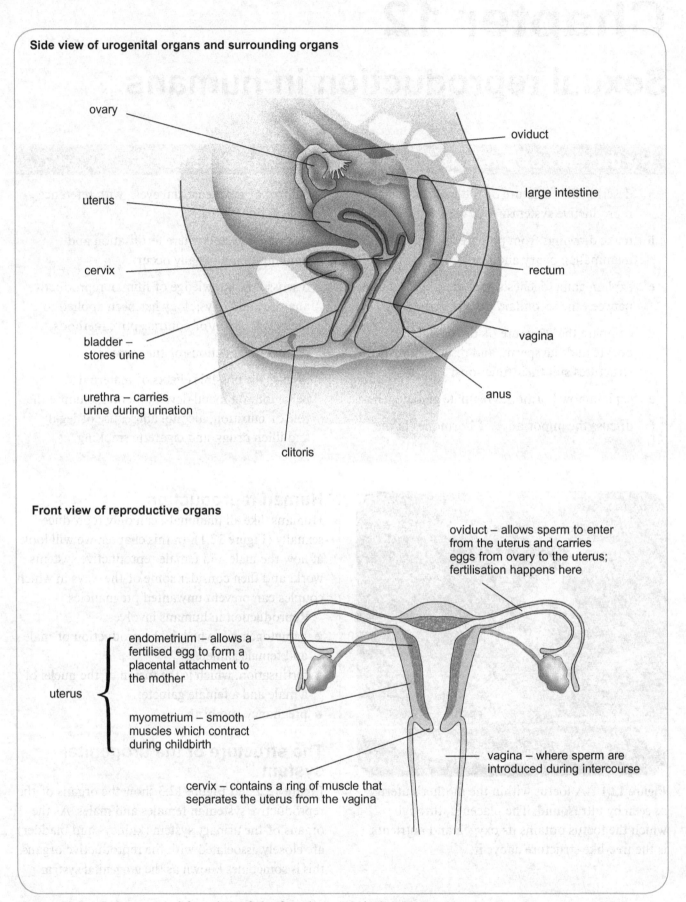

Side view of urogenital organs and surrounding organs

ovary

oviduct

uterus

large intestine

cervix

rectum

bladder – stores urine

vagina

urethra – carries urine during urination

anus

clitoris

Front view of reproductive organs

oviduct – allows sperm to enter from the uterus and carries eggs from ovary to the uterus; fertilisation happens here

endometrium – allows a fertilised egg to form a placental attachment to the mother

uterus

myometrium – smooth muscles which contract during childbirth

vagina – where sperm are introduced during intercourse

cervix – contains a ring of muscle that separates the uterus from the vagina

Figure 12.2 Structure and function of the female urogenital system.

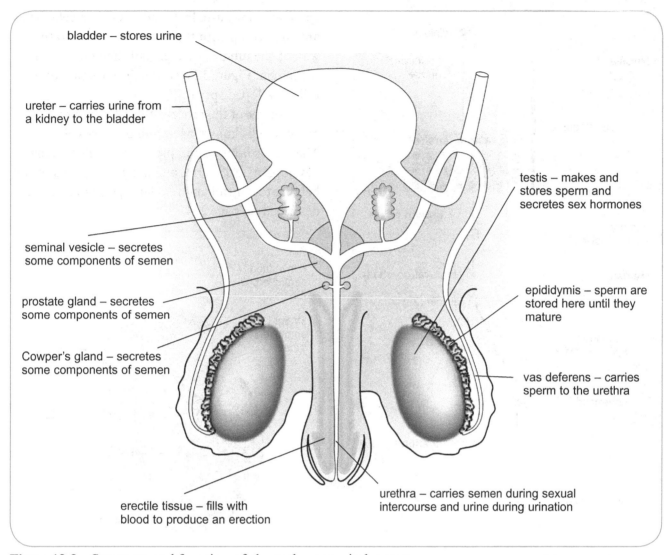

Figure 12.3 Structure and function of the male urogenital system.

Gametogenesis

Gametogenesis is the production of gametes. This happens in the testes and ovaries, where diploid cells divide by meiosis to produce haploid cells.

Spermatogenesis

Figure 12.4 shows how sperm are produced inside the testes. This process is called **spermatogenesis**. Sperm production begins in a boy around the age of 11, and then continues through the rest of his life. In most men, around 100 to 200 million sperm are made each day.

The testes are made up of many **seminiferous tubules** (Figure 12.5) and it is in the walls of these that spermatogenesis takes place. The process begins at the outer edge in the **germinal epithelium** and the new cells that are produced form towards

the inner edge. So by looking at a cross-section of a tubule you can see all the stages in sequence.

The cells that begin it all are called **spermatogonia** (singular: spermatogonium). These are diploid cells, and they divide by mitosis to form more diploid cells. Some of these grow into new spermatogonia, whilst others grow much larger and are called **primary spermatocytes**.

Now meiosis begins. The two new cells that are formed by the first division are called **secondary spermatocytes**. They are, of course, haploid. After a few days, the secondary spermatocytes go through the second division, each forming two **spermatids**. These are also haploid.

That is the end of the cell divisions, and all that is now required is for the spermatids to differentiate into **spermatozoa** (singular:

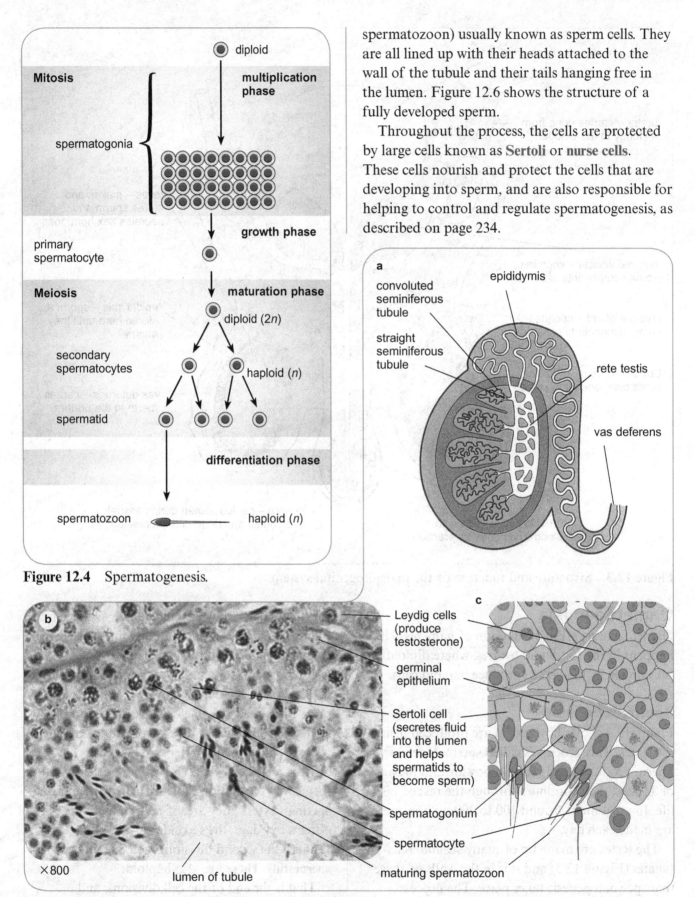

Figure 12.4 Spermatogenesis.

spermatozoon) usually known as sperm cells. They are all lined up with their heads attached to the wall of the tubule and their tails hanging free in the lumen. Figure 12.6 shows the structure of a fully developed sperm.

Throughout the process, the cells are protected by large cells known as **Sertoli** or **nurse cells**. These cells nourish and protect the cells that are developing into sperm, and are also responsible for helping to control and regulate spermatogenesis, as described on page 234.

Figure 12.5 The histology of the testis: **a** section through a testis; **b** light micrograph of part of a seminiferous tubule; **c** drawing from the micrograph.

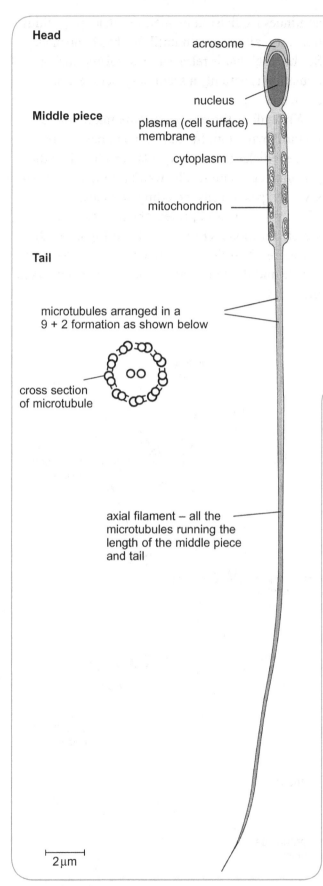

Head

acrosome

The acrosome contains hydrolytic enzymes which are released during fertilisation and are used to digest a path into the female gamete.

nucleus

Middle piece

plasma (cell surface) membrane

cytoplasm

The nucleus contains the haploid number of chromosomes. This is the only part of the spermatozoon that enters the cytoplasm of the female gamete.

mitochondrion

The numerous mitochondria are carrying out respiration to provide the large amounts of ATP required by the spermatozoon for the long distance it may have to swim to reach a female gamete.

Tail

microtubules arranged in a 9 + 2 formation as shown below

Microtubules use ATP as a source of energy and they can move relative to each other. This movement causes the tail to bend, producing the swimming movements used to take the spermatozoon to the female gamete.

cross section of microtubule

axial filament – all the microtubules running the length of the middle piece and tail

2 µm

Figure 12.6 The structure of a spermatozoon.

Oogenesis

Figure 12.7 shows how ova are produced. The first stages of this process take place in the ovary, but the ovum is not actually formed until fertilisation happens.

While a girl is still an embryo in her mother's uterus, germinal epithelial cells in her developing ovaries divide to form diploid **oogonia**. Within a few weeks, these oogonia begin to divide by meiosis. However, they don't get very far, only reaching prophase I. They are called **primary oocytes**. A lot of them disappear, but at birth the ovaries of a baby girl usually contain about 400 000 primary oocytes.

At puberty, some of these primary oocytes get a little further with their division by meiosis. They move from prophase I to the end of the first meiotic division, forming two haploid cells. However, one of these, the **secondary oocyte**, is much bigger than the other, the **polar body**. The polar body has no further role to play in reproduction. The secondary oocyte continues into the second division of meiosis, but gets no further than metaphase II.

Each month, one of the girl's secondary oocytes is released into the oviduct. If fertilised, it

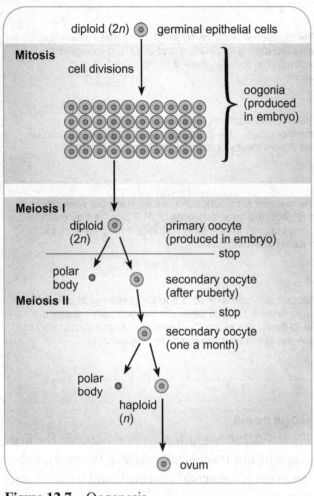

Figure 12.7 Oogenesis.

continues its division by meiosis. Strictly speaking, it is not really an ovum until this has happened. So the 'egg' that is released by an ovary during ovulation is actually a secondary oocyte, not an ovum.

While all of these processes are taking place, up until ovulation, the developing oocytes are inside follicles in the ovary, also produced by the germinal epithelium. The wall of a follicle contains several types of cells including **granulosa cells** which surround and protect the oocyte, and secrete hormones (Figure 12.8 and Figure 12.9).

Figure 12.10 shows the structure of a secondary oocyte and the cells that surround it when it leaves the ovary.

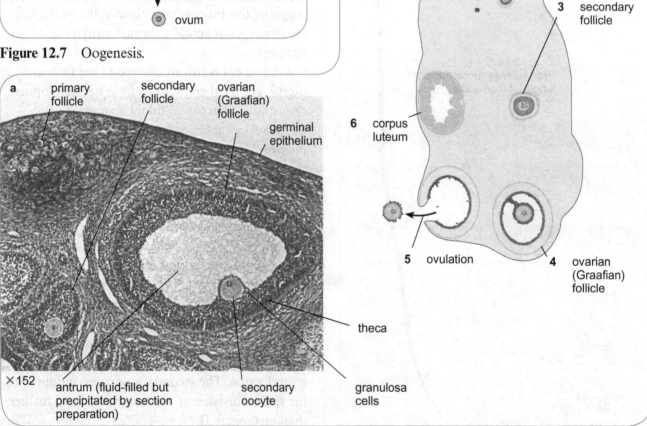

Figure 12.8 The histology of the ovary: **a** light micrograph of ovary; **b** stages leading up to ovulation in the ovary.

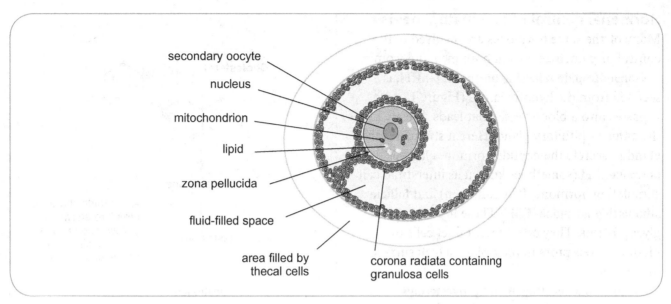

Figure 12.9 An ovarian follicle.

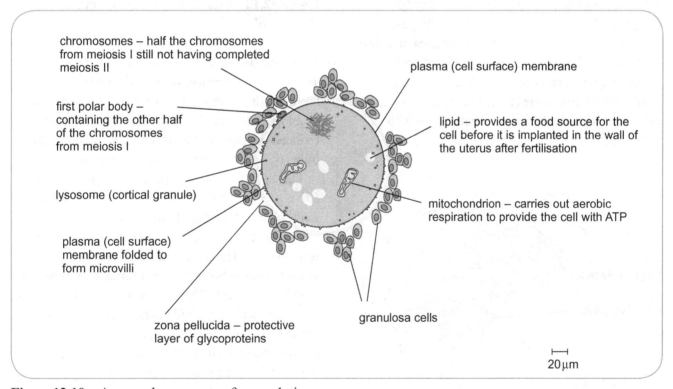

Figure 12.10 A secondary oocyte after ovulation.

1 Construct a table to compare the processes of spermatogenesis and oogenesis. List similarities as well as differences between them.

2 Explain how the structure of a sperm and the structure of an egg (secondary oocyte) adapt each for its functions.

Hormonal control of spermatogenesis

Many of the same hormones are involved in the control of gametogenesis in both men and women.

Gonadotrophin releasing hormone, GnRH, is secreted from the hypothalamus (Figure 12.11). It passes into a blood vessel that leads directly to the anterior pituitary gland. Here it stimulates the gland to secrete the peptide hormones **luteinising hormone, LH** (sometimes known as interstitial cell stimulating hormone, ICSH, in men) and **follicle stimulating hormone, FSH**. These hormones are glycoproteins. They affect their target cells by binding to receptors in their plasma (cell surface) membranes.

Testosterone, oestrogen and **progesterone** are secreted from the testes and ovaries. They are steroids (Figure 12.12) and so are soluble in lipids. This means that they can easily pass through plasma (cell surface) membranes and into cytoplasm. They bind to receptors inside cells, which triggers a response from their target cells.

In men, testosterone is secreted by the **Leydig cells** (sometimes called interstitial cells) which

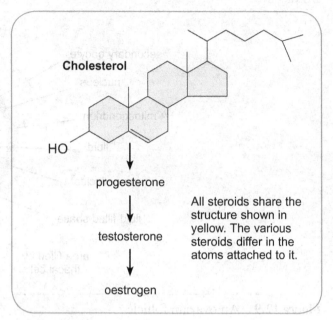

All steroids share the structure shown in yellow. The various steroids differ in the atoms attached to it.

Figure 12.12 Steroid hormones are made from cholesterol.

are found between the seminiferous tubules. Testosterone secretion begins when a boy is still a foetus in the uterus, and it causes the embryo to develop male sexual organs. Its secretion is greatly reduced during childhood, but then increases at puberty, when sperm production begins.

This secretion of testosterone happens because LH is secreted. LH travels in the blood from the anterior pituitary gland, and binds to receptors on the Leydig cells. This causes them to secrete testosterone. The testosterone passes into the seminiferous tubules, where it binds with receptors in the Sertoli cells. This stimulates spermatogenesis.

At the same time, FSH is also secreted from the anterior pituitary gland, and this binds to the plasma (cell surface) membranes of the Sertoli cells. This makes them more receptive to testosterone.

The level of all these hormones in the blood is controlled by negative feedback. If the concentration of testosterone in the blood rises too high, this inhibits the production of LH, which in turn reduces the secretion of testosterone. The level of testosterone in the blood remains fairly constant througout a man's life, so sperm production is a continuous process. As the activity of Sertoli cells increases, they release inhibin – a hormone which inhibits FSH secretion.

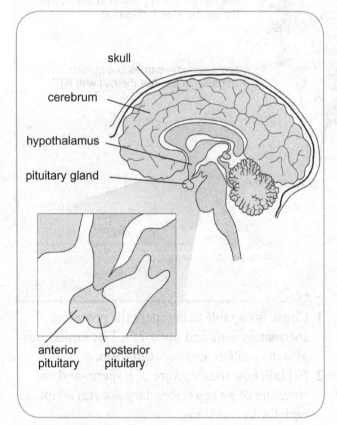

Figure 12.11 Hypothalamus and pituitary gland.

Hormonal control of oogenesis

In women there is an approximately 28-day cycle, the **menstrual cycle**, of activity in the ovaries and uterus, in contrast to the steady production of sperm that occurs in men. The menstrual cycle (Figure 12.13) usually begins between the ages of 10 and 14, and continues until the woman reaches the menopause at around 50 years old.

The hormones that control spermatogenesis in men also control the production of eggs in women. Figure 12.14 shows the changes in these hormones during the menstrual cycle.

The cycle is considered to begin with the onset of menstruation. Menstruation usually lasts for about 4 to 8 days. During this time, the anterior pituitary gland secretes LH and FSH, and their levels increase over the next few days.

In the ovary, one follicle becomes the 'dominant' one. The presence of LH and FSH stimulates it to secrete oestrogen from the theca surrounding the follicle. The presence of oestrogen in the blood has a negative feedback effect on the production of LH and FSH, so the levels of these two hormones fall. The oestrogen stimulates the endometrium (lining of the uterus) to proliferate – that is, to thicken and develop numerous blood capillaries.

When the oestrogen concentration of the blood has reached a level of around twice to four times its level at the beginning of the cycle, it stimulates a surge in the secretion of LH and, to a lesser extent, of FSH. The surge of LH causes the dominant follicle to burst and to shed its secondary oocyte into the oviduct. This usually happens about 14 to 36 hours after the LH surge. The follicle then collapses to form the **corpus luteum** ('yellow body'), formed from granulosa cells, which secretes progesterone. This maintains the lining of the uterus, making it ready to receive the embryo if fertilisation occurs.

If fertilisation does not occur, the corpus luteum degenerates and so the secretion of progesterone and oestrogen falls. As the levels

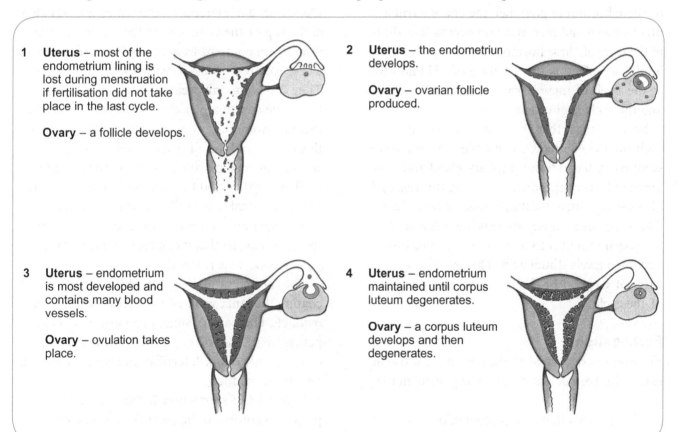

1 **Uterus** – most of the endometrium lining is lost during menstruation if fertilisation did not take place in the last cycle.

  Ovary – a follicle develops.

2 **Uterus** – the endometriun develops.

  Ovary – ovarian follicle produced.

3 **Uterus** – endometrium is most developed and contains many blood vessels.

  Ovary – ovulation takes place.

4 **Uterus** – endometrium maintained until corpus luteum degenerates.

  Ovary – a corpus luteum develops and then degenerates.

Figure 12.13   The menstrual cycle.

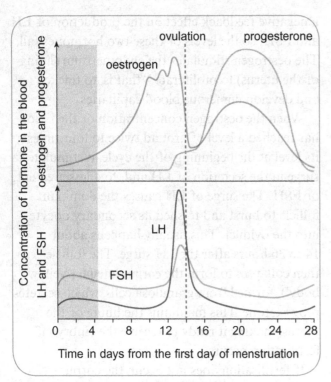

Figure 12.14 The menstrual cycle.

of these hormones plummet, the endometrium breaks down and menstruation occurs. The drop in the level of these hormones also removes the inhibitory effect on the secretion of LH and FSH, so the levels of these hormones begin to rise again and the cycle begins once more.

So, as in men, there is a negative feedback cycle involving interaction between the hormones secreted by the anterior pituitary gland and those secreted by the reproductive organs. But this cycle also has a positive feedback stage involved in it. This happens because, whereas lower levels of oestrogen inhibit the secretion of LH and FSH, very high levels stimulate it. This is why the surge of LH and FSH happens around day 12 of the cycle.

Fertilisation

Fertilisation occurs when the nucleus of a sperm fuses with the nucleus of a female gamete in the oviduct.

We have seen that sperm production is a continuous process that takes place in the seminiferous tubules of the testis. The fully formed sperm move from these tubules into the **epididymis**, carried in fluid secreted by the Sertoli cells.

During sexual activity, stimulation of various parts of the body, and especially of the head of the penis, causes impulses to be sent from the brain through parasympathetic nerve cells to an artery and its arteriole branches in the penis. They dilate, allowing blood to fill the spaces within the erectile tissue, so that the penis becomes hard and erect. Further stimulation causes impulses via sympathetic nerve cells to the vasa deferentia. Muscles in their walls contract and push the sperm within them into the urethra. At the same time, muscles around the prostate gland and in the seminal vesicles also contract, and this forces out fluid which mixes with the sperm to form semen. Semen contains citrate and calcium ions, and also fructose, which can provide the sperm with energy for their marathon swim towards a female gamete in an oviduct.

The muscles then contract in strong, rhythmic waves that force the semen along the vas deferens and out through the urethra. This is ejaculation. During sexual intercourse, the semen is ejaculated at the top of the vagina, near the cervix. A single ejaculation contains as many as 400 million sperm. These sperm are not yet ready to fertilise an egg. They can swim only weakly. Over the next few hours they gradually become **capacitated**, enabling them to swim more strongly and rapidly. During this process, a layer of glycoprotein around the sperm, and also plasma proteins in the seminal fluid, are hydrolysed by enzymes in the uterus. This appears to enable the tail to lash more strongly.

Changes also happen in the membrane around the **acrosome**, so that it can release its enzymes once an oocyte is reached.

The sperm gradually swim up through the fluid coating the inner walls of the uterus, towards the oviducts. Only a very small proportion of them, perhaps 0.025%, will complete the journey. And of these only one will fertilise an oocyte, if there is one in the oviduct.

Figure 12.15 shows how fertilisation takes place. Receptors on the sperm's plasma (cell surface) membrane bind to proteins in the zona pellucida surrounding the secondary oocyte, and this stimulates release of the enzymes in the acrosome. They digest the zona pellucida, forging

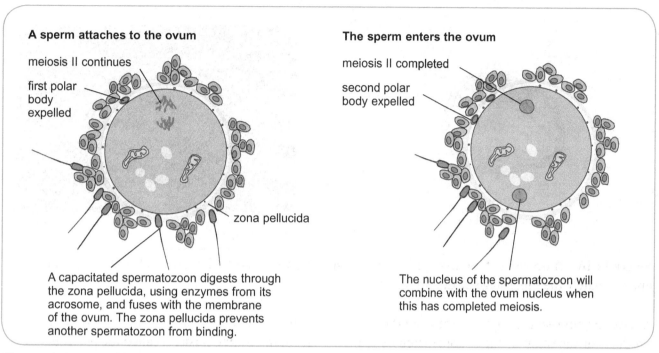

A sperm attaches to the ovum

meiosis II continues

first polar body expelled

zona pellucida

A capacitated spermatozoon digests through the zona pellucida, using enzymes from its acrosome, and fuses with the membrane of the ovum. The zona pellucida prevents another spermatozoon from binding.

The sperm enters the ovum

meiosis II completed

second polar body expelled

The nucleus of the spermatozoon will combine with the ovum nucleus when this has completed meiosis.

Figure 12.15 Fertilisation.

a pathway towards the membrane of the oocyte. As the membrane of the sperm makes contact with the membrane of the oocyte, the oocyte releases lysosomes (cortical granules) which rapidly pass into the zona pellucida and change the proteins there so that no more sperm can bind with them. This is known as the **cortical reaction**. The zona pellucida becomes an impenetrable barrier called the **fertilisation membrane**.

Contact of the sperm with its plasma (cell surface) membrane stimulates the secondary oocyte to complete meiosis. A second polar body is formed, effectively acting as a 'dustbin' for the disposal of one set of chromatids. The nucleus of the successful sperm can now enter the egg, and the two nuclei fuse together to form a diploid nucleus.

The cell is now a diploid zygote. It has the full complement of chromosomes – two complete sets, one from the father and one from the mother. This mixture of chromosomes from two parents will make the child that grows from this single cell unlike either of them. Taken together with the variation in the genotypes of gametes formed as a result of meiosis, you can see that there is tremendous scope for variation in the children of any two parents.

Implantation and the placenta

The first sign that fertilisation has occurred is the secretion of a glycoprotein hormone called **human chorionic gonadotrophin, hCG**. This is produced by the little ball of cells that has been formed by the repeated division of the zygote, and is known as a **blastocyst** (Figure 12.16). hCG is very similar to LH and has similar effects. It stimulates the corpus luteum to keep on secreting oestrogen and progesterone, which between them stimulate the endometrium so that its rich supply of blood vessels is maintained. This stops menstruation from taking place.

As the blastocyst is forming, it slowly moves along the oviduct towards the uterus. Its movement is caused by cilia on the epithelium lining the oviduct wall, and also by slow, steady contractions of muscles in the oviduct wall. It reaches the uterus about three to four days after fertilisation. Here, it sinks into the endometrium (the lining of the uterus), in a process called **implantation**.

Implantation depends on special cells on the surface of the blastocyst, called **trophoblasts**, which secrete enzymes that partially digest cells in the endometrium. The tiny blastocyst 'burrows' into the endometrium, eventually becoming completely surrounded by it (Figure 12.17). The embryo

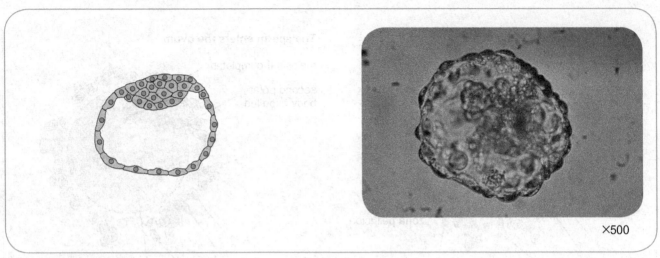

×500

Figure 12.16   A drawing of a cross-section of a blastocyst; and a light micrograph of a blastocyst at approximately the same stage of development.

receives its nourishment at this time from the products of digestion of the endometrium by the trophoblasts.

The trophoblast cells continue to divide, producing many tiny projections called **villi** which project into the endometrium. Within the villi, blood capillaries form. At the same time, the endometrium also develops, forming spaces around the villi called sinuses, which are filled with the mother's blood. The whole structure, made up partly of the mother's tissues from the

endometrium and partly of tissues produced from the trophoblasts, is called the **placenta**. Figure 12.18 shows the structure of a fully formed placenta. By this stage, the embryo has developed most of its organs and is now called a **foetus**.

As it develops in the uterus, the foetus is surrounded by a number of membranes – extra-embryonic membranes – the **yolk sac**, **allantois**, **chorion** and **amnion** (Figure 12.19).

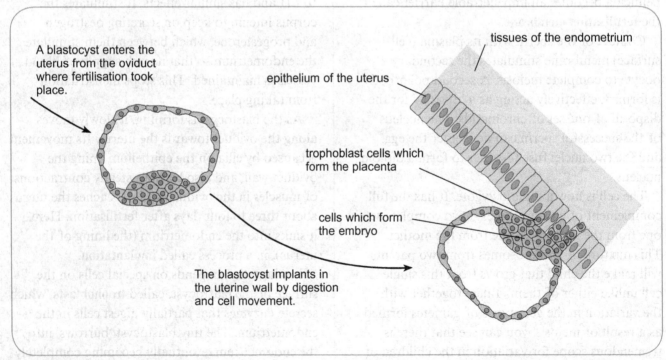

A blastocyst enters the uterus from the oviduct where fertilisation took place.

epithelium of the uterus

tissues of the endometrium

trophoblast cells which form the placenta

cells which form the embryo

The blastocyst implants in the uterine wall by digestion and cell movement.

Figure 12.17   Implantation.

The placenta and gas exchange

Oxygenated blood is brought to the sinuses through the mother's arteries, and taken back to her heart in veins. On the foetus's side, deoxygenated blood flows from the foetus through two **umbilical arteries** to capillaries in the villi. Here, the foetus's blood is brought very close to the mother's blood in the sinuses, although there is no direct contact between them. Oxygen is released from the mother's haemoglobin, diffuses across the thin barriers separating it from the foetus's blood, and combines with the foetus's haemoglobin in its red blood cells. The oxygenated blood flows to the foetus in the umbilical vein.

The large surface area provided by the villi enables this exchange of oxygen between the mother and the foetus to take place rapidly. Carbon dioxide is exchanged in the opposite direction, also by diffusion.

The placenta and exchange of nutrients and antibodies

Many other substances can pass between the mother's and the foetus's blood, across the cells

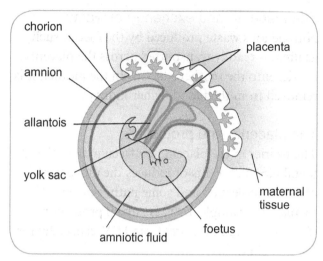

Figure 12.19 Extra-embryonic membranes.

that separate them, and these substances move by a variety of transport methods. For example, glucose moves by facilitated diffusion and amino acids by active transport. Antibodies, inorganic ions, vitamins and water also cross the placenta from mother to foetus, while urea diffuses from the foetus to the mother. Most drugs, such as nicotine and alcohol, also freely cross from the mother's blood to the foetus's blood.

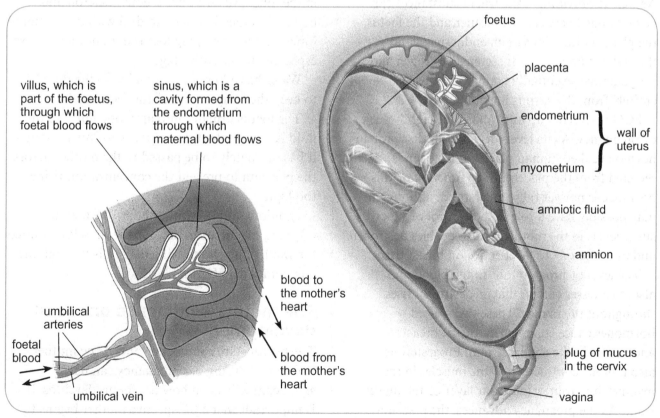

Figure 12.18 The structure of the placenta.

The placenta and exchange of waste

Nitrogenous waste produced by the foetus, such as urea, is readily transported across the placental barrier into the mother's blood. In the mother, it is removed from the blood by her kidneys.

The placenta and protection

The barrier in the placenta between maternal and foetal blood prevents or reduces the movement of larger molecules and some pathogens into the foetus. For example, there is partial protection (50% protection or more) from HIV, especially for the first child.

The barrier also prevents some large molecules from passing into the foetus. Some of these are maternal hormones which would disturb the metabolism of the foetus.

Because maternal and foetal blood are quite separate, the foetus maintains its own blood pressure, isolated from the pressure changes in the mother's blood, which can become dangerously high for the foetus.

The placenta as an endocrine organ

As well as providing a site where substances can be exchanged between the mother and the foetus, the placenta functions as an endocrine gland throughout pregnancy. It secretes oestrogen and progesterone, eventually taking over this role entirely from the corpus luteum.

hCG is secreted up until about the eighth week of pregnancy. As its levels decline, another protein hormone called human placental lactogen is secreted from the placenta. The function of this hormone is not yet fully understood, although it may possibly have a role in the control of blood sugar levels in the mother. It also enables oestrogen and progesterone to be effective on breast tissue.

The level of progesterone in the blood and also, to a lesser extent, that of oestrogen rises throughout pregnancy. These two steroid hormones cause enlargement of the breasts and maintain the endometrium. Progesterone also inhibits contraction of the muscles in the myometrium (outer, muscular layer of the uterus wall). As the pregnancy advances, the placenta increases its secretion of oestrogen.

Near the end of pregnancy, the peptide hormone oxytocin is produced by the hypothalamus and released from the mother's posterior pituitary gland. This hormone causes contraction of the uterine muscles, the myometrium, so that birth takes place.

Another steroid hormone, prolactin, is also secreted during pregnancy, from the mother's anterior pituitary gland. However, its actions are inhibited by the high levels of oestrogen and progesterone. After birth, when the placenta has been removed, the oestrogen and progesterone levels in the mother's blood plummet, and this allows prolactin to stimulate the breasts to secrete milk. The production and secretion of milk from the breasts is called **lactation**.

The amnion

The amnion secretes and contains amniotic fluid, which helps to support the foetus and protect it from mechanical damage – it acts as a shock-absorber, while allowing it to grow and move.

The amniotic fluid surrounding the foetus isolates the foetus from gravity. The foetus 'floats' in the fluid. An adult has a bony skeleton and highly developed muscles to deal with gravitational forces and a developing foetus does not have these, especially in the early stages.

Water has a high heat capacity, so the fluid helps to keep the temperature of the foetus stable.

The foetus urinates into amniotic fluid, so this fluid is, to an extent, a receptacle for urea, though it has ultimately to be passed to the mother across the placenta to prevent the concentration rising too high.

Amniotic fluid is swallowed by the foetus, allowing the foetus to develop the muscles required for swallowing and moving materials through the alimentary canal before birth.

The mother's influence on foetal development

The health status and behaviour of the mother before and during her pregnancy can have significant effects on how her foetus develops. It is important that a prospective mother begins to think about this even before she gets pregnant.

Most women will register with an antenatal clinic when they know that they are pregnant (Figure 12.20). 'Antenatal' means 'before birth'. The care that a prospective mother should take of herself, and the care that an antenatal clinic gives to her and her growing foetus, is known as **antenatal care**.

Preconceptual care

The time for the prospective mother to begin thinking about antenatal care is even before the baby is conceived. This is known as **preconceptual care**. The baby will be conceived inside her body, and will grow and develop there for nine months. So the health of her body can have a direct effect on the health of the baby.

Two particular steps are especially important. They involve the woman's immunity to rubella, and making sure she has plenty of folic acid in her diet.

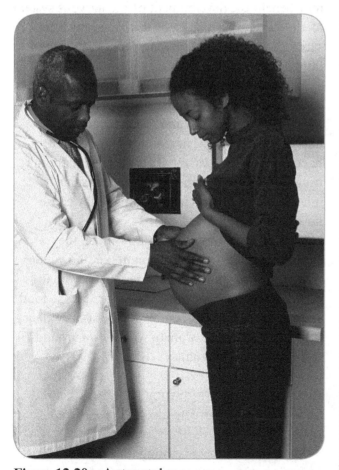

Figure 12.20 Antenatal care.

Rubella

Rubella, sometimes known as German measles, is an illness caused by a virus. It is not usually a dangerous disease for children or adults. They suffer a raised temperature, tiredness, and red spots all over the skin, but normally recover within ten days or so. But for a developing foetus it is a very different story. If the mother has rubella, the virus can cross the placenta and get into the foetus's blood. If this happens in the early stages of pregnancy, within the first three months, the baby's heart, brain, ears and eyes may fail to develop properly. Babies which have been exposed to the rubella virus may be born with heart and brain defects, deafness and sometimes cataracts.

So a woman who is trying to become pregnant needs to find out if she is immune to rubella. A simple blood test, to check for antibodies to rubella, will give the answer. If she is immune, she doesn't need to do anything else. If she is not, then she can have a rubella vaccination. But it is really important to do this at least three months before getting pregnant, because the vaccine itself could harm the developing foetus.

Folic acid

Another very important thing that the woman should do is begin to take supplements of **folic acid**. This is a B vitamin which is found in foods such as dark green vegetables, some breakfast cereals, milk products, oranges and bananas. A link has been found between lack of folic acid and the birth of a baby with neural tube defects. The neural tube is the part of the embryo which develops into the spinal cord and brain. It forms and grows during the very early stages of pregnancy – probably even before the woman knows that she is pregnant. If the neural tube does not develop properly, the baby may be born with spina bifida or other serious defects of the central nervous system. So, even if she eats a good diet, she should think about taking folic acid supplements. Taking folic acid pills each day before getting pregnant reduces the risk of having a baby with spina bifida by about 70%.

Postconceptual care

Once a woman knows that she is pregnant, she will be showered with advice about what she should and should not do. Friends, neighbours, relatives and magazines will all have their particular tips to give her. Some of it will be nonsense but a lot of it could be very helpful. The advice that she should be most ready to take will come from healthcare professionals.

The Rhesus factor

We have seen that red blood cells can carry A or B antigens in their cell surface membranes (page 136). These are not the only antigens they carry. Another important one is the Rhesus antigen. People who have this antigen are said to be Rhesus positive, while those who don't are Rhesus negative. Problems can arise when the mother is Rhesus negative and her partner is Rhesus positive. If this is the case, there is a chance that the baby might be Rhesus positive. Then, if during pregnancy or childbirth Rhesus positive baby's blood contacts the mother's blood, the mother becomes sensitised and makes anti-Rhesus antibodies. Should this occur during pregnancy, the antibodies will attack the baby. If it occurs at childbirth, it affects the next Rhesus positive foetus in her uterus.

To avoid these problems, Rhesus negative women are injected with anti-Rhesus antibodies after any potentially sensitising incidents in pregnancy, and routinely after childbirth. This stops the mother becoming sensitised.

Diet during pregnancy

Growing a complete, new human being from a starting point of just one cell is quite an achievement. The developing foetus needs a very wide range of different nutrients in order for all the organs to form and develop as they should. All of these nutrients come from the mother's blood. During pregnancy, it is important for her to eat a diet which not only supplies her foetus with these nutrients, but also supplies her own body with what it needs.

In the early stages of pregnancy, some women feel so ill with morning sickness that they don't really want to eat anything. But many women don't suffer in this way, and even those that do usually feel much better after some weeks.

A pregnant woman needs to think very carefully about what she is feeding to her unborn baby. Table 12.1 on page 244 lists some important nutrients that she should ensure she has in her diet. Overall, she should try to make sure that the food she eats each day contains:

- fresh fruit and vegetables, including ones which contain folic acid;
- complex carbohydrates (starch), found in bread, rice, pasta, potatoes;
- proteins, found in dairy products, fish and meat and pulses;
- vitamins and minerals, especially iron and calcium.

There is some evidence that increasing dietary intake of certain fatty acids, known as omega 3 fatty acids (one group of the essential fatty acids) may benefit the development of the nervous system of the foetus. Oily fish, such as tilapia, mackerel, sardines and herring, are rich sources of these fatty acids.

Pregnant women are usually recommended to avoid soft cheeses, smoked fish, precooked meat (such as chicken) and foods made with unpasteurised milk. These foods may contain a bacterium called *Listeria*. This bacterium doesn't usually cause people much harm, but even a mild infection in a pregnant woman may cause miscarriage.

Women often worry about how much they should eat once they are pregnant, and how much weight they are putting on. A pregnant woman certainly does need to eat a little more, as the food she eats must supply her growing foetus as well as herself. In general, a woman will put on around 10 to 13 kg during pregnancy.

Alcohol during pregnancy

Should a woman give up alcohol once she is pregnant?

Moderate drinking is thought to pose some risk of having a low-birthweight baby. One study has shown that women drinking four alcoholic drinks a week in the early stages of pregnancy had babies

Figure 12.21 Heavy drinking can damage the health of a baby.

which were on average 155 g lighter than babies born to non-drinking mothers. This is a small difference, but it does need to be borne in mind.

However, regular heavy drinking most certainly can cause major problems for the unborn child. Alcohol easily crosses the placenta. When a pregnant woman drinks alcohol, it goes into her blood and can then diffuse from her blood into her baby's blood (Figure 12.21).

A child born to a mother who is a heavy drinker, and who often has blood alcohol levels of more than 80 mg 100 cm^{-3} of blood, may have foetal alcohol syndrome. Alcohol affects the development of the nervous system, so the child's brain does not grow as it should. There is reduced growth both before and after birth. Other symptoms of this syndrome include poor muscle tone, abnormal limbs and heart defects. There is an increased risk of the child having a cleft palate. A baby born with foetal alcohol syndrome may have learning difficulties throughout life.

It is often recommended that a pregnant woman should limit her drinking to no more than two units a week. A unit is a small glass of wine or half a pint of beer.

Smoking and pregnancy

Everyone who smokes knows that it is bad for their health. They know they should give up, but many don't have the willpower or motivation to do it. However, smoking during pregnancy is a different matter. It isn't only the mother's health that is affected, but also her unborn baby's. Moreover, the number of miscarriages is higher in women who smoke.

Every time a pregnant woman smokes, harmful chemicals from the cigarette smoke enter her blood. They readily pass through the placenta and enter the foetus's blood. It is as though the foetus is smoking cigarettes, too.

There are many different harmful chemicals in tobacco smoke. Of these, perhaps carbon monoxide and nicotine are the most dangerous for the foetus.

Carbon monoxide combines with haemoglobin (Hb) in the foetus's red blood cells. If the Hb is combined with carbon monoxide then it cannot combine with oxygen. So when a mother smokes, she reduces the amount of oxygen being carried in her own blood and also the baby's blood.

Nicotine reduces the diameter of the foetus's blood vessels. This reduces the volume of blood that can flow through them. This, too, reduces the amount of oxygen reaching the foetus's developing tissues. Nicotine also appears to affect the development of the nervous system.

In addition, it has been found that the combination of carbon monoxide and nicotine can reduce the size of the baby's lungs by as much as 30%.

These effects of carbon monoxide and nicotine result in retardation of the foetus's growth in the uterus. They increase the chance of the baby being born prematurely, of respiratory problems in the newborn child, and also of perinatal mortality – that is, death of the child before, during or shortly after birth.

There are also suggestions that the damaging impact of the mother's smoking in pregnancy affects the child's development in the years following birth, even if the mother is no longer smoking then.

Nutrient	Function	Good food sources	Notes
carbohydrates	Providing energy, which is released from them by respiration inside body cells.	Bread, potatoes, rice, pasta, breakfast cereals, pulses.	Complex carbohydrates (starches) are better than sugars, because the energy in them is released steadily.
proteins	Forming new cells and tissues. Also for formation of haemoglobin, plasma proteins, collagen (in skin and bones), enzymes.	Meat, eggs, fish, dairy products, pulses.	Proteins contain 20 different amino acids, of which 8 are essential in the diet as the body cannot make them from other amino acids.
lipids	Making cell membranes. Formation of nerve cells. Providing energy when broken down in respiration.	Dairy products, red meat, oily fish.	Lipids contain several different fatty acids, of which two are essential.
vitamin A (retinol)	Making the pigment rhodopsin, needed for vision in dim light.	Meat, egg yolks, carrots.	Daily doses at around 100 times the recommended daily intake are toxic. Pregnant women should not eat too much of this vitamin.
vitamin D	Formation of bones and teeth.	Dairy foods, oily fish, egg yolks.	This vitamin is made in the skin when it is exposed to sunlight.
folic acid	Formation of the neural tube.	Dark green vegetables, some breakfast cereals, milk products, oranges and bananas.	Folic acid supplements are often recommended before pregnancy and up to about the 12th week, after which the neural tube will have formed.
iron	Formation of haemoglobin.	Meat, beans, chocolate, shellfish, eggs.	Shortage of iron can cause anaemia in the mother.
phosphorus and calcium	Bone formation.	Dairy products, fish (especially with bones, e.g. sardines).	

Table 12.1 Nutrients important for maintaining foetal growth.

3 The table shows some statistics about smoking by women before and during pregnancy in England in 2000.

Age of mother	Percentage who smoked in the year before or during pregnancy	Percentage who smoked throughout pregnancy	Percentage of smokers who gave up smoking in the year before or during pregnancy
under 20	64	39	38
20–24	52	29	44
25–29	36	19	45
30–34	25	12	50
35 and over	23	12	48

a i Describe the relationship between smoking in the year before or during pregnancy and age.

ii Suggest reasons for this relationship.

iii Do the data about the percentages of women who smoked throughout pregnancy support your answer to **ii**? Explain your answer.

b The number of women between 25 and 29 who took part in this survey was 1397. Calculate the number of smokers in this age category who:

i smoked in the year before or during pregnancy;

ii gave up in the year before or during pregnancy.

c Explain why health professionals always strongly advise a woman smoker to give up while she is pregnant.

Other drugs and pregnancy

We have seen that most drugs freely cross the placenta and enter the foetus's circulation. Whatever drug a mother takes, her foetus takes it too.

It almost goes without saying that any illicit drug will harm the developing foetus. The reason that a drug is illicit is that it has the potential to cause harm to a person's health, and the risk of this happening to a tiny foetus is even greater than for an adult, because the drug will affect its entire development. For example, there is considerable evidence that if a mother smokes marijuana (cannabis) during pregnancy, her baby is more likely to be underweight at birth. Addictive drugs such as cocaine or heroin will also cause addiction in the baby. Starting life as a drug addict does not give an individual the best opportunities for success.

In general, a woman should not take any drugs – even commonly used legal drugs – during pregnancy that have not been agreed with her doctor or healthcare worker. There are several drugs that can safely help a pregnant woman with any health problems, but care should be taken not to unwittingly harm the developing baby.

Birth control

The use of birth control gives people control over the number of children that they have. Contraception means 'against conception', and specifically refers to methods that can be used to prevent fertilisation when sexual intercourse takes place. There are also several methods of birth control that do not prevent conception but instead prevent the tiny embryo from implanting into the lining of the uterus. These can be termed anti-implantation methods and they include the use of IUDs and the so-called 'morning-after pill'.

The term 'family planning' refers to the ways in which a couple can control when to have their children and how many they have. It may involve contraception or anti-implantation methods.

Methods of contraception

The birth control pill

At the moment, the only birth control pill is prescribed for women and was developed in the 1960s. Its introduction had a huge impact on the freedom of women to have sexual intercourse without the risk of pregnancy. While most would consider that this has been a great advance, it has also contributed to the rise in the incidence of sexually transmitted diseases, including HIV/AIDS, as more women have had unprotected sex with more than one partner.

The pill contains steroid hormones that suppress ovulation. When a woman becomes pregnant, the levels of hormones in her body change. This prevents her ovaries from producing any ova while she is pregnant and for a period of time after birth. In simple terms, the birth control pills are adding progesterone and oestrogen hormones to a woman's metabolism, mimicking the normal hormone concentrations during pregnancy. However, without a real pregnancy, the cycle of hormones in the pill also has to allow menstruation so that the endometrium remains healthy.

Usually, synthetic hormones rather than natural ones are used, because they are not broken down so rapidly in the body and therefore act for longer.

The **mini birth control pill** contains progesterone only. This works by thickening the cervical mucus, stopping sperm from meeting an egg. More effective than this is the **combined birth control pill** containing both progesterone and oestrogen. There are many different types of combined birth control pill, with slightly different ratios of these two hormones, as women are not all alike in the way their bodies respond to the pill.

With most types of oral contraceptive, the woman takes one pill daily for 21 days, and then stops for 7 days during which time menstruation occurs. With some types, she takes a differently coloured, inactive, pill for these 7 days.

4 The graph shows part of a woman's 28-day oral contraceptive cycle. The top row shows the days on which she took a progesterone and oestrogen pill. The part of the graph below this illustrates the changes in levels of progesterone and oestrogen (steroids) in her blood. The bottom graph shows the activity of the follicles in her ovaries.

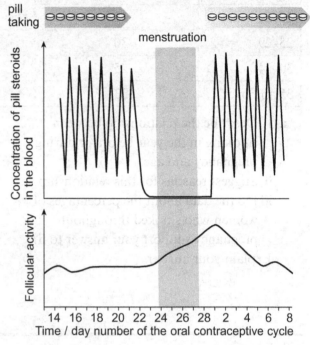

a How many days of the cycle are shown in these graphs?

b Describe the patterns shown by the level of steroids in the woman's blood, and relate these to her pill-taking schedule.

c Describe the patterns shown by the level of follicular activity. Explain how the levels of steroids in the blood can cause the patterns you describe.

d A medical textbook states: 'Towards the end of the 7-day pill-free period, some women can come dangerously close to ovulating.' Using the data on the graph, and also your knowledge of the roles of hormones in follicle development and ovulation, discuss this statement.

e One way around the problem described in **d** would be for the women to take the pill all the time, without a 7-day break. Suggest why this is not the normal way in which oral contraceptives are used.

Condoms

Condoms are a very widely used method of contraception. The use of a condom is a mechanical or barrier method of contraception, in which a physical barrier is placed between sperm and ova. Many people like to be able to use a contraceptive method as and when they need it, rather than altering their body physiology long-term as is done with oral contraceptives or sterilisation and also, to a lesser extent, using an IUD (Figure 12.22).

Condoms are cheap and readily available. They are made from a material that does not allow the passage of sperm, nor of bacteria or viruses. So, apart from their contraceptive effect, they also help to prevent the transmission of HIV/AIDS and other sexually transmitted infections. With careful use, they have a very high success rate in preventing conception.

Condoms are the only method of contraception, apart from sterilisation, that is used by men. However, many women prefer to be in control of preventing their own pregnancy, and choose to use a femidom. Like a condom, this is an impermeable barrier, but it is placed inside the vagina instead of over the penis.

The effectiveness of both condoms and femidoms is increased even further if they are used with spermicidal cream inside the vagina.

Diaphragm

The diaphragm or cap is a flexible device which a woman can insert into the vagina so that it sits over the cervix. Like a condom, it provides a physical barrier preventing sperm from reaching an egg. Also like a condom, it should be used with spermicidal cream to be certain that no sperm can get past (Figure 12.23).

The diaphragm should not be worn all the time, but can be inserted some time before it is expected to be needed. It must be left in place for at least six hours after intercourse. A woman will need to be 'fitted' with a diaphragm initially, as it needs to be a snug but comfortable fit, but from then on she can easily put it into position herself. Nevertheless, many women find it a clumsy method and prefer to use a different method of contraception.

Implanted contraceptives

Contraceptive hormones can be delivered via a device which is implanted into a woman's body and gives her protection against conception over

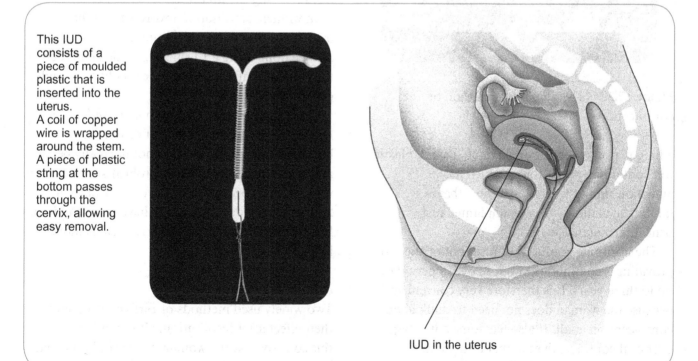

This IUD consists of a piece of moulded plastic that is inserted into the uterus.
A coil of copper wire is wrapped around the stem. A piece of plastic string at the bottom passes through the cervix, allowing easy removal.

IUD in the uterus

Figure 12.22   The IUD method of contraception.

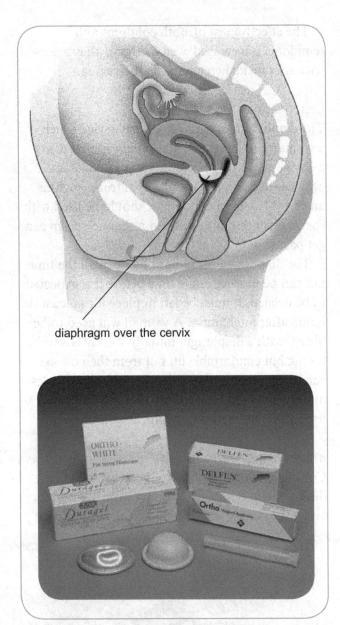

diaphragm over the cervix

Figure 12.23 The diaphragm method of contraception.

a long period of time. For example, a small plastic rod containing progestogen, which has an action similar to progesterone – that is, it inhibits ovulation – may be implanted just beneath her skin.

The hormone gradually diffuses out of the rod, providing a continuous low dose over a period of up to three years. It is therefore very convenient because the woman does not need to think about contraception at all. It also has some advantage over oral contraceptives in that it maintains a steady concentration in the blood that is lower than that seen soon after each pill is taken.

On the down side, as in all the non-barrier methods, it does not have any protective effect against the transmission of bacteria or viruses. If the woman decides that she wants to become pregnant, the implant can be removed.

DMPA (Depo-Provera®)

Depo-Provera® is the registered trademark of a contraceptive that is injected into a woman's body about once every 12 weeks. It provides continuous protection against conception throughout this time.

Depo-Provera® contains a synthetic hormone similar to progesterone called non-cyclic medroxyprogesterone acetate or DMPA. It prevents ovulation. To maintain contraception, injections must be given every 12 weeks.

Sterilisation

When a couple have had as many children as they want, they may opt for either the man or the woman being sterilised.

In men, sterilisation is achieved by vasectomy. In a relatively simple operation, usually done under local anaesthesia, the vasa deferentia leading from both testes are cut and tied, so preventing sperm from entering the urethra.

In women, sterilisation involves tying the oviducts, a procedure called tubal ligation. It can also be done using a Filshie clip, which clips across the oviduct. Tubal ligation is normally done under general anaesthesia.

If properly carried out, both male and female sterilisation are 100% effective. The big disadvantage is that it cannot normally be reversed, and this may cause problems for a person who, perhaps because they have a new partner, finds that they do still want to have more children. If a Filshie clip has been used, reversal is more likely to be successful.

Anti-implantation methods

Two widely used methods of birth control have their effect after fertilisation, although they do this so early that the woman will never know that conception has taken place.

IUD

IUD stands for intra-uterine device. It is a small, folded piece of plastic and copper that fits inside the uterus (Figure 12.22, page 247).

The uterus responds to the presence of the IUD in the same way that it would respond to a slight bacterial infection. Leucocytes congregate in the uterus lining, and cytokines are secreted that bring about a low-level immune response. This helps to prevent sperm from passing through (and therefore it can also be considered to be a contraceptive) and prevents a blastocyst from implanting into the uterus lining. Moreover, the copper in the IUD is toxic to both sperm and the young embryo.

An IUD has to be placed in the uterus by a doctor or specialist nurse, and once in place it is left there. For some women, it is not suitable because it causes discomfort and also because it carries a small risk of infection. IUDs are therefore not usually recommended for young people who wish to have children later.

The morning-after pill

This form of birth control is intended to be taken after a woman has had unprotected sexual intercourse and thinks that she might be pregnant. It might be taken by a woman who forgot to take her oral contraceptive pill, or if a condom broke, or by someone who was raped, as well as by a woman who simply did not take any precautions to prevent pregnancy. It works up to 72 hours afterwards, not just the 'morning after'.

The pill contains a synthetic progesterone-like hormone. If taken early enough, it reduces the chances of a sperm reaching and fertilising an egg. However, in most cases it probably prevents a pregnancy by stopping the embryo implanting into the uterus.

Ethical issues arising from birth control

People have used methods of birth control for thousands of years, but the last few decades have seen a huge increase in their efficacy and in the ease with which people can obtain and use them. This has led to disagreement between groups of people with strongly held views.

Some of the arguments against the widespread use of contraceptives include the following.

- The easy availability of a wide range of highly effective contraceptives appears to have led to an increase in promiscuity among people of all ages, but especially the young. Some people take the view that young people should wait until they are married before having sex, or at least should limit their sexual activities to a single partner.
- Parents are concerned that their children may be sexually active without their knowledge, because young people can get contraceptives easily and in privacy.
- Some people see the use of anti-implantation pills as being equivalent to 'unsupervised abortion on demand'.

Opinions on the other side of the debate are equally strongly voiced. Some of the advantages put forward include the following.

- It is likely that easy access to contraception prevents many unwanted pregnancies and many terminations that could cause great distress to young people and their families.
- Contraceptives have given increased control over their bodies and their lives to women in circumstances where their partners have no interest in preventing them from becoming pregnant.
- There is an urgent need to try to reduce world population growth.

Summary

- Spermatogenesis takes place in seminiferous tubules in the testes. Diploid spermatogonia grow into diploid primary spermatocytes, which then undergo the first division of meiosis to form haploid secondary spermatocytes, and then the second division to form haploid spermatids, which differentiate into spermatozoa.

- Oogenesis takes place in the ovaries, where diploid oogonia go through the early stages of meiosis I to become diploid primary oocytes. Some of these complete meiosis I, becoming haploid secondary oocytes. Each month, one of these is released into the oviduct, where it may be fertilised. This triggers the second division of meiosis, and the nucleus produced by this fuses with the nucleus of a sperm cell to form a diploid zygote.

- Gametogenesis and the menstrual cycle are controlled by hormones. In a woman, this involves secretion of FSH and LH from the anterior pituitary gland, and oestrogen and progesterone from the ovaries. The cycle involves negative feedback by oestrogen, which reduces the secretion of FSH and LH.

- Fertilisation happens in the oviduct, and results in the formation of a zygote which divides by mitosis to form a blastocyst. This travels to the uterus, where it implants and develops into first an embryo and then a foetus.

- The placenta develops from a combination of foetal and maternal tissues. It allows easy exchange of solutes between the mother's blood and that of her foetus. It also secretes progesterone. The amnion surrounds the foetus, secreting amniotic fluid that provides support and protection.

- A woman who is thinking of becoming pregnant should check that she is immune to rubella, and should take care to eat a well-balanced diet, including folic acid. Once pregnant, her diet should contain plenty of iron and calcium. Alcohol, nicotine and all illicit drugs should be avoided during pregnancy.

- Contraceptive methods can be used to reduce the risk of a woman becoming pregnant. Many of the most effective methods involve the use of hormones that mimic those involved in the normal control of reproduction, and prevent either ovulation or implantation. Condoms are especially useful in also preventing the transfer of pathogens, such as the HIV/AIDS virus, between partners.

Questions

Multiple choice questions

1 Into which structure is the secondary oocyte released at ovulation?
 A uterus **B** oviduct **C** cervix **D** ovary

2 When does the secondary oocyte complete meiosis? When the:
 A sperm enters the oocyte.
 B luteinising hormone triggers ovulation.
 C acrosomal enzyme digests the zona pellucida.
 D sperm approaches the oocyte.

continued …

3 The male reproductive system is shown below.

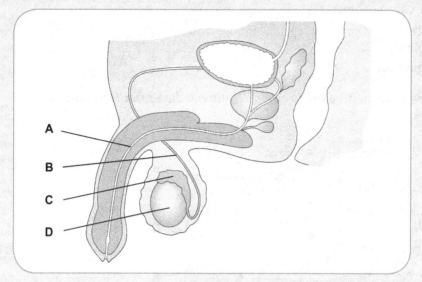

In which tract are sperm stored to complete maturation before they leave the body?

4 Which of the following **correctly** identifies the structural changes which occur to the ovary during the human menstrual cycle?
  A ovulation → corpus luteum development → follicle development
  B follicle development → ovulation → corpus luteum development
  C corpus luteum development → follicle development → ovulation
  D follicle development → corpus luteum development → ovulation

5 Ovulation and production of hormones by the ovary are controlled by pituitary hormones. Which of the following correctly identifies these hormones?
  A GnRH and FSH
  B FSH and LH
  C oestrogen and FSH
  D progesterone and oestrogen

6 The combined birth control pill contains high doses of progesterone and oestrogen. What is the biological basis of this birth control pill?
  A It stops ovulation by inhibiting the secretion of FSH and LH.
  B It causes the lining of the uterus to become thin.
  C It makes the mucus around the cervix become thinner.
  D It causes the corpus luteum to degenerate.

7 In which part of the female reproductive system does fertilisation of the ovum occur?
  A oviduct (fallopian tube) **B** uterus **C** corpus luteum **D** vagina

8 Which of the following comparisons between oogenesis and spermatogenesis is correct?
  A Egg development is promoted by LH while sperm development is stimulated by FSH.
  B The development of the primary spermatocytes is continuous while that of the oocytes is not.
  C Both processes produce four gametes per germ cell.
  D Both processes start in the foetal period.

continued ...

9 What do both the sperm and ovum have in common?

   A a haploid set of chromosomes

   B cortical granules

   C very little cytoplasm

   D a large mitochondrion

10 The diagram below shows a developing embryo fourteen days after fertilisation.

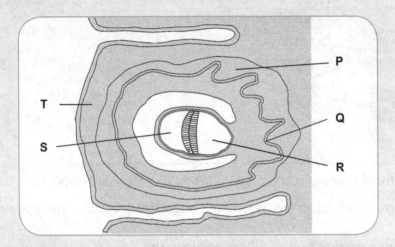

Which of the labelled tissues are foetal and which are maternal?

	P	Q	R	S	T
A	maternal	maternal	foetal	foetal	foetal
B	foetal	foetal	maternal	foetal	maternal
C	maternal	foetal	foetal	foetal	maternal
D	foetal	foetal	foetal	maternal	maternal

11 The diagram below shows a section of a human secondary oocyte.

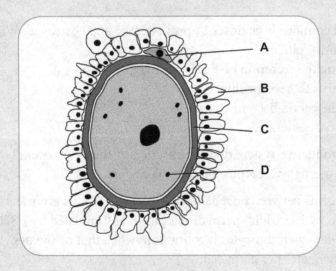

Which of the labelled structures represents the first polar body?

continued …

Structured questions

12 The micrograph below shows a Graafian follicle of a mammal.

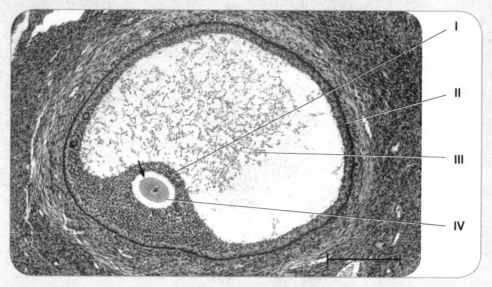

a i What are the visible characteristics of a Graafian follicle? [2 marks]
   ii Identify the structures labelled **I** to **IV**. [4 marks]
   iii The scale bar represents 50 μm. Calculate the magnification of the micrograph. [2 marks]
b Make a fully labelled drawing to show the detailed structure of the secondary oocyte that is shown in the micrograph. [4 marks]
c The structure shown by the arrow in the micrograph is a secondary oocyte. Explain why the term 'ovum' is not used to describe it. [2 marks]
d The development of the primordial follicle, ovulation and formation of the corpus luteum are controlled by hormones produced by the pituitary gland and the ovary. The diagram below shows a graph of these hormonal changes.

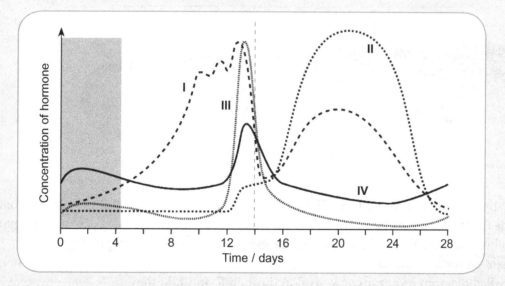

Identify the hormones labelled **I** to **IV**. [2 marks]

continued …

13 The diagram below shows the foetus 10 weeks after fertilisation.

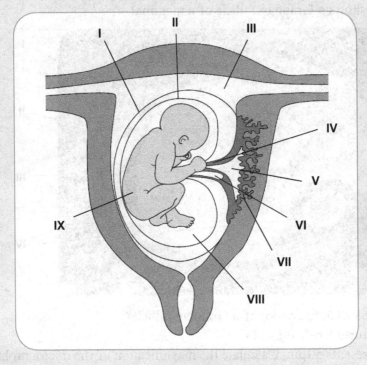

a Identify the structures labelled **I–IX** in the diagram.   [5 marks]
b Identify the structures which are termed 'extra-embryonic membranes'.   [2 marks]
c Give **three** functions of structure **II** during pregnancy.   [3 marks]
d The actions of the mother during pregnancy can affect the development of the foetus.
   What are the possible effects of smoking and drinking excessive alcohol on
   the foetus?   [5 marks]

14 The diagram below shows a mature human sperm.

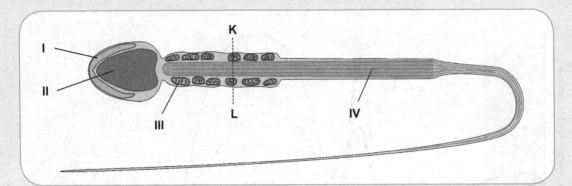

a i   Identify the structures labelled **I, II, III** and **IV**.   [2 marks]
   ii   State **one** function each of **I** and **II**.   [2 marks]
   iii Explain how the structures labelled **III** and **IV** assist the sperm in moving through
        the female reproductive tract.   [2 marks]
b Draw a labelled diagram of a transverse section through **K** and **L** in the
   diagram above.   [4 marks]

continued ...

c The table below shows the results of an experiment investigating factors affecting sperm quality in the Asian elephant, *Elephas maximus*. One of the factors was age and the quality of the sperm was measured in terms of viable sperm and normal morphology.

Age range / years	% viable sperm	% normal morphology
10–19	46.50	73.12
23–43	33.70	84.48
51–70	27.10	66.98

   i Plot a histogram to show the relationship between age and percentage viable sperm and percentage normal morphology. [4 marks]

   ii Suggest one possible conclusion from the results of this experiment. [1 mark]

Essay questions

15 a Hormones which control the menstrual cycle are produced by the pituitary gland and the ovary.

   i Identify these hormones. [2 marks]

   ii State the main site of secretion of the hormones identified in **a i** and describe their effects on the endometrium of the uterus and the ovary. [4 marks]

b Describe the roles of the hormone testosterone in spermatogenesis. [3 marks]

c Discuss the biological basis of the following contraceptive methods:

   i IUD

   ii combined pill

   iii vasectomy [6 marks]

16 a The placenta has a number of important roles to play in reproduction.

   i Identify **three** functions of the placenta. [3 marks]

   ii Draw an annotated diagram to show how the placenta is adapted for efficient transport of substances. [4 marks]

b Some substances are able to pass across the barrier separating maternal and foetal blood in the placenta while others cannot.

   i Describe **two** processes by which named substances can cross the placental barrier. [2 marks]

   ii Give **one** reason why some substances are able to cross the placenta. [1 mark]

c Discuss the possible effects of maternal behaviour on foetal development. (Make reference to cigarette smoking, alcohol abuse, role of nutrition). [5 marks]

17 a Sperm needs to spend about five hours in the female reproductive tract before they are capable of fertilising a secondary oocyte.

   i Explain the changes which occur to the sperm before it can fertilise a secondary oocyte. [3 marks]

   ii Describe the events of fertilisation and explain how penetration by more than one sperm is prevented. [6 marks]

b Compare the processes of spermatogenesis and oogenesis. [5 marks]

SAQ answers

Chapter 1

1. **a** Water requires a relatively large amount of heat energy to evaporate – that is, water has a high heat of vaporisation. Heat energy which is transferred to water molecules in sweat allows them to evaporate from the skin, which cools down, helping to prevent the body from overheating. A relatively large amount of heat can be lost with mimimal loss of water from the body.

   b Water moves up each xylem vessel as a tall column, by mass flow. Cohesion between the water molecules helps to hold the column together. If the column broke, then the pulling force exerted by transpiration in the leaves would not be transmitted to the whole of the column, and the water would not move up the plant in this way.

   c Although the atoms in glucose are covalently bonded, there is a small electrical charge on the hydrogen atom and oxygen atom of each hydroxyl group in the molecule. These dipoles weakly bond with water molecules. This is known as hydrogen bonding. The bonding of water and glucose is stronger than the tendency for glucose molecules to associate in a solid lattice, so glucose is soluble in water.

   d In solution, particles can move around, allowing reactants to meet, which they would otherwise not be able to do so easily as solids. Some reactions may require the presence of hydrogen ions, which water provides.

   e The mineral ions in sea water, mainly sodium and chloride ions, bind water molecules through hydrogen bonding. This makes it less likely that water undergoes sufficient hydrogen bonding with itself to form ice until the temperature falls well below $0\,°C$.

2. **a** $C_3H_6O_3$
   b $C_5H_{10}O_5$

3. Prepare a standard solution of glucose. This could be at a concentration of $0.1\,mol\,dm^3$, made by dissolving $18\,g$ of glucose and making it up accurately to $1\,dm^3$. Put a volume of the unknown solution into one tube and the same volume of standard solution into another. Add equal volumes of Benedict's reagent to each tube. Heat both. Weigh each of two filter papers. Filter to collect the precipitate from each tube. Dry the filter papers and precipitates to constant weight. Find the mass of each precipitate by subtraction. Determine the mass of precipitate you get from the known mass of glucose from the standard solution. Use this value to determine the mass of reducing sugar in the unknown solution.

4. Add Benedict's reagent to a sample of each and heat, then stop heating. The tube that does not produce a red precipitate contains the non-reducing sugar. Filter the contents of the two tubes that produced a precipitate. Add a few drops of dilute hydrochloric acid to the filtrate, boil for a minute and then add Benedict's reagent. The solution that contained a mixture of reducing sugar and non-reducing sugar will produce much more red precipitate.

5.

	Amylose	Cellulose	Glycogen
monosaccharide from which it is formed	α-glucose	β-glucose	α-glucose
type(s) of glycosidic bond	α1–4	β1–4	α1–4, α1–6
overall shape of molecule	linear, spiral	linear	linear, spiral, branching
hydrogen bonding within or between molecules	within	within and between	within
solubility in water	very low / insoluble	insoluble	very low / insoluble
function	energy reserve	structural	energy reserve

6. **a** the amino acid sequence:
   glycine – proline – alanine – glycine

b the helix in **a** or **b**

c the shape of one strand in **b**

d a triple helix in **b** or **c**

7 hydrolysis; lipase

8 **a** For example, $C_{45}H_{86}O_6$

b A glucose molecule has twice as many hydrogens as carbons, and the triglyceride also has very nearly twice as many hydrogens as carbons. But while the glucose molecule has the same number of carbon atoms as oxygen atoms, the triglyceride has more than seven times as many carbons as oxygens.

Chapter 2

1 maximum diameter of cell in micrograph
$$= 40 \, mm$$
$$\text{so magnification} = \frac{40 \times 1000}{50}$$
$$= \times 800$$

2 $$\text{magnification} = \frac{\text{size of image}}{\text{real size of object}}$$
$$= \frac{12}{5}$$
$$= \times 2.4$$

3 length of scale bar $= 24.0 \, mm$
$$\text{so magnification} = \frac{24.0}{0.1}$$
$$= \times 240$$

4 thickness of leaf on micrograph
$$= 50.5 \, mm$$
$$= 50.5 \times 1000 \, \mu m$$
$$\text{magnification} = \times 240$$
$$\text{real thickness} = \frac{50.5 \times 1000}{240}$$
$$= 210 \, \mu m$$

5 It is a TEM. It is not three-dimensional.

6 Ribosomes, details of mitochondria, centrioles, endoplasmic reticulum, nuclear pores, Golgi body, lysosomes. (It is just possible to see Golgi bodies and centrioles using a light microscope, but no detail of their structure.)

Chapter 3

1 Some amino acids within the polypeptide chain have hydrophilic side chains (e.g. containing amino groups) or hydrophobic side chains (e.g. containing hydrocarbon only).

2 The outer surface of the membrane has chains of sugars – carbohydrate – that are part of the glycolipid and glycoprotein molecules.

3 Photosynthesis in the palisade cell uses carbon dioxide, maintaining a very low concentration inside the cell. The concentration of carbon dioxide in the air outside the leaf, and in the air spaces inside the leaf, is higher than inside the cell, so carbon dioxide diffuses into the cell down its concentration gradient. The cell wall and plasma membrane of the cell are permeable to carbon dioxide.

4 The hydrophilic parts lining the channel through the membrane allow water-soluble ions to diffuse through the channel. The hydrophobic parts of the protein are found next to the lipid bilayer and keep the protein bound within the membrane.

5 The source of energy for the diffusion of a substance which moves by facilitated diffusion is from the kinetic energy of the particles diffusing and is not provided by respiration within the cells.

6 This space is outside the plasma membrane and only bounded by the cell wall. The cell wall is fully permeable to the external solution.

Chapter 4

1 6×10^8

2 There is a risk of inaccuracy in a single measurement at 30 seconds. The shape of the curve is more likely to give an accurate value, because it is based on many readings taken over a period of time, rather than just one.

3

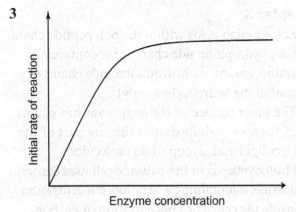

4 a

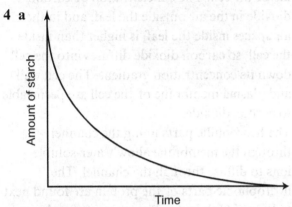

b Calculate the slope of the curve right at the beginning of the reaction.

5 Measure the volume of oxygen given off over time for several catalase–hydrogen peroxide reactions at different temperatures. In each case, all conditions other than temperature must remain constant. In particular, the same volume of hydrogen peroxide and of catalase solution should be used each time. Plot the total volume of oxygen against time for each reaction. Calculate the slope of the line at the beginning of the reaction in each case to give the initial reaction rate. Then plot initial reaction rate against temperature.

Chapter 5

1 At the 5' end, the covalent bond to the next nucleotide is with carbon number 5 on the deoxyribose. At the 3' end, it is with carbon number 3.

2

	DNA	RNA
Sugar	deoxyribose	ribose
Bases	A, C, G, T	A, C, G, U
Number of strands	two	usually one
Where it is found	in the nucleus	both nucleus and cytoplasm
Function	stores the genetic code	copies the code from DNA and uses it to make proteins

3 a 64

b In some cases, more than one triplet codes for the same amino acid. There are also triplets indicating 'start' and 'stop'.

c A two-letter code could only code for 16 different amino acids.

4 a AAA would be formed from a TTT triplet on DNA, which codes for lysine.

b ACG would be TGC on DNA, which codes for threonine.

c GUG would be CAC on DNA, which codes for valine.

d CGC would be GCG on DNA, which codes for arginine.

e UAG would be ATC on DNA, which codes for 'stop'.

5 a 4

b Tyrosine, serine, asparagine, glycine.

c UAU UCU AAC GGG

d i AUA

ii UUG

6 a Amino acids are not 'made' during protein synthesis. They are already there, in the cytoplasm. During protein synthesis they are simply linked together. A better statement would be: 'The sequence of bases in a DNA molecule determines the sequence in which amino acids are linked together during protein synthesis.'

b DNA does not contain amino acids. DNA is a polynucleotide, made of many nucleotides linked together. Amino acids are found in proteins. A better statement would be: 'The sequence of bases in a DNA molecule determines what kinds of proteins are formed in a cell.'

c The names of the bases are spelt incorrectly. The student has confused their names with those of other substances. For example 'thiamine' is a vitamin, not a base. The correct names of the bases are adenine, cytosine, thymine and guanine.

d During transcription, an mRNA molecule is built up. It is not already made, so it cannot 'come and lie' next to the DNA strand. A better statement would be: 'During transcription, a complementary mRNA molecule is built up against part of a DNA molecule.'

7 The diagram should show the mRNA moved to the left, so that the codon UCC is now on the left inside the ribosome, and the next codon (UGC) is on the right. The tRNA molecule with the anticodon AGG is sitting directly above the UCC codon, and its amino acid (serine) has now joined the chain. Another tRNA molecule, with the anticodon ACG and carrying the amino acid cysteine is about to bond with the UGC codon on the mRNA.

Chapter 6

1 4

2 a 32
b 64

3 meiosis I

4 a Possible. Any cell with chromosomes in its nucleus can divide by mitosis, as chromosomes do not pair up.
b Possible. Chromosomes can pair up.
c Possible, as for **a**.
d Not possible. Each chromosome needs to be able to pair up with its homologous partner in order to undergo meiosis, and in a haploid cell there is only one copy of each chromosome.

5 a prophase I
b prophase I
c anaphase I
d anaphase II
e telophase I

Chapter 7

1 a We have no idea how many children they will have – they may have none or a large number. Nor can we tell how many will have cystic fibrosis and how many will not. The genetic diagram tells us only the *probability* of any one child having cystic fibrosis. All we can say is that, each time they have a child, there is a one in four chance that it will have cystic fibrosis.
b If their first child has cystic fibrosis, this does not affect the chances of the next one having it as well. For that second child, the chances are still one in four that it will have cystic fibrosis.

2 The gametes from the father are F and f. The genotypes of the offspring are FF and Ff, in equal proportions (you could write this as 1:1). Therefore the chance of a child with cystic fibrosis is zero.

3 All of the gametes from the female parent are the same. If we show the same one twice over, we just end up doing unnecessary work in doubling up everything in the diagram.

4 You should find that one in four children born to this couple would be expected to have blood group O. This can be expressed as a probability of 0.25 or 25%.

5 A son is conceived when a sperm carrying a Y chromosome fuses with an egg (all eggs contain an X chromosome). Therefore none of the genes on a man's X chromosome, including the haemophilia gene, can be passed on to his son.

6 a The condition appears to be dominant, as it seems to appear in the sons and daughters of everyone who has it. A mutation may have occurred in the ovaries or testes of one of the two original parents; they did not show the condition themselves as the mutant allele was not present in their body cells, but it was passed on to some of their children in their gametes. It would be possible for this pedigree to arise if the allele was recessive, but this is unlikely.
b Equal numbers of males and females have brachydactyly, so it does not appear to be sex-linked. However, as it is a dominant

allele, it could possibly produce most of this pattern even if it was on the X chromosome. (Try this out on one or two parts of the pedigree.) The piece of evidence against this possibility is that a man in the middle row has a son who has the condition, even though the man's wife did not have it. As we have seen, it is not possible for a man to pass on an allele on his X chromosome to his son.

7   a   Male cats cannot be tortoiseshell because a tortoiseshell cat must have two alleles of this gene. As the gene is on the X chromosome, and male cats have one X chromosome and one Y chromosome, they can only have one allele of the gene.

   b
   C^O is the allele for orange fur; C^B is the allele for black fur

phenotypes of parents:	male $\times$ orange fur	female tortoiseshell fur
genotypes of parents:	$X^{C^O}Y$	$X^{C^O}X^{C^B}$
genotypes of gametes:	X^{C^O} and Y	X^{C^O} and X^{C^B}

   genotypes and phenotypes of offspring:

gametes from mother	gametes from father X^{C^O}	Y
X^{C^O}	$X^{C^O}X^{C^O}$ female orange fur	$X^{C^O}Y$ male orange fur
X^{C^B}	$X^{C^O}X^{C^B}$ female tortoiseshell fur	$X^{C^B}Y$ male black fur

   The kittens would therefore be expected to be in the ratio of 1 orange female : 1 tortoiseshell female : 1 orange male : 1 black male.

8   Each cell should be shown with two chromosomes (each chromosome will be a single thread, not with two chromatids), one long with one of the A/a alleles and one short with one of the B/b alleles.

9   all Ab

10 AB and Ab

11 a   She is ffI^AI^O. He is FFI^OI^O.
   b   The woman's gametes will have genotypes fI^A and fI^O in equal proportions.
   The man's gametes will all be FI^O.
   The possible genotypes of the children are therefore FfI^AI^O and FfI^OI^O, with an equal chance of each combination arising.

The phenotypes corresponding to these two genotypes are 'no cystic fibrosis with blood group A', and 'no cystic fibrosis with blood group O'.

12 a   The expected offspring would be 1 purple cut : 1 purple potato : 1 green cut : 1 green potato.
   b   The expected offspring would be 9 purple cut : 3 purple potato : 3 green cut : 1 green potato.

13 a   Albino could be AAcc, Aacc or aacc. Black could be aaCc or aaCC. Agouti could be AACc, AaCc, AACC or AaCC.
   b   The table of offspring genotypes and phenotypes will look like this:

   genotypes and phenotypes of offspring:

gametes from mother	gametes from father AC	Ac	aC	ac
AC	AACC agouti	AACc agouti	AaCC agouti	AaCc agouti
Ac	AACc agouti	AAcc albino	AaCc agouti	Aacc albino
aC	AaCC agouti	AaCc agouti	aaCC black	aaCc black
ac	AaCc agouti	Aacc albino	aaCc black	aacc albino

   You would therefore expect a ratio of 9 agouti : 3 black : 4 albino.
   c   The mouse could have any genotype with both an A allele and a C allele. We are not interested in the A allele. To find out if the mouse is CC or Cc, cross it with an albino mouse. We know that an albino mouse must have the genotype cc. If we get any albino mice in the offspring, then both parents must have contributed a c allele, so our unknown mouse must have been Cc. If we do not get any albino offspring, then the unknown mouse is probably CC – but we cannot be sure, as it is possible that just by chance none of its c gametes was involved in fertilisation.

14 White could be AABB, AABb, AAbb, AaBB, AaBb or Aabb. Black could be aaBB or aaBb. Brown can only be aabb.

genotypes and phenotypes of offspring:

gametes from male

	(AB)	(Ab)	(aB)	(ab)
(AB)	AABB white	AABb white	AaBB white	AaBb white
gametes from female **(Ab)**	AABb white	AAbb white	AaBb white	Aabb white
(aB)	AaBB white	AaBb white	aaBB black	aaBb black
(ab)	AaBb white	Aabb white	aaBb black	aabb brown

White, black and brown cats would be expected to be produced in the ratio of 12 : 3 : 1.

15 a i GGC^DC^P or GgC^DC^P

ii ggC^DC^P

b The genotypes of the parents are GgC^DC^P or GGC^DC^P (dark green) and ggC^DC^P (cobalt blue).

genotypes and phenotypes of offspring:

gametes from father

	(GC^D)	(GC^P)
gametes from mother (gC^D)	GgC^DC^D olive green	GgC^DC^P dark green
(gC^P)	GgC^DC^P dark green	GgC^PC^P light green

or

genotypes and phenotypes of offspring:

gametes from father

	(GC^D)	(GC^P)	(gC^D)	(gC^P)
gametes from mother (gC^D)	GgC^DC^D olive green	GgC^DC^P dark green	ggC^DC^D mauve	ggC^DC^P cobalt blue
(gC^P)	GgC^DC^P dark green	GgC^PC^P light green	ggC^DC^P cobalt blue	ggC^PC^P sky blue

16 Red peppers can be AABB, AABb, AaBB or AaBb. Orange peppers can be AAbb or AAbb. Yellow peppers can be aaBB or aaBb. Lemon yellow peppers can only be aabb.

Phenotypes of parent AABB is red and aabb is lemon yellow.

Genotype of parent AABB gametes is all AB.

Genotype of parent aabb gametes is all ab.

Genotype of the F_1 is AaBb.

Phenotype of the F_1 is red.

Genotypes and phenotypes of the F_2 are:

genotypes and phenotypes of offspring:

gametes from male

	(AB)	(Ab)	(aB)	(ab)
(AB)	AABB red	AABb red	AaBB red	AaBb red
gametes from female **(Ab)**	AABb red	AAbb orange	AaBb red	Aabb orange
(aB)	AaBB red	AaBb red	aaBB yellow	aaBb yellow
(ab)	AaBb red	Aabb orange	aaBb yellow	aabb lemon yellow

Red, orange, yellow and lemon yellow fruits are expected in the ratio of 9 : 3 : 3 : 1.

17 a The offspring would all have genotype GgTt, and phenotype grey fur and long tail.

b Grey long, grey short, white long, white short in a ratio of 9 : 3 : 3 : 1.

c The total number of offspring is 80, so we would expect $\frac{9}{16}$ of these to be grey long and so on.

Expected numbers:

$\frac{9}{16 \times 80} = 45$ grey, long

$\frac{3}{16 \times 80} = 15$ grey, short

$\frac{3}{16 \times 80} = 15$ white, long

$\frac{1}{16 \times 80} = 5$ white, short

	Grey long	Grey short	White long	White short
Observed number, O	54	4	4	18
Expected number, E	45	15	15	5
$O - E$	+9	−11	−11	+13
$(O - E)^2$	81	121	121	169
$\dfrac{(O - E)^2}{E}$	1.8	8.1	8.1	33.8
$\sum \dfrac{(O - E)^2}{E} = 51.8$				
$\chi^2 = 51.8$				

This gives a huge value for χ^2.
Now look at the Table 7.1 on page 144.

We have four classes of data, so there are 3 degrees of freedom. Looking along this line, we can see that our value for χ^2 is much greater than any of the numbers there, and certainly well above the value of 7.82, which is the one indicating a probability of 0.05 that the difference between the observed and expected results is due to chance.

Our value is way off the right-hand end of the table, so we can be certain that there is a significant difference between our observed and expected results. Something must be going on that we had not predicted.

(In fact, these results suggest that these two genes are linked.)

Chapter 8

1 The cuts all produce sticky ends with four unpaired bases.

2 **a** Restriction enzymes are used to cut DNA into shorter lengths. There are many different kinds, and each cuts DNA at a particular sequence of bases. They therefore cut each strand at a different point, leaving sticky ends. If the same kind of restriction enzyme is used to cut another piece of DNA, then the sticky ends will be complementary and will form hydrogen bonds with the first piece.

 b DNA ligase is used to link together different lengths of DNA. It catalyses the formation of bonds between the phosphate and deoxyribose groups in the backbone of the DNA molecule.

 c Reverse transcriptase is used to synthesise DNA molecules using mRNA as a template.

3, 4, 5 These questions do not have definitive answers. They would be best used as starting points for class discussion, during which sets of bullet points could be developed.

Chapter 9

1 Sex is the clearest example. Tongue rolling and shape of earlobes are other possibilities.

2 **a** Both genes and environment.

 b Probably genes only.

 c Both genes and environment.

 d Environment only. (It depends on how many ovules were fertilised.)

3 **a** Deaths from *S. aureus* that are not specified as resistant have remained fairly constant, at between 350 and 500 per year over this time period. However, deaths from MRSA have increased from less than 100 in 1993 to around 1600 in 2005. This is an increase of over 16 times.

 b The figures are for deaths, not for infections. It is likely that there have been many more infections than deaths, especially for non-resistant bacteria, which are easily treated with antibiotics. (The constant death rate from the non-resistant bacteria could possibly mask an increase in actual infections – but we cannot know this.) The figures suggest that there has been a steady increase in the number of people being infected with MRSA, and this could be because MRSA is becoming more common. The resistance arises as the result of selection – if antibiotics are used, then bacteria that have resistance to them are more likely to survive. They pass on their resistance genes to their offspring, producing a population of resistant bacteria. The numbers have increased because antibiotics continue to be used and so populations of resistant bacteria continue to breed more successfully than those that are not resistant.

4 **a** Costs remained relatively constant between 1980 and 1991, varying around a value of about $40 per acre. There was a sharp rise in 1992, when costs more than doubled, to about $100 per acre. Between 1992 and 2000, they varied around a value of about $85 per acre.

 b There may have been more infestations with insect pests, so farmers had to spray more often. The pests may have become resistant to insecticides, so newer and more expensive ones may have been required.

5 Each time the insecticide is used, it provides a selection pressure. Boll worms that by chance are resistant are more likely to survive and breed. Their offspring are more likely

to survive and breed than the offspring of non-resistant parents, and this continues over many generations. Over time, if the insecticide continues to be used, more and more of the boll worms in the population have the gene that confers resistance.

Chapter 10

1 a • offspring identical and so remain well adapted to the environment
   • a population can grow from one individual and quickly colonise a new habitat (in animals and plants)
   • offspring can sometimes use their parent as a source of nutrients until they become independent, so more likely to survive (in some plants)
   b • reduces the amount of variation on which natural selection can act to produce evolutionary change in the species (in animals and plants)
   • offspring may remain close to parents and compete with them (in plants)
   c • offspring not identical so promoting variation for the evolution of the species (in animals and plants)
   • the reproduction may be associated with special structures designed for spreading the organism and allowing it to colonise new habitats (in plants)
   d • within the offspring, some individuals are less well adapted to survive than others (in animals and plants)
   • usually two individuals are needed and they need to be to be able to fertilise each other (in animals and many plants)

2 Students may think of these:
advantages of asexual propagation:
   • getting many plant or animals with identical features to those already present – useful if these features are desired; e.g. flavour, ability to grow well in particular conditions, tendency for a crop to ripen at the same time (making harvesting easier), uniform characteristics of crop (making marketing easier).

   • in plants, may be able to get very large numbers of plants relatively quickly from a single parent or (if tissue culture is used) a very small amount of original material.
disadvantage of asexual reproduction:
   • lack of variation among the offspring, which could lead to vulnerability to diseases.
advantage of sexual reproduction:
   • variation among the offspring, which ensures that some members of the population may be more resistant to certain diseases than others.
disadvantage of sexual reproduction:
   • significant time and energy consumed.
Other valid points may be suggested. This would be a good class discussion exercise, building up a list of bullet points together.

Chapter 11

1 microsporangial cell ($2n$) → pollen mother cell ($2n$) → microsporocyte (n) → immature pollen grain with generative nucleus (n) → mature pollen grain with generative nucleus (n)

2 megasporangial cell ($2n$) → embryo sac (megaspore) mother cell ($2n$) → megaspore (n) → embryo sac containing egg cell (n)

3 Similarities: Both involve meiosis and mitosis. In both, the haploid products of meiosis undergo mitosis and some of the products never contribute to tissues in the seed. In male gamete production this applies to the pollen tube nucleus; in female gamete production this applies to the synergid and antipodal cells of the embryo sac.
In both, two gametes are produced; one male and female gamete combine to form the diploid embryo, the other pair forms the endosperm.
Differences: In female gamete production, only one of the four meiotic products (the megaspores) contributes to tissues that form a seed. In male gamete production all meiotic products are used to produce pollen.
In female gamete production one of the two gametes (the primary endosperm nucleus) is the product of the fusion of two haploid nuclei. This does not occur in male gamete production.

Chapter 12

1

Spermatogenesis	Oogenesis
takes place in the testis	takes place in the ovary
involves meiosis	involves meiosis
meiosis completed in days	meiosis takes many years
involves mitosis	involves mitosis
gamete is haploid	gamete is haploid
mature gamete is small and motile	mature gamete is large and non-motile
mature gamete contains no food reserves	mature gamete contains large amounts of food reserves

2 Sperm:
- motile – to reach the ovum
- small – increases motility
- contains many mitochondria – for respiration providing energy for movement
- contains microtubules – for movement
- streamlined shape – for movement
- contains an acrosome – stores enzymes needed for fertilisation

Secondary oocyte (egg):
- contains lipid – food reserve to help it survive till it can embed in the uterus wall
- surrounded by zona pellucida – prevents more than one sperm fertilising the egg
- contains mitochondria – provides energy to allow survival and completion of meiosis
- large – to store food and slow down its transport in the fallopian tube

3 a i The younger the mother, the more likely she is to have smoked in the year before or during her pregnancy. Women who are 35 and over are almost three times less likely to have smoked than women under 20.

ii Without further information, it is not possible to put forward reasons with any confidence. It is possible that, as a woman gets older, she becomes more aware of the potential harm that smoking could do to her baby, and so finds the motivation to give up smoking. Another possibility is that older women are more likely to have planned to get pregnant, and therefore give up smoking before their pregnancy begins; younger women are perhaps more likely to become pregnant without planning it, and therefore have not predicted the need to give up smoking. Yet another possibility is that the type of young women who smoke are perhaps more likely to become pregnant without planning this to happen. More data would be needed before any of these hypotheses could be considered further.

iii The trend shown by the data about the percentages of women who smoked throughout pregnancy quite closely matches that for the percentages of women who smoked in the year before or during pregnancy; the younger the woman, the more likely she is to smoke throughout pregnancy. This supports the suggestion that as a woman gets older, she becomes more aware of the potential harm that smoking could do to her baby, and so finds the motivation to give up smoking. It also makes the 'unplanned pregnancy' hypothesis in **ii** less likely. However, the numbers in all age groups are now much smaller. About half of women aged 35 and over who smoked before pregnancy found the willpower to give up once they knew they were pregnant. But in under-20-year-olds the proportion giving up is smaller, at about 40%.

b i 36% of these women smoked in the year before or during pregnancy
$(36 \div 100) \times 1397 = 503$

ii 45% of smokers gave up smoking. The number of smokers was 503.
$(45 \div 100) \times 503 = 226$

c Smoking decreases the supply of oxygen to the growing foetus, both by reducing the amount of oxygen which can be carried by haemoglobin (because carbon monoxide combines irreversibly with Hb) and by inhibiting the development of blood vessels. Babies born to smokers are, on average, of lower weight than those born to non-smokers. Smokers are also more likely to have a miscarriage than non-smokers. There is also some evidence of increased ill health in babies born to women who have smoked – even if the mother remains a non-smoker.

4 **a** 23 days.

  b Each day, steroid levels rise to a peak and then fall. They rise just after a pill is taken and then fall. During the seven days when no pills are taken, days 21 to 28, the steroid concentrations remain very low.

  c Follicular activity remains low and relatively constant throughout the days when a pill is taken. During days 21 to 28, when no pill is taken, follicular activity gradually increases. The progesterone and oestrogen in the pill inhibit the secretion of LH and FSH from the anterior pituitary gland. Their disappearance allows these hormones to be secreted, and they stimulate the development of a follicle in the ovary.

  d From the appearance of the graph, the 'follicular activity' does seem to rise rapidly throughout the pill-free seven days, but we cannot tell how great this 'activity' is compared with the normal activity that would be associated with ovulation. Certainly, in the normal menstrual cycle, the presence of FSH and LH first cause a follicle to develop and then a surge in LH causes ovulation to take place. This follicle development takes place over roughly days 5 to 12 of the cycle, but in the pill-taking woman it won't begin until she stops taking her pills on day 21. So it is possibly not a long enough time to allow a follicle to develop and be ready for ovulation to take place.

  e Most women like to be reassured that they are still fertile, and they mistakenly associate this with having a period (menstruating). Menstruation may also help the endometrium to remain healthy. They also like to be sure that they are not pregnant, and having a period confirms to them that this is so.

Glossary

α 1–4 glycosidic bond a covalent bond joining two sugar molecules between carbon atom 1 of one sugar and carbon atom 4 of the other and where the hydroxyl group at carbon atom 1 is in the alpha orientation

α-helix a secondary structure of a polypeptide chain; a specific spiral stabilised by hydrogen bonding

β carotene an orange-yellow chemical found in the chloroplasts of plant cells with a role in photosynthesis

β-strand a secondary structure of a polypeptide chain which has a tendency to join to neighbouring β-strands to form a quaternary structure called a β-pleated sheet (β-sheet)

χ^2 (chi-square) test a statistical test that can be used to determine whether differences between observed and expected results are statistically significant or could be due to chance

abiotic factor a non-living component of the environment that affects the distribution and abundance of a species

acrosome a vesicle containing hydrolytic enzymes, in the head of a sperm

activation energy the initial energy that must be given to a substrate in order for it to change into a product

active site the part of an enzyme molecule to which the substrate binds

active transport the movement of molecules or ions through transport proteins across a cell membrane, against their concentration gradient or electrochemical gradient, involving the use of energy from ATP

adenine one of the two purine nitrogenous bases found in DNA and RNA

adenosine triphosphate (ATP) an energy-containing substance that acts as the energy currency of a cell, supplying an instantly available energy source that the cell can use

adipose tissue a tissue made up of cells containing large lipid droplets

aerobic respiration the sequence of reactions – including glycolysis, the link reaction, the Krebs cycle and the electron transport chain – that result in the complete oxidation of glucose in a cell

allantois one of the four extra-embryonic membranes of a foetus

allele one of two or more alternative forms of a gene

allopatric speciation the production of new species from populations that are geographically separated from one another

allopolyploid a polyploid cell or organism which contains chromosome sets from two different species

allosteric site a binding site on an enzyme, other than the active site, at which a chemical can bind and reduce the enzyme's activity

amino group $-NH_2$; this group readily ionises to form $-NH_3^+$, so it is basic

amnion one of the four extra-embryonic membranes of a foetus

amphipathic molecule a molecule, such as a protein, in which there are hydrophilic and hydrophobic portions

amylopectin one of the two polysaccharides in starch, containing α-glucose molecules linked together and branching chains

amylose one of the two polysaccharides in starch, containing α-glucose molecules linked together

anabolic reaction a metabolic reaction in which smaller molecules are combined to make larger ones

anaphase the stage in cell division when chromatids (or chromosomes in meiosis I) are pulled apart and travel to opposite ends of the dividing cell

androecium the male reproductive organs of a flower – the stamens

aneuploidy possessing more or fewer copies of one of the chromosomes within the normal complement of chromosomes

antenatal care (ante – before; natal – relating to birth) the care of pregnant women

anther the part of the male organ of a flower which produces pollen

anti-parallel running in opposite directions; the two polynucleotide strands in a DNA molecule are anti-parallel

anticodon a sequence of three bases on a tRNA molecule that determines the specific amino acid it can pick up, and the mRNA codon with which it can bind

antipodal cells three cells within the embryo sac of a flower, at the opposite end of the sac to the synergid cells

antiport a transmembrane transporter through which two particles are transported simultaneously but in opposite directions

asexual reproduction reproduction in which genetically identical individuals are produced by mitosis; no gametes or fertilisation are involved

aster the microtubules radiating out in all directions, star-like, from the centriole pairs at the poles of an animal cell during cell division but not connecting to chromosomes (the aster is not present in plant cells)

ATP (adenosine triphosphate) a substance that acts as the energy currency of a cell, supplying an instantly available energy source that the cell can use

ATPase an enzyme that can catalyse the conversion of ADP and P_i into ATP or the reverse reaction; sometimes known as ATP synthase

autopolyploid a polyploid cell or organism which contains multiple sets of chromosomes from the same species

autosome any chromosome other than a sex chromosome

autotetraploid a polyploid cell or organism which contains four sets of chromosomes from the same species

auxin a hormone produced by young plant tissues that affects growth – synthetic auxin is used in plant tissue culture

basal cell the cell which attaches a developing embryo to the wall of the ovule in plants

Benedict's reagent a reagent used to detect reducing sugars; the sugars reduce copper ions in the reagent resulting in colour changes from transparent blue through cloudy yellow precipitate to heavy red precipitate

bilayer a structure made up of two layers of molecules, e.g. the phospholipid bilayer in a cell membrane

binary fission the principal method by which bacteria reproduce, splitting into two; it is not the same as mitosis because bacteria do not have linear chromosomes

biological species concept the idea of a species as being a group of organisms with similar morphology and physiology, which are unable to breed successfully with other species

biotic factor a living component of the environment that affects the distribution and abundance of a species

blastocyst the tiny ball of cells formed by a zygote as its cells repeatedly divide

buffer an ionic or polar substance that resists changes in its pH when acidic or alkaline substances are added to it

calyx the sepals of a flower

capacitated a fully matured spermatozoon after having spent some time in the uterus

carboxyl group $-COOH$; this group readily ionises to form $-COO^-$ and H^+, so it is acidic

Casparian strip a band of suberin in the walls of cells making up the endodermis of plant roots, preventing water from passing between the cells

catabolic a metabolic reaction or pathway in which larger molecules are broken down into smaller ones

cell cycle the sequence of events that takes place from one cell division until the next; it is made up of interphase, mitosis and cytokinesis

cell wall a structure that is present outside the plasma membrane in plant cells, fungal cells and bacterial cells; in plants, it is made of criss-crossing layers of cellulose fibres

cellulose a polysaccharide made of many β-glucose molecules linked together

centriole a pair of organelles found in animal cells, which build the microtubules to form the spindle during cell division

centromere the place where two chromatids are held together, and where the microtubules of the spindle attach during cell division

chalaza the end of the ovule in a flower which is furthest from the micropyle and nearest to the funicle

channel protein a protein that forms a pathway through a membrane, through which molecules that would not be able to move through the phospholipid bilayer can pass

chiasma (pl. chiasmata) a point at which a chromatid of one of a pair of homologous chromosomes breaks and rejoins to the chromatid of the other one of the pair, swapping genes between them

chloroplast an organelle found only in plant cells, where photosynthesis takes place; it is surrounded by an envelope of two membranes, and contains chlorophyll

cholesterol a lipid (or a lipid-like substance) that helps to maintain the fluidity of cell membranes

chorion one of the four extra-embryonic membranes of a foetus

chromatid one of two identical parts of a chromosome, held together by a centromere, formed during interphase by the replication of the DNA molecule

chromatin the DNA in a nucleus when the chromosomes have not condensed

chromosome a structure made of DNA and histones, found in the nucleus of a eukaryotic cell; the term bacterial chromosome is now commonly used for the circular strand of DNA present in a prokaryotic cell

chromosome mutation an unpredictable permanent change in the structure or number of chromosomes

cisterna (pl. cisternae) a space enclosed by the membranes of the endoplasmic reticulum

clone a group of genetically identical cells – a result of mitosis or binary fission; or a group of genetically identical organisms – a result of asexual reproduction

cloning producing genetically identical copies of an organism

coding strand (reference strand) the strand of a DNA molecule against which mRNA is built up during transcription

codominance when two alleles at a locus both have an effect on the phenotype of a heterozygous organism

codon a sequence of three bases in mRNA that codes for one amino acid

cohesion the attractive force that holds water molecules together by hydrogen bonding

collagen a fibrous protein found in skin, bones and tendons

combined birth control pill birth control pill containing both progesterone and oestrogen

competitive inhibition inhibition in which an enzyme inhibitor has a similar shape to the substrate molecule, and competes with it for the enzyme's active site

complementary base pairing the pattern of pairing between the nitrogenous bases in a polynucleotide; in DNA, A pairs with T and C with G

concentration gradient a difference in concentration of a substance between one area and another

conidia asexual, non-motile spores of a fungus

conidiophore an aerial fungal hypha producing conidia

continuous variation variation in which a feature of an individual does not fit into a definite category, but can have any value between two extremes; it is likely to be caused partly by the environment and/or by polygenes or multiple alleles

corolla the petals of a flower

corpus luteum the tissue of a Graafian follicle after ovulation that secretes hormones

cortex the area of a stem or root between the surface layers and the centre; it is made up of parenchyma tissue

cortical reaction the changes that take place when a spermatozoon unites with an oocyte plasma membrane, preventing another sperm binding and entering

cotransport a form of active transport in which the active movement of one ion provides a gradient which can provide energy for the movement of another ion or molecule up its concentration gradient

cotyledon the first leaf or pair of leaves produced by an embryo plant. In some seeds, the cotyledons contain food reserves

crenated the shape that red blood cells become when they have lost a lot of water by osmosis

crista (pl. cristae) one of the folds in the inner membrane of a mitochondrion, on which the electron transport chain is found

cross-pollination transfer of pollen from one plant to another

crossing over the exchange of alleles between chromatids of homologous chromosomes as a result of chiasma formation during prophase of meiosis I

cutting a section of a plant organ that is removed from a plant and then allowed to grow roots when placed in soil or another medium

cystic fibrosis a genetic disease caused by the recessive allele of a gene that codes for a membrane protein responsible for the passage of chloride ions out of cells

cytokinin a plant hormone that stimulates cell division

cytosine one of the two pyrimidine nitrogenous bases found in DNA and RNA

degenerate a term used to describe the genetic code, in which more than one triplet of bases codes for the same amino acid

degrees of freedom the number of values in a statistical calculation that are free to vary. In the chi-squared test it relates to the number of comparisons of observed and expected data

dehiscence the splitting open of a plant organ e.g. anthers to release pollen

deletion the removal of one base pair from a DNA molecule

denatured when a protein molecule, for example an enzyme, has lost its molecular shape, so that its function can no longer be carried out; this is usually permanent

deoxyribonucleic acid (DNA) the genetic material contained in chromosomes; a polynucleotide in which the five-carbon sugar is deoxyribose

deoxyribose a five-carbon sugar found in DNA

differentiation the development of a cell to become specialised for a particular function

diffusion the net movement, as a result of random motion of its molecules or ions, of a substance from an area of high concentration to an area of low concentration

dihybrid inheritance the study of the inheritance of two different genes

dioecious male and female organs present on the same plant

diploid containing two sets of chromosomes

dipole a partial separation of electrical charge in a molecule e.g. in water where, as a result, the oxygen atom has a partial negative charge and the hydrogen atoms a partial positive charge

direct active transport transport of a substance across a membrane where ATP is directly used for that transport

directional (evolutionary) selection natural selection in which a change in environment or a change in the alleles present in the gene pool selects for a different feature than in the past, so that this feature becomes more common in successive generations

disaccharide a sugar whose molecules are made of two sugar units

discontinuous variation variation in which a feature of an individual fits into one of a few definite categories; it is likely to be caused by a small number of genes with a small number of alleles

disruptive selection a type of selection which reduces the number of individuals of a species with intermediate values of a particular character, thus increasing the numbers with more extreme values of that character

DNA (deoxyribonucleic acid) the genetic material contained in chromosomes; a polynucleotide in which the five-carbon sugar is deoxyribose

DNA helicase an enzyme that unwinds and separates the two strands of a DNA molecule

DNA ligase an enzyme that links nucleotides together by catalysing the formation of covalent bonds between the deoxyribose and phosphate groups

DNA polymerase an enzyme that makes complementary copies of DNA

dominant allele an allele having an effect on the phenotype even when a recessive allele is also present

dominant epistasis a type of epistasis when a gene is not expressed because of the effect of a dominant gene at a different locus

double fertilisation fertilisation in a plant which uses two male gametes, one of which fuses with the female gamete to form the zygote and the other fuses with the primary endosperm nucleus to produce endosperm

Down's syndrome a genetic disease resulting from trisomy 21, which is an aneuploidy where each cell has three copies of chromosome 21

egg cell female gamete in an embryo sac in a flower

electrophoresis the separation of fragments of DNA according to their lengths, by applying a voltage across them; the DNA fragments are pulled towards the positive end, smallest fastest

embryo the structure produced by repeated mitotic division of the zygote in plants or animals in which the earliest stages of development take place

embryo sac the multicellular structure formed by three mitotic divisions of one of the haploid products of meiosis of a megaspore mother cell in a flower

emulsion a stable suspension of tiny globules of one liquid in another liquid with which it does not mix, e.g. a suspension of oil in water

end-product inhibition where the end product of a metabolic pathway inhibits the first enzyme of the pathway

endocytosis the movement of bulk liquids or solids into a cell, by the indentation of the plasma membrane to form vesicles containing the substance; it is an active process requiring ATP

endodermis the outer layer of the stele in a plant root

endoplasmic reticulum (ER) a network of membranes within a eukaryotic cell, where various metabolic reactions take place

endosperm a tissue made of triploid cells that stores food in some seeds, made by the mitotic division of the endosperm nucleus

endosperm nucleus the triploid nucleus in a flower's embryo sac, formed from the union of a male gamete with the diploid primary endosperm nucleus

endosymbiont an organism living inside another where both organisms gain benefit from the association

envelope a pair of membranes surrounding some organelles; that is, chloroplast, mitochondrion and nucleus

environmental factor a feature of the environment that has an effect on an organism

enzyme–substrate complex the temporary association between an enzyme and its substrate during an enzyme-catalysed reaction

epidermal cell a cell in the epidermis

epidermis the tissue forming the outer surface of the body of a plant, composed of a single layer of cells

epididymis a long, tightly folded tube outside the testis in which sperm is stored

epistasis the interaction of two or more different genes at different loci to produce a particular phenotype

equator the mid-line of a cell between the two poles, where the chromosomes are lined up during metaphase of mitosis or meiosis

ester bond a linkage between a fatty acid and glycerol in a lipid molecule

euchromatin the less dense form of chromatin within a nucleus, containing DNA that is or can easily be transcribed

eukaryote an organism made up of cells with organelles enclosed within internal membranes i.e. eukaryotic cells

evaporation the change of liquid to gas, below boiling point, at the surface of a liquid in contact with the air

evolution a directional change in the characteristics of a population over time

exine the pitted outer layer of the cell wall of a mature pollen grain

exocytosis the movement of bulk liquids or solids out of a cell, by the fusion of the vesicles containing the substance with the plasma membrane; it is an active process requiring ATP

exon one of a number of sections within a gene that codes for part of the sequence of amino acids in a protein

explant a small group of cells taken from a parent plant to be used in tissue culture

extracellular outside a cell

extrinsic (peripheral) protein a protein that is attached to a membrane but can be removed easily

facilitated diffusion the diffusion of a substance through protein channels in a cell membrane; the proteins provide hydrophilic areas that allow the molecules or ions to pass through a membrane that would otherwise be less permeable to them

factor VIII one of several substances in blood required for clotting to occur

fatty acid a molecule containing a hydrocarbon chain with a carboxyl group at one end

fertilisation fusion of the nuclei of two gametes to form a zygote

fertilisation membrane the barrier which is impenetrable to sperm, formed from the zona pellucida after fertilisation has occurred

filament the stalk of a stamen supporting the anther

flaccid a plant cell is said to be flaccid if it has lost a lot of water, and its volume has shrunk so much that the cell no longer pushes outwards on the cell wall

fluid mosaic model the universally accepted model of membrane structure, in which proteins float in a phospholipid bilayer

foetus the stage in the development of a mammal that follows the embryonic stage and that ends in birth

folic acid a vitamin which results in anaemia if deficient in the diet and is important for the development of a foetus

follicle stimulating hormone (FSH) a mammalian hormone that stimulates the growth of a follicle in the ovary

frame shift a shift in the genetic code reading frame

fruit the structure produced from an ovary in a flower that has been fertilised

funicle the stalk that attaches an ovule or seed to the ovary wall

gamete a haploid cell specialised for reproduction; the nuclei of two gametes fuse together at fertilisation to form a diploid zygote

gene a sequence of DNA nucleotides that codes for a polypeptide

gene mutation an unpredictable change in the structure of DNA

gene technology the manipulation of an organism's DNA to produce an organism or a product that can be made use of in some way

gene therapy changing the DNA in some of a person's cells – for example, to attempt to cure a disease caused by faulty genes

genetic diagram a conventional way of showing the genotypes of parents, their gametes and the genotypes and phenotypes of the offspring they would be expected to produce

genetic drift a change in the characteristics of a population, or the proportions of different alleles in the population, as a result of chance; it is most likely to happen in small, isolated populations

genetic engineering using technology to change the genetic material of an organism

genetically modified organism (GMO) an organism whose DNA has been modified using gene technology – for example, by the removal of some of its genes or by the addition of DNA from another organism

genome the complete DNA of an organism, or of a species

genotype the alleles of a particular gene or genes possessed by an organism

geographical isolation the separation of two populations by a geographical barrier

germ line the DNA contained in cells that can be passed on to the next generation, e.g. in eggs and sperm or in the cells from which these will be made

germinal epithelium in a male, a layer in the wall of a seminiferous tubule producing spermatogonia that eventually mature into sperm; in a female, the outer layer of cells of the ovary producing oogonia that eventually may release oocytes

globular protein a protein with a roughly spherical three-dimensional shape; many are metabolically active and soluble in water

glucose a hexose sugar; the form in which carbohydrate is transported in the blood of mammals

glyceraldehyde a triose sugar i.e. one containing three carbon atoms

glycocalyx a layer on the outside of a cell made up of a mixture of large molecules, including many polysaccharides

glycogen a storage polysaccharide found in liver cells and muscle cells, made of many α-glucose units linked by glycosidic bonds

glycolipid a molecule made of a lipid to which sugars are attached

glycoprotein a molecule made of a protein to which sugars are attached

glycosidic bond the linkage that joins sugar molecules together in a disaccharide or polysaccharide

Golden Rice™ a variety of rice that has been genetically engineered to produce large amounts of β carotene

Golgi body a stack of curved membranes inside a cell, in which protein molecules are packaged and modified – for example, by adding sugars to them to produce glycoproteins

gonadotrophin releasing hormone (GnRH) a mammalian hormone produced by the hypothalamus which affects the release of luteinising hormone, follicle stimulating hormone, oestrogen and progesterone from the ovaries or placenta

granulosa cells cells that form a loose layer around a secondary oocyte in a Graafian follicle in the ovary

granum (pl. grana) stack of thylakoid membranes inside a chloroplast, containing chlorophyll

guanine one of the two purine nitrogenous bases found in DNA and RNA

gynoecium the female reproductive organs of a flower, which include the ovules

haem a group of atoms with an Fe^{2+} ion at the centre, found in several proteins including haemoglobin and cytochrome

haemophilia a sex-linked genetic disease caused by a recessive allele of the gene that encodes factor VIII

haploid containing one set of chromosomes

heterochromatin the more dense form of chromatin within a nucleus, containing highly condensed DNA that is not being transcribed

heterostyly a mechanism preventing self-pollination in flowers whereby plants have either one of two forms of flower in which the positions of anthers and stigma are switched e.g. in *Primula*

heterozygous possessing two different alleles of a gene

histone a protein that forms a part of chromosomes; it binds to DNA and can regulate whether genes are transcribed

homologous having similar base sequences (in genes); homologous chromosomes pair during meiosis

homozygous possessing two identical alleles of a gene

human chorionic gonadotrophin (hCG) a mammalian hormone made by the developing embryo, and then the placenta, maintaining the corpus luteum and its hormone secretion

hydrogen bond an attractive force between a slight negative charge on one atom, for example oxygen, and a slight positive charge on another, for example hydrogen

hydrolase an enzyme that catalyses a hydrolysis reaction

hydrolysis a reaction in which two molecules are separated from each other, involving the combination of water

hydrophilic attracted to water; hydrophilic substances have small electric charges that are attracted to the small electric charges (dipoles) on water molecules

hydrophobic repelled by water; hydrophobic substances do not have electric charges, and are not attracted to water molecules

hypertonic solution a solution with a lower water potential than another due to its higher concentration of solute

hypha (pl. hyphae) one of the long, thin threads that make up the body of a fungus, which is a single cell wide

hypostatic a type of epistasis when a gene is not expressed because of the effect of genes at a different locus

hypotonic solution a solution with a higher water potential than another due to its lower concentration of solute

immune response the way in which lymphocytes respond to infection by pathogens

implantation the attachment and embedding of a fertilised ovum in the wall of the uterus

inbreeding sexual reproduction between closely related individuals of a species

incipient plasmolysis the state of a plant cell in which any further loss of water from the cell would cause plasmolysis

independent assortment the result of the random orientation of each pair of homologous chromosomes on the equator of the spindle in metaphase of meiosis I, ensuring that either one of a pair of homologous chromosomes can be found with either one of another pair

indirect active transport transport of a substance across a membrane where ATP is used for the transport of another substance, which sets up a gradient down which the first substance diffuses passively

induced fit the change of shape of an enzyme when it binds with its substrate, caused by contact with the substrate

initial rate of reaction the rate at which substrate is converted to product right at the beginning of a reaction

initiation the first stage of transcription and translation; in transcription, it involves the binding of RNA polymerase to the promoter region of DNA, while in translation, it involves the binding of the first ribosomes to mRNA

insertion the addition of one base pair within a DNA molecule

integuments the outer layers of tissue of an ovule in a flower

interphase the longest stage in the cell cycle; this is when the replication of DNA takes place

intine the inner layer of the cell wall of a mature pollen grain

intracellular within cells

intraspecific variation variation within a species

intrinsic (integral) protein a protein that forms part of the structure of a membrane and cannot easily be removed

intron one of several sections of a gene that does not code for the sequence of amino acids in a protein

irreversible inhibition inhibition of an enzyme that is permanent, and that is not affected by the addition of more substrate

isolation (geographical isolation) the separation of two populations by a geographical barrier

isotonic two solutions can be described as being isotonic if they have the same water potential

lactation the production of milk for the newborn by a female mammal

Leydig cell one of many cells surrounding the seminiferous tubules in the testis, secreting testosterone

ligand-gated channel a transport channel in a membrane in which transport is affected by the binding of a regulator chemical to the protein of the channel

ligase (DNA ligase) an enzyme that links nucleotides together by catalysing the formation of covalent bonds between the deoxyribose and phosphate groups

light microscope a microscope that uses light rays

lipid a fat, oil or wax

liposome a tiny ball of lipid, which contains other substances such as DNA or protein

locus the position on a chromosome at which a particular gene is found

luteinising hormone (LH) a mammalian hormone produced by the anterior pituitary gland, which helps regulate the ovarian cycle

lysosome small vesicle containing hydrolytic enzymes, surrounded by a membrane

magnification the number of times greater that an image is than the actual object:

$$\text{magnification} = \frac{\text{image size}}{\text{object size}}$$

male gamete nuclei the two male gametes in a pollen tube which are used for double fertilisation in plants

maltase an enzyme that catalyses the hydrolysis of maltose to glucose

matrix the 'background material' inside a mitochondrion, where the link reaction and the Krebs cycle take place

mechanically gated channel a transport channel in a membrane in which transport is affected by a mechanical change e.g. pressure

megaspore mother cell the cell that divides by meiosis in male and female organs in a flower; in the male organs they can be called pollen mother cells and, in the female, embryo sac mother cells

meiosis reduction division; a type of nuclear division in which the nucleus divides twice to form four genetically different daughter cells from one parent cell, each containing half the number of chromosomes of the parent cell

menstrual cycle the cycle of changes in the wall of the uterus, preparing it for implantation, which is coordinated with the ovarian cycle in an ovary

meristem an area of a plant in which there is a collection of meristematic cells e.g. at the tips of stems and roots

meristematic cell a plant cell that is able to divide by mitosis

messenger RNA (mRNA) a type of RNA which is a copy of the genetic code for a gene, made during transcription and carrying the code into the cytoplasm for translation into polypeptide

metabolic poison a substance that prevents a metabolic reaction from taking place – for example, a heavy metal that inhibits enzymes

metabolism (metabolic reactions) the chemical reactions that take place in living organisms

metaphase the stage of cell division just after the disappearance of the nuclear envelope during which chromosome centromeres are aligned to the equator

micrometre (μm) one thousandth of a millimetre; 1×10^{-6} metres

micropyle a small hole in the integuments of an ovule which is retained in a seed; the seed absorbs water through it for germination

microtubule long, thin tubes of protein that help to make up the cytoskeleton in a eukaryotic cell

middle lamella a layer found between adjacent cell walls in plant tissues holding the cells together, containing pectin

mini birth control pill birth control pill containing only progesterone

mitochondrion (pl. mitochondria) organelle found in most eukaryotic cells, in which aerobic respiration takes place; it is surrounded by an envelope, of which the inner membrane is folded to form cristae

mitosis the division of a nucleus such that the two daughter cells acquire exactly the same number and type of chromosomes as the parent cell

monoecious male and female organs on different plants

monomer one unit of a polymer

monosaccharide a sugar whose molecules are made of a single sugar unit

morphology study of the structure of organisms

mRNA (messenger RNA) a type of RNA which is a copy of the genetic code for a gene, made during transcription and carrying the code into the cytoplasm for translation into a polypeptide

multicellular made of many cells

mutation an unpredictable change in the structure of DNA, or in the structure and number of chromosomes

mycelium all the hyphae of a fungus in the substrate in which the fungus is growing

nanometre (nm) one millionth of a millimetre; 1×10^{-9} metres

natural selection the way in which individuals with particular characteristics have a greater chance of survival than individuals without those characteristics; they are therefore more likely to reproduce and pass on genes for those characteristics to their offspring

non-competitive inhibition inhibition in which an enzyme inhibitor does not resemble the substrate, and binds with the enzyme molecule at a place other than its active site

non-disjunction an error during cell division in which there is a failure of chromosomes to separate after metaphase

non-overlapping the genetic code is non-overlapping, which means that the moving of the reading frame by one or two bases results in the code not producing useable protein

non-reducing sugar a sugar, such as sucrose, that is not capable of reducing Benedict's reagent and changing its colour

nonpolar molecule a molecule without full (ionic) or partial (dipolar) electric charges

nucellus a layer of tissue in the anthers of a flower which supplies nutrients to the pollen developing in the pollen sacs

nuclear envelope two membranes that surround the nucleus and separate it from the cytoplasm

nuclear pore pore in the nuclear envelope through which messenger RNA can pass

nucleolus a darkly staining region of the nucleus which contains DNA coding for rRNA

nucleotide a molecule consisting of a five-carbon sugar, a phosphate group and a nitrogenous base

nucleus a large organelle found in eukaryotic cells; it is surrounded by an envelope and contains the chromosomes

null hypothesis the hypothesis applied to the results of any investigation which states that the results arise by chance alone

oestrogen a mammalian hormone produced by developing follicles, the corpus luteum and placenta. It is important in ovarian and menstrual cycles

oogonium (pl. oogonia) a cell representing the earliest stage of oogenesis in an ovary of a mammal

optimum temperature the temperature at which a reaction occurs most rapidly

organ a structure within a multicellular organism that is made up of different types of tissues working together to perform a particular function, e.g. the stomach in a human or a leaf in a plant

organelle a functionally and structurally distinct part of a cell – for example, a ribosome or mitochondrion

osmosis the net movement of water molecules from a region of high water potential to a region of low water potential, through a partially permeable membrane, as a result of their random motion

outbreeding sexual reproduction between distantly related individuals in a species

ovary female organ in mammals and plants

ovule the part of the ovary in a plant which gives rise to the cells used in sexual reproduction

parenchyma tissue a relatively unspecialised plant tissue – in the root, its roles are mainly storage and support; in a leaf or stem, mainly photosynthesis and support

partially permeable a partially permeable membrane allows some solutes to pass through it and others not, but is permeable to water and will therefore allow osmosis to take place across it

passive in passive transport, the movement of particles is driven by the energy in the particles themselves and is derived from concentration gradients with no energy provided by the metabolism of the cell in the form of ATP

pectin the major component of the middle lamella helping to hold adjacent plant cells together

peptide bond the −CO−N− linkage between two amino acids

peptidoglycan molecule made of a chain of amino acids and sugars; it is found in bacterial cell walls

pericarp the wall of the fruit

pericycle the layer of cells inside the endodermis of a root

petal part of a flower, often used to attract or guide insects to a flower or act as a suitable landing stage for them

phenotype the characteristics of an organism

phloem sieve element a living cell found in phloem tissue; the cells have perforated ends, called sieve plates, and are connected to form long tubes for the transport of substances such as sucrose

phloem tissue tissue which transports substances such as sucrose throughout the plant

phospholipid a lipid that contains a phosphate group and is a major component of the bilayer of plasma membranes

photosynthesis the manufacture of carbohydrates from inorganic substances (carbon dioxide and water) using energy from light; the light is transformed to chemical energy

phylogenetic (evolutionary) species concept the idea of a species as being a group of organisms that is geographically separated and morphologically or behaviourally distinct from other species, even if they may still be able to interbreed with individuals of another species; also called evolutionary species concept

physiology study of the processes within the body of an organism

pinocytosis an example of endocytosis in which the vesicles that are formed are small and contain solutions rather than solid objects

pit a thinner section of a plant cell wall, in xylem allowing sap to pass through, in the exine of pollen allowing the exit of a pollen tube

plan drawing a drawing in which only the boundaries of tissues are shown and no individual cells are drawn

plasma (cell surface) membrane the outer membrane of a cell

plasmid a small, single-stranded molecule of DNA found in bacteria

plasmodesma (pl. plasmodesmata) a direct connection between the cytoplasm of one plant cell and an adjacent cell, made up of plasma membrane endoplasmic reticulum running between the two cells

plasmolysis when a plant cell loses so much water that the plasma membrane is torn away from the cell wall

plumule the first stem produced by an embryo plant

polar having full or partial electrical charges

polar body one of the two small, unused cells produced at each of the two stages of meiosis during oogenesis in mammals

pollen grain a structure containing a number of nuclei, such as the male gametes, used in plant sexual reproduction to carry the gametes to the female organs

pollen mother cell a cell that divides by meiosis in the production of pollen

pollen sac a structure in an anther of a flower containing cells that form pollen, which contains male gametes

pollen tube a cellular outgrowth from a pollen grain which passes from the stigma, through the style to the ovule of a flower, to bring the male gametes to the embryo sac for fertilisation

pollination the transfer of pollen from a stamen to a stigma by wind or insect

polymer a large molecule made up of many monomers joined together

polymerase chain reaction (PCR) a method of making a large number of copies of a DNA molecule in a relatively short time; it involves the denaturation of DNA (i.e. the separation of its two strands), the attachment of a primer and then the construction of a complementary DNA strand against each exposed strand; this sequence is carried out repeatedly

polymorphism the presence of distinct inheritable forms of a species e.g. the different forms within a bee species

polynucleotide a substance made of many nucleotides linked together in a chain; RNA and DNA are polynucleotides

polypeptide a chain of amino acids

preconceptual care care of a woman before conception to ensure a healthy pregnancy

pressure potential the pressure of the cell wall of a plant cell that is directed inwards on the contents of the cell

primary endosperm nucleus the diploid nucleus resulting from the fusion of two haploid nuclei in an embryo sac in a flower

primary oocyte the cell formed from an oogonium in oogenesis in a mammal

primary spermatocyte the cell formed from a spermatogonium in spermatogenesis in a mammal

primary structure the sequence of amino acids linked together in a polypeptide or protein

primer a short section of DNA which contains code that acts as the starting point for DNA synthesis

probe a short length of DNA that is labelled – for example, by being radioactive; it binds to complementary lengths of DNA and so indicates their position

product a new substance that is made by a chemical reaction

progesterone a mammalian hormone produced by the ovary and, during pregnancy, the corpus luteum and placenta, necessary for the ovarian cycle and pregnancy

prokaryote an organism made up of prokaryotic cells – in which there is no nucleus surrounded by an envelope or any other membrane-bound organelles

promoter a length of DNA that is needed for a gene to be transcribed; in a prokaryotic cell, part of an operon to which RNA polymerase binds in order to initiate transcription of the gene

prophase the stage of cell division during which the chromosomes first appear

prosthetic group a non-protein group that is tightly bound to a protein, e.g. the haem in haemoglobin

protandry a mechanism to achieve cross-pollination in plants in which the stamens shed pollen before the female organs are receptive to pollen in that flower

protein a substance whose molecules are made of chains of amino acids

protogyny a mechanism to achieve cross-pollination in plants in which the ovum is mature and fertilised before the stamens in that flower shed pollen

purine base a nitrogenous base whose molecules contain two carbon–nitrogen rings; adenine and guanine are purines

pyrimidine base a nitrogenous base whose molecules contain one carbon–nitrogen ring; thymine, cytosine and uracil are pyrimidines

Q_{10} **temperature coefficient** the number of times a reaction or process is speeded up following a 10 °C rise in temperature

quaternary structure the overall shape of a protein molecule that is made up of two or more intertwined polypeptides

radicle embryo root of a plant

receptor site the part of a protein to which a regulatory molecule, such as a hormone, binds

recessive allele an allele having an effect on the phenotype only when a dominant allele is not present

recessive epistasis a type of epistasis when a gene is not expressed because of the effect of the homozygous recessive state at a different locus

recombinant DNA DNA that has had DNA from a different source (often from a different species) inserted into it

recombinant organism an organism containing DNA with a new combination of genes, either naturally, as a result of sexual reproduction, or artificially, as a result of adding DNA to it

reducing sugar a sugar, such as glucose, that is capable of reducing Benedict's reagent and changing its colour from blue

replica plating a technique used to determine which of several colonies of bacteria have successfully taken up a plasmid containing the desired gene, by testing for antibiotic resistance

reproductive isolation the prevention of interbreeding between two populations because of problems in courtship, mating, fertilisation or development, or because of hybrid sterility

resolution the ability to distinguish between two objects very close together; the higher the resolution of an image, the greater the detail that can be seen

restriction enzyme an enzyme produced by a bacterium to destroy viral DNA entering the cell; they are widely used in gene technology to cut DNA into smaller lengths

retinol vitamin A, necessary for bone growth and vision

retrovirus a virus that contains RNA – for example, HIV

reverse transcriptase an enzyme that uses RNA to make a single-stranded molecule of complementary DNA

reversible inhibition inhibition of an enzyme that lasts for only a short time, or that can be reduced by the addition of more substrate

rhizome a plant stem that is found underground, generally running horizontally, and from which roots and aerial stems can arise

ribonucleic acid (RNA) a polynucleotide made of nucleotides containing ribose

ribose a five-carbon sugar found in RNA

ribosome one of many thousands of tiny organelles, sometimes free in the cytoplasm and sometimes attached to rough endoplasmic reticulum, where protein synthesis takes place

RNA (ribonucleic acid) a polynucleotide made of nucleotides containing ribose

RNA polymerase an enzyme that links together RNA nucleotides during transcription

rough endoplasmic reticulum (RER) endoplasmic reticulum that has ribosomes attached to it, where protein synthesis takes place

saturated fatty acid a fatty acid in which there are no double bonds between carbon atoms; a component of saturated fats

scanning electron microscope a microscope that forms three-dimensional images of surfaces, using electron beams

secondary oocyte the cell formed from a primary oocyte in the ovary of a mammal as a result of starting to divide by meiosis

secondary spermatocyte a haploid cell formed from a primary spermatocyte by meiosis

secondary structure a regular, repeating pattern or shape in a polypeptide chain – for example, an α-helix or β-sheet

secretion the production and release of a useful substance – for example, salivary glands secrete saliva

selection pressure an environmental factor that decreases or increases the chance of survival of organisms with particular variations

self-incompatibility a state of a flower in which the pollen from that flower is incompatible with the female organs, so preventing fertilisation

self-pollination when pollen from a flower pollinates the same flower resulting in successful sexual reproduction

self-sterility a state of a flower where one of a number of possible mechanisms are used to prevent viable seed production from self-pollination

semiconservative replication the method by which a DNA molecule is copied to form two identical molecules, each containing one strand from the original molecule and one newly synthesised strand

seminiferous tubules small tubes in the testis in which spermatozoa are produced

sepal one of a ring of structures underneath the ring of petals in a flower

Sertoli (nurse cell) a cell supporting and nurturing developing spermatozoa in the seminiferous tubules

sex chromosomes the chromosomes that determine gender; in mammals there are two, X and Y, and males are XY and females are XX

sex-linked a characteristic caused by a gene that is found on the non-homologous portion of the X chromosome

sexual reproduction reproduction in which two gametes (usually but not necessarily from two different parents) fuse to form a zygote; the offspring are genetically different from each other and their parent or parents

silent mutation an unpredictable change in the structure of DNA which has no observable effect, that is, it does not affect the phenotype

smooth endoplasmic reticulum (SER) endoplasmic reticulum that does not have ribosomes associated with it; it carries out various metabolic reactions, such as the synthesis of steroids

sodium–potassium pump protein molecules in a plasma membrane that use ATP to move sodium ions out of a cell and potassium into it, against their concentration gradients

solute a substance that can dissolve in a solvent

solution a homogeneous mixture of a solute and solvent

solvent the liquid that can dissolve a solute

somatic gene therapy gene therapy in which body cells are transformed, but with no effects on cells that will become gametes

spermatid an immature spermatozoon

spermatogenesis the production of spermatozoa

spermatogonium (pl. spermatogonia) the cells that divide by meiosis to form primary spermatocytes during spermatogenesis

spermatozoon (pl. spermatozoa) motile male gamete

spindle a structure made up of microtubules that is formed during cell division and manoeuvres the chromosomes into position

stabilising selection a type of natural selection in which the characteristics of the population are kept constant by selection against extreme variants

stain a substance that adds colour or contrast to a microscope section

stamen the male organ of a flower

starch a polysaccharide containing a mixture of amylose and amylopectin, both of which are made of many α-glucose molecules linked together; it is the storage polysaccharide in plants

stele the central area of a root, containing xylem and phloem tissues

steroid a lipid-like substance derived from cholesterol; many hormones, including testosterone, are steroids

sticky end a short stretch of unpaired nucleotides at the end of a DNA molecule

stigma the part of the female organ of a flower which is receptive to pollen

stroma the 'background material' in a chloroplast, in which the light-independent stage of photosynthesis takes place

style the column that links the stigma and ovary in the female part of a flower

substitution the replacement of one base pair from a DNA molecule with another

substrate the substance that is altered by an enzyme during an enzyme-catalysed reaction

sucrose a carbohydrate made up of two sugar units, α-glucose and β-fructose, linked between carbons 1 and 2

surface tension an effect produced by the attractive forces between water molecules that pull downwards on the molecules at the surface, resulting in the surface molecules packing more tightly together

sympatric speciation the production of a new species from populations that are living in the same place

symport a transmembrane transporter through which two particles are transported simultaneously and in the same direction

synergids two cells within the embryo sac of a plant flower which are at the opposite end of the sac to the antipodal cells

tapetum the ring of tissue outside a pollen sac in an anther of a flower that supplies nutrients to the developing pollen in the sac

telophase the stage in cell division where chromosomes have arrived at opposite ends of the dividing cell; they decondense and become surrounded by nuclear membranes

termination the stopping of transcription of a gene at the end of the gene and the releasing of the synthesised mRNA

tertiary structure the overall three-dimensional shape of a protein molecule

testa the seed coat, derived from the integuments of the ovule

testosterone a mammalian hormone secreted by testes and ovaries, necessary for sexual reproduction

tetrad the four cells produced by the meiotic division of one cell

thylakoid one membrane disc within a stack of discs (granum) inside a chloroplast, where chlorophyll is found and the light-dependent reactions of photosynthesis take place

thymine one of the two pyrimidine nitrogenous bases found in DNA (replaced by uracil in RNA)

tissue a layer or group of cells of similar type, which together perform a particular function

tissue culture the growth of many genetically identical plants from a small group of cells; the cells are grown in sterile nutrient medium before transfer to agar gel where they are stimulated to grow roots and shoots; the technique is also known as micropropagation

tonoplast the membrane that surrounds a vacuole in a plant cell

totipotent a cell is said to be totipotent if it is able to divide to form any of the different types of specialised cell in the body – for example, meristematic cells in plants and stem cells in mammals

transcription producing an mRNA molecule with a complementary base sequence to one strand of a length of DNA

transcription factor a substance that can bind with a particular region of DNA and either initiate or prevent transcription

transfer RNA (tRNA) a type of RNA found in the cytoplasm, made of a single strand looped back on itself; each tRNA molecule has a particular anticodon that pairs with a codon on mRNA, and also determines the type of amino acid with which the tRNA will bind

transformed (transgenic) organism an organism that has had foreign DNA inserted into its genome

translation the synthesis of proteins on a ribosome; the sequence of amino acids is determined by the sequence of bases in the mRNA

translocation a chromosomal mutation in which a section of one chromosome is broken off and joined to the end of a different chromosome

transmembrane protein a protein spanning the width of a plasma membrane from one surface to the other

transmission electron microscope a microscope that forms images of thin specimens, using electron beams

transporter protein a protein molecule in a plasma membrane that uses ATP to move ions or molecules across the membrane against their concentration gradient

triglyceride a lipid whose molecules are made from three fatty acids linked to a glycerol molecule

triplet a group of three bases in a DNA molecule, coding for one amino acid

triploid a cell or organism with three sets of chromosomes; in plants, the endosperm is a triploid tissue

trisomy 21 the genetic defect in Down's syndrome – an aneuploidy with three copies of chromosome 21

tRNA (transfer RNA) a type of RNA found in the cytoplasm, made of a single strand looped back on itself; each tRNA molecule has a particular anticodon that pairs with a codon on mRNA, and also determines the type of amino acid with which the tRNA will bind

trophoblast the outer layer of cells of a blastocyst which is responsible for attaching a blastocyst to the wall of the uterus and invading it to form part of the placenta

tube nucleus one of the two nuclei that migrate down a pollen tube, but is only used to direct tube growth

tubulin a globular protein that makes up microtubules

turgor the outward pressure of a plant cell's contents on its cell wall resulting from the osmotic uptake of water and the resistance of the wall to further expansion

turnover number the maximum number of molecules of substrate that an enzyme can convert to product per second

ultrastructure detailed structure, including very small things; electron microscopes reveal the ultrastructure of cells

umbilical artery an artery in the umbilical cord which carries foetal deoxygenated blood from the foetus to the placenta

unicellular made up of only one cell

unsaturated fatty acid a fatty acid in which there is at least one double bond between carbon atoms; a component of unsaturated fats

uracil one of the two pyrimidine nitrogenous bases found in RNA (replaced by thymine in DNA)

vascular tissue the transporting tissues of plants: xylem and phloem

vector in gene technology, anything that is used to transfer DNA into the organisms to be genetically transformed; plasmids, viruses and liposomes can be used as vectors

vesicle a tiny 'space' inside a cell, surrounded by a single membrane and containing substances such as enzymes

vessel element that part of a xylem vessel which is formed from one cell that has lost its end walls and cytoplasm; it is a dead cell with thickened walls

villus (pl. villi) a part of the placenta in which foetal and maternal blood are brought close together for exchange of substances between them

voltage-gated channel an ion channel in a plasma membrane that responds to a change in voltage (potential difference) across the membrane by opening or closing

water potential the tendency of a solution to lose water; water moves from a solution with high water potential to a solution with low water potential; water potential is decreased by the addition of solute, and increased by the application of pressure; symbol is ψ (psi).

water potential gradient a difference in water potential between one place and another; water moves down a water potential gradient

whorls the rings of flower parts; in a typical flower, for example, the outer whorl contains the sepals and the next one the petals

xylem tissue the vascular tissue in plants which transports water and minerals from roots to leaves; it contains xylem vessels, xylem fibres and parenchyma cells

yolk sac one of the four extra-embryonic membranes of a foetus

zygote a diploid cell formed by the fusion of the nuclei of two haploid gametes

Index

Acknowledgements

The author and publisher are grateful for the permissions granted to reproduce copyright materials. While every effort has been made, it has not always been possible to identify the sources of all the materials used, or to trace all the copyright holders. If any omissions are brought to our notice, we will be happy to include the appropriate acknowledgements on reprinting.

Cover image: Banana Pancake/Alamy; p. 4 Richard Wehr/Custom Medical Stock Photo/Science Photo Library; p. 7 Hermann Eisenbeiss/Science Photo Library; p. 20 Prof. P. Motta/Dept of Anatomy/ University 'La Sapienza', Rome/Science Photo Library; p. 21 J. Gross, Biozentrum/Science Photo Library; p. 32*l* Richard Kirby/Timeframe Productions Ltd/PhotoLibrary Group Ltd; pp. 32*r*, 33, 69, 199, 200*r*, 220*l*, 220*r*, 225 Geoff Jones; p. 33r Herve Conge/Phototake; p. 34 Dr. Jeremy Burgess/ Science Photo Library; p. 35*t* Pascal Goetgheluck/ Science Photo Library; p. 35*b* Power and Syred/ Science Photo Library; p. 37 Carolina Biological Supply Company/PhotoLibrary Group Ltd; p. 38 Herve Conge/Phototake; pp. 40, 43*t*, 173 Dr Gopal Murti/Science Photo Library; p. 41 A.M. Page at Royal Holloway College, University of London; p. 42*l* Dennis Kunkel Microscopy, Inc/Phototake; p. 42*rt*, 42*rb*, Dr Don Fawcett/Science Photo Library; p. 43*rb* Biology Media/Science Photo Library; p. 45*l* Dr Kari Lounatmaa/Science Photo Library;p. 45*r*, CNRI/Science Photo Library; p. 46*t* Don W. Fawcett / Science Photo Library; p. 46*b* Biophoto Associates/Science Photo Library; p. 47 Marilyn Schaller/Science Photo Library; p. 50*l* Dr Jeremy Burgess/Science Photo Library; p. 50*r* Dr Keith Wheeler/Science Photo Library; p. 51 Herve Conge/ Science Photo Library; pp. 55t, 132*b* Peter Arnold/ PhotoLibrary Group Ltd; p. 55*b* Dr Jeremy Burgess/ Science Photo Library; p. 85 Simon Fraser/Science Photo Library; pp. 86, 157 James King-Holmes/ Science Photo Library; p. 107*l* CNRI/Science Photo Library; p. 107*r* Steve Gschmeissner/Science Photo Library; p. 120 M. I. Walker/Science Photo Library; p. 121*l* Jerome Wexler/Science Photo Library; p. 121*r* Phototake Inc/Alamy; pp. 122, 230, 232 John Adds (The Coachman's House, Rectory Lane, Latchington, Chelmsford CM3 6HB); p. 128*t* Kaj R. Svensson/ Science Photo Library; p. 128*b* Greg Balfour Evans/ Alamy; pp. 131, 132*t* Melba Photo Agency/Alamy; p. 141*t* Vario Images GmbH & Co.KG/Alamy; p. 141*m* Redmond Durrell/Alamy; p. 141*b* Arco Images GmbH/Alamy; p. 142 Petra Wegner/Alamy; p. 156 Sinclair Stammers/Science Photo Library; p. 161 Golden Rice Humanitarian Board; p. 174 Kevin Beebe, Custom Medical Stock/Science Photo Library; p. 176 Wendy Lee; p. 178*l* John Carnemolla/ Corbis; p. 178*rt* NHPA/Stephen Dalton;p. 178*rb* John Howard; Corday Photo Library Ltd/Corbis; p. 179 John Durham/Science Photo Library; p. 180 PhotoLibrary Group Ltd; pp. 181, 204*b* Nigel Cattlin/Alamy; pp. 183*t*, 183*b* John Mason/Ardrea. com p. 188*l*, 188*r* Michael Redmer/Science Photo Library; p. 189*lt* FotoFlora/Alamy; p. 189*rt* Paul Collis/Alamy; p. 189*rb* Mikes Powles/PhotoLibrary Group Ltd;p. 189*lb* Suzanne Long/Alamy; p. 200*l* Brandon Cole Marine Photography/Alamy; pp. 201*m*, 216, 218 Biodisc/Visuals Unlimited/Alamy; p. 201*b* Biology Media/Science Photo Library; p. 201*t* Dr Keith Wheeler/Science Photo Library; p. 202 John Swithinbank/Alamy; p. 203 Grant Heilman Photography/Alamy; p. 204*t* Meriel Jones, University of Liverpool, UK; p. 204*m* Phototake Inc/Alamy; p. 206 Jason Friend Photography Ltd/Alamy; pp. 212*l*, 212*r* Blickwinkel/Alamy; p. 214 Scenics and Science/Alamy; p. 217 Tim Gainey/Alamy; p. 219 Vicki Wagner/Alamy; p. 227 © Chad Ehlers - Stock Connection/Science Faction/Corbis; p. 238 Andy Walker, Midland Fertility Services/Science Photo Library; p. 241 Tony Cordoza/Alamy; p. 243 Angela Hampton Picture Library/Alamy; p. 247 Garry Watson/ Science Photo Library; p. 248 Wellcome Photo Library.p. 253 Alain Gougeon/ISM/Science Photo Library.

l = left, *r* = right, *t* = top, *b* = bottom, *m* = middle

Typesetting and illustration by Greenhill Wood Studios www.greenhillwoodstudios.com